Mathematics and Visualization

The series *Mathematics and Visualization* is intended to further the fruitful relationship between mathematics and visualization. It covers applications of visualization techniques in mathematics, as well as mathematical theory and methods that are used for visualization. In particular, it emphasizes visualization in geometry, topology, and dynamical systems; geometric algorithms; visualization algorithms; visualization environments; computer aided geometric design; computational geometry; image processing; information visualization; and scientific visualization. Three types of books will appear in the series: research monographs, graduate textbooks, and conference proceedings.

More information about this series at http://www.springer.com/series/4562

Ingrid Hotz · Talha Bin Masood ·
Filip Sadlo · Julien Tierny
Editors

Topological Methods in Data Analysis and Visualization VI

Theory, Applications, and Software

Springer

Editors
Ingrid Hotz
Department of Science and Technology
Linköping University
Norrköping, Sweden

Talha Bin Masood
Department of Science and Technology
Linköping University
Norrköping, Sweden

Filip Sadlo
Heidelberg University, IWR
Heidelberg, Germany

Julien Tierny
CNRS, Department of Scientific Computing
Sorbonne Université
Paris, France

ISSN 1612-3786 ISSN 2197-666X (electronic)
Mathematics and Visualization
ISBN 978-3-030-83499-9 ISBN 978-3-030-83500-2 (eBook)
https://doi.org/10.1007/978-3-030-83500-2

Mathematics Subject Classification: 76M24, 53A45, 62-07, 62H35, 65D18, 65U05, 68U10

Cover illustration from the chapter entitled “Using Contour Trees in the Analysis and Visualization of Radio Astronomy Data Cubes” kindly provided by Paul Rosen, University of South Florida.

This Springer imprint is published by the registered company Springer Nature Switzerland AG
The registered company address is: Gewerbestrasse 11, 6330 Cham, Switzerland

Preface

Efficient analysis of large and complex data is playing a rapidly increasing role in essentially all scientific disciplines. Large amounts of data originate from diverse sources, e.g., simulations and experiments, and contain combinations of scalar, vector, or tensor data. Therefore, topology-based methods have gained increasing recognition also in context with visualization due to their robustness and rigorous mathematical guarantees. However, while reaching more and more applications, also novel challenges are developing continuously. These include increasing data complexity with respect to size and structures, multi-scale features, noise, and uncertainty. Developing efficient and robust numerical methods for specific applications and making them available to a large user base is another urgent demand.

The book is the 8th in a series based on the biannual TopoInVis workshops that aim for scientific exchange between researchers working in this field in an open atmosphere. The 8th workshop was held in Nyköping, Sweden, in June 2019 with a specific focus on software for topological data analysis and its applications. Most contributions in the book are related to or based on work that has been presented at the workshop. The book is structured into four parts. Part I focuses on topological methods for scalar fields and Part II focuses on more complex fields as vector, tensor, or multi-fields. Part III deals with topological methods for non-field data. Finally, Part IV reports the results from efforts trying to establish a community code bases for software development.

Part I focuses on theory and methods related to scalar field topology, followed by a couple of chapters on applications which demonstrate how theory translates into practical solutions to problems from diverse scientific domains. The first chapter discusses the W-structures in contour tree, algorithms for extraction of such structures, and their impact on the performance of distributed algorithms. It also demonstrates using an example that extended persistence is not equivalent to branch decomposition and leaf-pruning. The next chapter presents a novel interaction interface for exploring scalar fields using merge trees called *mergemaps*. The third chapter is an interesting discussion on application of graph theoretic concept of percolation analysis to scalar fields. With both Gaussian random fields and real data, it describes how the histogram or degree of structure influences the shape of the percolation function. Then a chapter focuses on a discussion on implementing contour-tree-guided volume rendering of very large scalar field datasets in the context of in situ and distributed computing. The fifth chapter demonstrates how ideas from scalar field topology can be used for robust extraction and tracking of features in climate data. Multi-center cyclones are modeled as a set of critical points, and the tracking is done based on Morse complex. The last chapter of this part presents another application of scalar field topology in solving domain-specific problems, here the domain being astrophysics. It deals with design of contour-tree-guided feature extraction and visualization tool for analysis of image data obtained by current generation of radio and millimeter telescopes which are complex both in its spatial and spectral structures.

Part II contains contributions concerned with the analysis of fields with more complex attributes. The first three chapters deal with vector data. In the first chapter the notions of saddles, sinks, and sources originally defined for instantaneous vector fields are generalized to the finite-time setting based on a flow categorization with respect to contraction or expansion. The second chapter addresses the extraction of vortex corelines of inertial particles. Therefore, 3D and 6D parallel vector operators are introduced resulting in straight and bent inertial vortex corelines, respectively. In the following chapter it is shown how implicit visualization of 2D vector field topology can be used for periodic orbit extraction in 2D vector fields. The next two papers investigate fields with multiple attributes that shall be analyzed at the same time. The fourth chapter of this part applies visualization to evaluating a new topological equivalence relation, called *topological B+-equivalence* for the study of bounded bivariate fields. Invariants are introduced that approximate the equivalence. It is shown that visualizing the Reeb space gives us a near-instant way of evaluating these invariants. Topological relationships between multiple scalar fields to approach the analysis of time-varying multi-fields are considered in the next contribution. A novel method of finding similarity between two multi-fields by comparing their respective fiber component distributions is proposed. The Part II concludes with a chapter that deals with automatic chart analysis from rasterized images applying ideas from tensor field topology. It is demonstrated how positive semidefinite second-order tensor fields can be used as an effective model for this purpose.

Part III deals with topology-based methods for analysis of non-field data. It starts with a chapter that describes an approximate solution to the tree reconstruction problem for any finite point cloud in a Euclidean space with theoretical guarantees. The second chapter presents a novel approach for the extraction of micro-structural features called fibers from 3D scans of wood-based insulation materials. It describes how splitting geometry and topology processing of the data allows for topological simplification while still preserving the geometry of the scanned objects.

Part IV focuses on community efforts for developing the impact of topological methods in practice. It will particularly interest readers coming from the applications. It is composed of four chapters. The first chapter is a thorough introduction to vector field topology, which is an excellent entry point to any new comer to the field. The following three chapters are dedicated to the Topology ToolKit (TTK), an open-source library for topological data analysis and visualization, which is a community-powered effort for making the research results of our community more accessible to the applications. The second chapter provides a global overview of the algorithms currently supported by TTK, including methods for scalar, bivariate, uncertain, and time-varying data. The following chapter describes with an example how TTK can be extended with the addition of a new module, dedicated in this illustration to the persistence-driven clustering of high-dimensional point cloud data. Finally, the last chapter provides experience feedback regarding the TTK hackathon organized in conjunction with TopoInVis 2019. It describes its organization and main results as well as reflections which spontaneously emerged then. It also provides detailed organizational pieces of advice which will be useful to anyone willing to organize a hackathon.

Lastly, we would like to mention that the 2019 TopoInVis Workshop was organized by the division of Media and Information Technology at Linkoping University. Here we would like to particularly acknowledge the support from Gun-Britt Löfgern in organizing the workshop and the social event. Further, we would like to acknowledge the financial support from the Swedish e-Science Research Center (SeRC). Naturally, we thank all the participants of the workshop for a successful event and the contributors to this collection of manuscripts. We also thank the diligent reviewers who helped immensely in improving the quality of the manuscripts during the two-phase review process. Special thanks goes also to Leonie Kunz for her help in the production process of this volume.

March 2021

Ingrid Hotz
Talha Bin Masood
Filip Sadlo
Julien Tierny

Contents

Part I: Scalar Field Topology – Algorithms and Applications

W-Structures in Contour Trees

Petar Hristov and Hamish Carr

Abstract The contour tree is one of the principal tools in scientific visualisation. It captures the connectivity of level sets in scalar fields. In order to apply the contour tree to exascale data we need efficient shared memory and distributed algorithms. Recent work has revealed a parallel performance bottleneck caused by substructures of contour trees called W-structures. We report two novel algorithms that detect and extract the W-structures. We also use the W-structures to show that extended persistence is not equivalent to branch decomposition and leaf-pruning.

1 Introduction

Topology is the basis for persistent homology [17], a framework for extracting structural information from data, and Computational Topology [35], which studies how to compute and scale topological structures efficiently. Topological algorithms and data structures have been applied to various problems in structural biology [14, 34], computer vision [3, 23] medical imaging [32] and visualisation [5, 6, 10].

The Contour Tree (CT) is a data structure that captures the topological connectivity of a scalar field. In scientific visualisation it is used to identify features of more than local importance in large scale scientific and engineering simulations [22, 30]. As the size of data sets grows to exascale there is an increasing demand to develop scalable massively multicore and distributed algorithms and systems.

Electronic supplementary material The online version of this chapter (https://doi.org/10.1007/978-3-030-83500-2_1) contains supplementary material, which is available to authorized users.

P. Hristov (✉) · H. Carr
University of Leeds, Leeds LS2 9JT, UK
e-mail: mm16pgh@leeds.ac.uk

H. Carr
e-mail: h.carr@leeds.ac.uk

I. Hotz et al. (eds.), *Topological Methods in Data Analysis and Visualization VI*, Mathematics and Visualization, https://doi.org/10.1007/978-3-030-83500-2_1

The contributions of this paper are to extend the understanding of a pathological case, called the *W structure* [12], that emerges in one of the state of the art parallel contour tree algorithm (Subsect. 2.2); to demonstrate that W-structures have implications for parallel algorithmic efficiency (Subsect. 2.2) and that they can be detected and characterized (Subsect. 3.2); and finally that they can be used to show that contour tree simplification via persistent homology and branch decomposition [29] are equivalent (Sect. 5).

2 Background

The goals of this work are study how W-structures affect parallel contour tree computation and how branch decomposition relates to persistent homology. The relevant background is split between Sect. 2 and Sect. 5 because these two tasks have different prerequisites. In this section, we introduce methods to compute (Subsect. 2.2) and simplify (Subsect. 2.3) the Contour Tree (Subsect. 2.1) and leave the background relating to persistent homology to Subsect. 5.1.

2.1 Contour Trees

Given a scalar function $f : \mathbb{R}^n \to \mathbb{R}$, a *level set* is the set of all points with a given *isovalue* h: $f^{-1}(\{h\}) = \{x \in \mathbb{R}^n \mid f(x) = h\}$. We refer to individual connected components of a level sets as *contours*. As h varies, contours may appear, disappear, connect or disconnect at *critical points* where the gradient vector is zero.

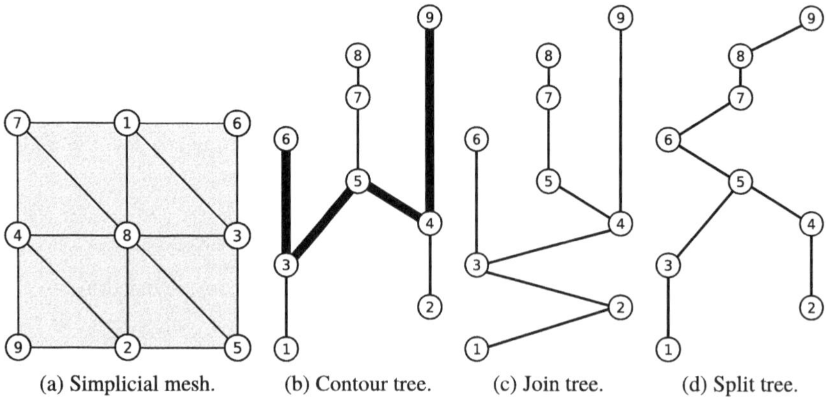

Fig. 1 A simplicial mesh **a** that generates a W-structure, the corresponding contour tree **b** with the W-structure shown as thicker edges and the corresponding join and split trees **c** and **d**. The vertices are labeled with their height value

The Reeb Graph of f is constructed by contracting the contours at every isovalue to a point. The resulting structure is a graph whose vertices are critical points and whose edges are families of contours with identical connectivity. Since $\mathbb{R}^n$ is simply connected, the Reeb Graph is connected and acyclic, also called the Contour Tree [4].

We extend the definition of a level set to a super-level set: the points with higher values than h, i.e. $\{x \in \mathbb{R}^3 \mid f(x) \geq h\}$, or a sub-level set with lower values, $\{x \in \mathbb{R}^3 \mid f(x) \leq h\}$. Contracting the connected components of the super-level and sub-level set gives two *merge trees*, also referred to as the join & split trees.

2.2 Contour Tree Algorithms

In practice, we assume that the domain is approximated by a simplicial mesh with a linear interpolant. Under these constraints, critical points occur at vertices, and we only need to process the graph composed of the vertices and edges of the mesh. Although these assumptions can be relaxed [7], they are the most common in practical data analysis, and they simplify our analysis without loss of generality.

The standard contour tree algorithm [8] is based on the idea of an isovalued sweep - i.e. processing the vertices of the mesh in sorted order from high to low. As each vertex u is processed, any edge (u, v) to a higher-valued vertex v is also processed. At each step there is a subgraph representing the super-level set, whose connectivity can be tracked with an incremental version of the union-find data structure [33]. In the first stage of the algorithm we construct the join tree, then repeat with a low-to-high sweep to compute the split tree. In the second stage, we construct the contour tree iteratively by transferring leaves and their adjacent edge from the merge trees, using induction on a simple invariant to guarantee correctness. As a result, this algorithm is sometimes referred to as the *sweep and merge* algorithm.

There are several approaches to scaling sweep and merge. Distributed algorithms [22, 26–28] adopt a divide and conquer approach where each node computes the contour/merge tree on parts of the data. The scalability of distributed methods relies not only on minimising node communication but also efficient utilisation of individual nodes. Efficient utilisation of nodes relies on parallel algorithms using vector [12], thread [19, 20] or hybrid [2, 24, 31] shared memory parallelism.

The parallel peak pruning algorithm [12] is the only shared memory algorithm which parallelises the merge phase. Other algorithms introduce a novel way of computing the join and split tree, but combine them in serial. Note that the merge phase has linear complexity and it is significantly faster to compute than the join and split trees. Nonetheless, parallelising the merge phase is important for resource utilisation and parallel speed up according to Amdahl's law [20].

In the merge phase of the serial contour tree algorithm [8] we transfer the leaves and their adjacent edge from the merge trees to the contour tree. Since this is a local operation all leaves can be batched and transferred in a single parallel step. The algorithm alternates between transferring leaves from the join and split tree until the contour tree is fully constructed. In the ideal case, each batch transfers at least half of the vertices, guaranteeing logarithmic performance. In a tree with no vertices with degree two, half of the vertices are leaves. Thus we can achieve logarithmic collapse if we remove degree two vertices in a post process for each batch.

Removing a degree two vertex is straightforward when its neighbouring vertices have values spanning the value of the vertex. This is the case when the vertex is connected by a chain of such vertices to a leaf. In effect, these vertices are regular at this stage (although they may not have been in earlier stages), and can be removed. For vertices of degree two whose value is smaller or bigger than the value of both neighbours, this is not so easy to perform. We call these vertices forks.

A *W-structure* consists of repeated forks zigzagging between upwards and downwards, as illustrated in Fig. 1. In order to collapse the W-structure completely we can only prune from an endpoint of the W-structure to a fork. The internal vertices cannot be process until we have pruned all forks. Therefore computation is effectively serialized along the largest W-structure in the contour tree. This prevents logarithmic collapse and complicates the parallel complexity analysis of the algorithm.

W-structures were first visible in Carr et al. [9], they have previously complicated proofs [11] and have caused issues with contour tree parallel algorithm design and analysis [12]. The contribution of this paper is to initiate a systematic study of these W-structures. A natural starting point is to describe them mathematically and develop algorithms to detect and extract them from contour trees. These algorithms will allow us to quantify the impact they have on computation and determine whether they are an issue in real life data sets. Finally we will show that W-structures lead to theoretical complications as well, by demonstrating that contour tree simplification via branch decomposition is not equivalent to persistent homology.

2.3 Contour Tree Simplification

The principal technique for contour tree simplification is branch decomposition [29]. The contour tree is partitioned into a set of disjoint monotone paths (branches). A branch decomposition is hierarchical when there is exactly one branch that connects two leaves called the master branch and every other branch connects a leaf to a saddle. Branches represent pairs of critical points that can be cancelled [25].

In order to decide the order of cancellation we use the persistence of the branches [29]. The persistence of a branch is the greater of the difference between the height value at its endpoints and the persistence of its children. Simplification consists of removing branches that do not disconnect the tree in order of their persistence. This produces a hierarchy of cancellations as shown in Fig. 2. In branch decomposition we repeatedly pair upper leafs to join saddles and lower leaves to split saddles in order of persistence until all vertices are paired.

As an example we consider branch decomposition of the contour tree from Fig. 1 (b). The first two candidate branches are $(6, 3)$ with persistence 3 and $(4, 9)$ with persistence 5. We take the branch with lower persistence $(6, 3)$. In the next step the candidates are $(1, 5)$ with persistence 4 and $(4, 9)$ with persistence 5. We take

$(1, 5)$. The remaining candidate branches are (4, 8) with persistence 4 and (4, 9) with persistence 5. After removing (4, 8) the only remaining branch is the master branch - $(2, 9)$. To conclude, the pairs of critical points produced by the branch decomposition of the contour tree are $(3, 6)$, $(1, 5)$, $(4, 8)$ and $(2, 9)$.

We can also compute the branch decomposition of the join and split trees. The candidate branches for the join tree are (3, 6), (4, 8) and (4, 9). We remove them in order of persistence. First (3, 6), then (4, 8) and finally the master branch (1, 9). In the split tree the two candidate branches are (1, 5) and (2, 5). We remove (2, 5) first because it has lower persistence and then the master branch $(1, 9)$.

3 W-Structures in Contour Trees

In this section we will develop the existing understanding about W-structures and introduce three algorithms to compute the largest W-structure in a contour tree. The value of these algorithms will be in that they will allow us to build our understanding of W-structures and allow us to detect W-structures in real life data. We adopt the following notation: in a contour tree T the set of vertices is V and the set of edges is E. We refer to paths in the contour tree by their first and last vertex because there is a unique path between any two vertices.

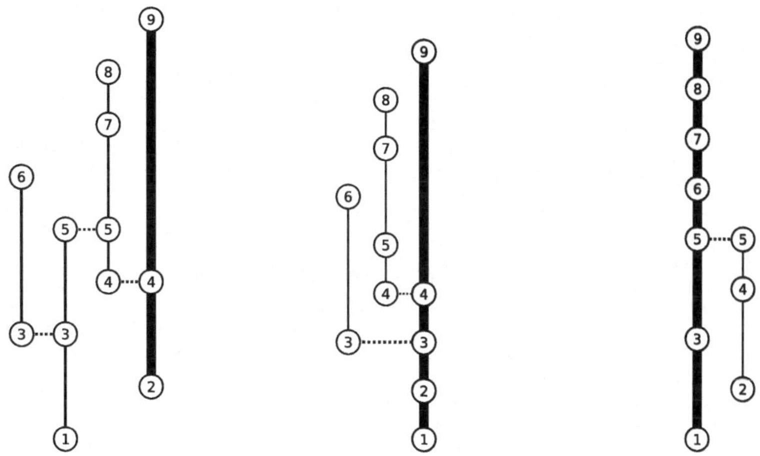

(a) Contour Tree (b) Join Tree (c) Split Tree

Fig. 2 Branch decomposition of the contour tree and the two merge trees from Fig. 1 with the edges of the master branches in thicker lines. Vertices are labeled with their height. In both merge trees the master branch is the monotone path from the global minimum 1 to the global maximum 9. Note that in the absence of a monotone path between 1 and 9 in the mesh and its contour tree they cannot be paired

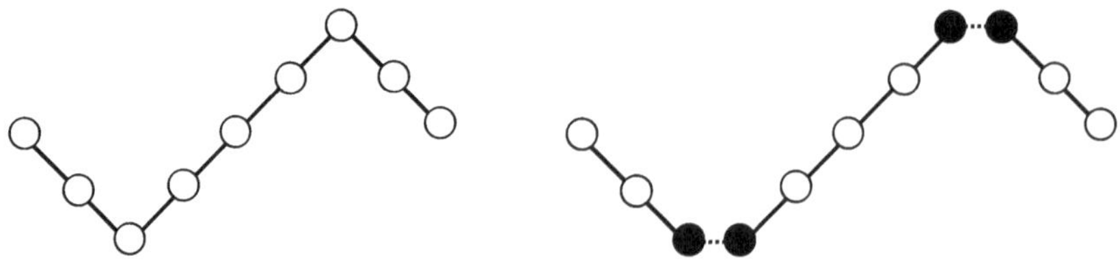

(a) Extracted W-structure. (b) Monotone path decomposition.

Fig. 3 A W-structure and its monotone path decomposition (forks in solid black)

3.1 *Spatial Characterization*

An important property of paths in contour trees is their monotone path decomposition. This is a sequence of monotone subpaths that share exactly one vertex and have alternating direction (Fig. 3). We can use the number of subpaths in the monotone path decomposition to characterize them. To simplify this characterization we note that the number of subpaths in the monotone path decomposition is one more than the number of vertices where an ascending subpath ends and a descending subpath begins (or vice versa). We will call these vertices forks.

We define the w-length of a path as the number of forks in that path and the length of a path as the number of edges. To avoid ambiguity between w-length and length we will use the term w-path to emphasize that we are referring to a path's w-length. Note that if two paths share a vertex it may be a fork in one of them, but not the other. For example vertex 5 from the contour tree in Fig. 1 is a fork in the path from 6 to 9, but not in the path from 1 to 8. This property is crucial in understanding how we develop algorithms for detecting W-structures.

In this terminology the largest W-structure in a contour tree is a path between two leaves with maximum w-length (or longest w-path). We will call this the *w-diameter* of the contour tree. This again is analogous to how the longest path in a tree is called the diameter of the tree. As there are efficient algorithms for computing the diameter of a tree a natural question to ask is whether we can adapt these algorithms to compute the w-diameter of a contour tree.

3.2 *W-Diameter Algorithms*

We will begin the development of w-diameter algorithms by first describing three tree diameter algorithms—brute force, endpoint search and root based search. The brute force algorithm computes the lengths of all paths in a tree and outputs the maximum one. The second algorithm relies on the fact that the most distant leaf from any vertex is the endpoint of the diameter [15]. It requires two breadth first searches, so it has linear running time. The third algorithm is defined for rooted trees where the longest path may or may not pass through the root. If it passes through the root then it must

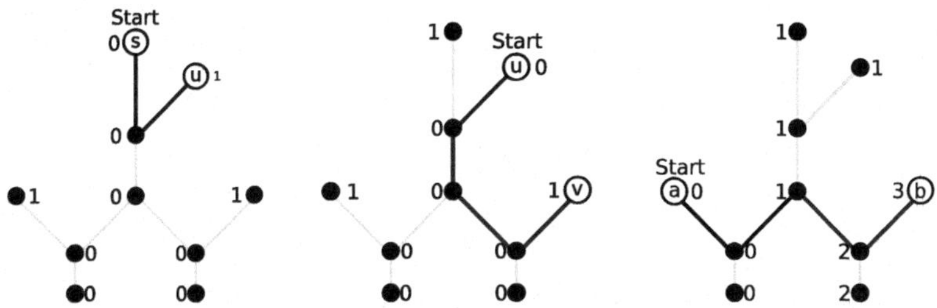

(a) Stage 1. (b) Stage 2. (c) Actual w-diameter.

Fig. 4 Execution of the Double BFS algorithm **a** and **b** on an example contour tree and the actual w-diameter **c**. Black edges indicate a path of maximum w-length. Numerical labels next to vertices indicate their w-distance from the start vertex. In stage 1 we start from an arbitrary vertex s and find the most w-distant vertex from it u. In stage 2 we find the most w-distant vertex from u and call it v. The w-length of the path from u to v is however suboptimal as demonstrated in the last figure. If our initial root was a then the algorithm would have obtained the w-diameter

start in a leaf in one of the subtrees of a child of the root, pass through the root and end in a leaf in another subtree. If it does not then it must be entirely contained in a subtree of a child of the root. With the use of dynamic programming this algorithm has linear complexity as well. In the next three subsections we will adapt each one of these algorithms to compute the w-diameter of a contour tree.

3.3 Algorithm 1—Multi BFS

The brute force approach to finding the w-diameter of a contour tree compares the w-lengths of all paths in the contour tree. To implement it we modify Breadth First Search (BFS) to traverse the tree and compute w-length (number of forks) instead of length (number of edges). We then run this modified BFS from every vertex in the tree and output the maximum value found. It has quadratic running time and we will refer to it as Naive BFS.

3.4 Algorithm 2—Double BFS

The algorithm works the same as the second tree diameter algorithm we described in Subsect. 3.2 except we measure w-distance instead of distance. Consider a contour tree T and an arbitrary start vertex s. First we find the most w-distant vertex from s and call it u. Then we find the most w-distant vertex from u and call it v. After the first search from s we are not guaranteed that v is an endpoint of a w-diameter. We are however guaranteed that v is an endpoint of a path whose w-length is at least that of the w-diameter minus two. We will demonstrate why that is true in the following two paragraphs.

We illustrate this with an example in Fig. 4. In this example the algorithm does not produce the w-diameter of the contour tree which is the path from a to b. The mismatch between the output of the algorithm and the actual w-diameter is due to vertices which are forks in one path, but not in others. First consider the black vertex on the path from s to u in Fig. 4 a); it is a fork on the path from s to u, but not on the path from u to $v = b$ (Fig. 4 b). Secondly, consider the midpoint of the path from a to b; it is a fork on the path from a to b, but not on the path from u to b or a.

Our argument for the general case is based on this example. In a contour tree T let s be the start vertex and v the most w-distant vertex from s. During the course of the algorithm there are two graph searches, one from s to identify v and one from v. There are at most two turning points which may or may not be forks during the two searches. One of the turning points is where the path from s to a (or symmetrically to b) diverges from the path from s to v. The other one is where the path from v to a (or symmetrically to b) diverges from the path from a to b (or symmetrically from b to a). This causes the w-length of the path we find to vary by at most two from the w-diameter of the contour tree.

To implement Double BFS we pick a starting vertex and then run the modified Breadth First Search twice. The first BFS from the root to find the farthest leaf from it and then a second BFS from that leaf. The algorithm consists of two consecutive Breadth First searches and therefore its running time is $O(|V|)$.

3.5 Algorithm 3—Dynamic

The third algorithm works by progressively combining paths from subtrees of the contour tree to obtain the longest w-path. For a contour tree T we pick an arbitrary start vertex s to be the root of the rooted tree T_s. Observe that the w-diameter of T either passes through s or it does not. If it does pass through s then it must also pass through two children of s and be contained in their subtrees. If it does not pass through s then it must be entirely contained in the subtree of one of the children of s. We can then extend this reasoning recursively to all the subtrees of T_s.

For every vertex u in T_s we define $T_{s,u}$ as the subtree of T_s with root u. We find the w-diameters of all subtrees of $T_{s,u}$ and use them to compute the w-diameter of $T_{s,u}$. We now demonstrate how to compute the w-diameter of $T_{s,u}$ assuming that the optimal solutions for all subtrees of $T_{s,u}$ have been found recursively. Note that the base case is at the leaves of the tree Fig. 5 a).

Case 1—the w-diameter of $T_{s,u}$ goes through u. To handle this case we must find two maximum paths contained in two subtrees whose roots are children of u, say a and b. As we have recursively found all such maximum paths we only have to determine how to combine them. When combining them *three* vertices can become forks. The first one is u and the other two are a and b which were previously endpoints of the maximum paths in their subtrees. To account for u we simply have to compare

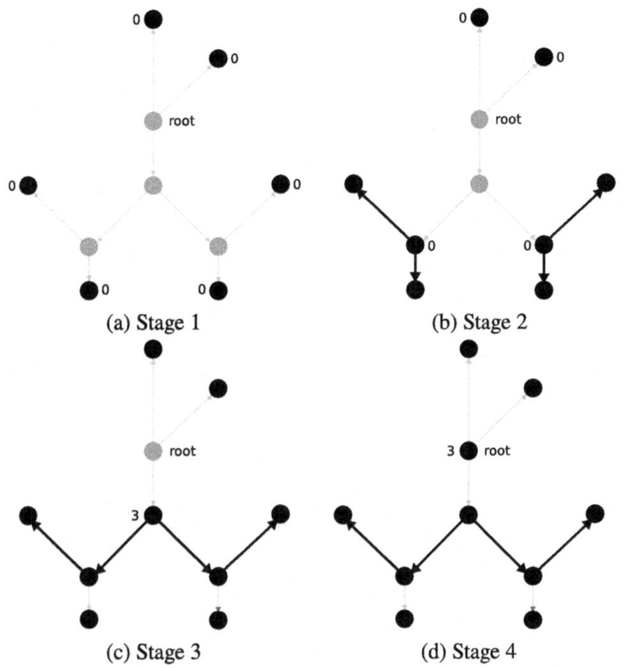

Fig. 5 Execution of the Dynamic algorithm on an sample contour tree. Black vertices have been processed, gray ones have not. The black edges are the w-diameters for all subtrees. The numerical label next to a vertex indicates that it is being processed in the current stage and the number is the w-diameter in its subtree. Each stage combines w-diameters of smaller subtrees to obtain the diameter of the whole tree

its height to a and b. To account for a and b we must compare their height with u and the previous vertex in the maximum w-path they are the endpoint of.

An example of where Case 1 holds is in Fig. 5 c). Note that this requires us to not only look at all children of u, but also to all children of children of u. In addition to this it may be the case that u is a leaf and has only one child say a. In this case u must be the endpoint of the w-diameter. In this case we find all maximum paths that end at a and account for whether a becomes a fork in them or not.

Case 2—the w-diameter of $T_{s,u}$ does not pass through u. In this case the w-diameter has to be entirely contained in one of the subtrees whose root is a child of u so we pick the maximum one. An example of where Case 1 holds is in Fig. 5 d).

Let us derive the time complexity of the algorithm. Firstly it takes $O(|V|)$ time to traverse the tree and root it via either BFS or DFS. Secondly, we iterate over the children of all children of all vertices. Since the algorithm operates on trees, every vertex has a unique grandparent. Therefore every vertex will be visited exactly once and contribute $O(|V|)$. Finally we pick all pairs of children of a vertex to find the maximum w-path that goes through the vertex. Computing this for all vertices in the graph yields $O(\sum_{u \in V} d(u)^2)$ where $d(u)$ is the degree of a vertex. To see how we can evaluate this consider that in a tree $d(u) + d(v) \leq |V|$ for any two vertices connected by an edge (otherwise we would have a cycle). If we sum over all edges we obtain that $\sum_{uv \in E(T)} d(u) + d(v) \leq |V|^2$. In the summation on the left hand side every term $d(u)$ is present $d(u)$ times and therefore $\sum_{uv \in E(T)} d(u) + d(v) = \sum_{u \in V(T)} d(u)^2 \leq |V|^2$. Therefore the overall complexity of the algorithm is $O(|V|^2)$. Note that this is formally more complex than the dynamic programming tree diameter algorithm in Sect. 3.2. The difference between the two is that the base algorithm does not need to look at all pairs of children, it can simply pick the child with the largest height.

We have shown that the algorithmic complexity of the Dynamic algorithm is no better than that of the Naive BFS algorithm. However, the running time of the Dynamic algorithm depends on the average degree of the vertices of the contour tree. If we assume the function is generic and PL Morse [16], then the degree of every vertex is bounded and the running time of the algorithm is linear.

4 Empirical Study

In this section we supplement the theoretical investigation of the W-structures with an empirical study. We verify the correctness and running time of the w-diameter algorithms and study the W-structures in contour trees of real life data sets.

We implemented all three w-diameter algorithms in C++. Their source code is in the supplementary materials. We avoided a recursive implementation of the Dynamic algorithm due to excessive overhead caused by recursive calls. Instead we traversed the tree once with a standard Breadth First Search to identify the children and parents of all vertices. We then put them in a list and starting from the leaves we processed all vertices in the tree making sure their children are processed beforehand.

The data sets we have used are taken from the Open SciVis Dataset [1]. The contour tree was computed using the open source VTK-m implementation. All tests were run on a Lenovo E550 Laptop with Intel(R) Core(TM) i5-5200U CPU at 2.20 GHz and 8 GB DDR3 RAM at 1600 MHz. The running time of the w-diameter programs was obtained from an average of five runs on each data set.

4.1 Results

The results from the empirical tests are shown in Table 1. The first column shows the name of the data set, the second column the number of vertices and the third the diameter of the contour tree (not w-diameter). The fourth column shows the number of iterations in the merge phase of the PPP algorithm. The following columns show the running times and output of the Double BFS, Dynamic and Naive BFS algorithms.

From previous work [12] we know that in a contour tree without W-structures the parallel step complexity of the merge phase is logarithmic. When W-structures are

Table 1 Analysis of the W-Diameter of real life data sets

Dataset	Vertices	Tree Diam.	Merge Iter.	Double BFS		Dynamic		Naive BFS	
				Time(s)	W-Diam	Time(s)	W-Diam	Time(s)	W-Diam
fuel	236	43	6	0.0001	2	0.0003	4	0.0133	4
marschner lobb	1604	371	9	0.0008	3	0.0022	3	0.3881	3
hydrogen atom	13038	3153	6	0.0039	2	0.0139	4	25.382	4
aneurism	65625	23701	10	0.0251	4	0.0671	4	723.39	4
engine	518780	92559	12	0.2183	6	0.5504	6	N/A	N/A
foot	870371	133655	14	0.4922	7	1.0342	7	N/A	N/A
skull	2199876	340611	14	1.0839	7	2.5048	7	N/A	N/A
backpack	7441922	1365783	18	4.2384	7	8.4635	7	N/A	N/A
bunny	13078906	1450364	18	8.1435	5	15.903	6	N/A	N/A
present	17006950	2349226	16	11.339	5	21.075	6	N/A	N/A
christmas tree	24643034	4866458	17	13.399	5	29.015	5	N/A	N/A
magnetic rec.	40321359	6401594	18	34.480	6	57.287	6	N/A	N/A

taken into account the best formal guarantee that could be given is the tree diameter. However it was noted that the merge phase usually takes less than a logarithmic number of iterations. This is reflected in our results as well - the number of iterations is always less than $\log_2(|V|)$ and substantially less than the diameter.

In this paper we investigate the case where W-structures are present in the contour tree. The key issue in doing so is to detect when they are present and to quantify their size. Based on the results of this empirical study we can confirm that W-structures do appear in real world data. Fortunately, the size of the W-structures is relatively small compared to the size of the data. Furthermore larger data sets do not seem to have a proportionally larger w-diameter.

A w-diameter of more than $2\log_2(|V|)$ can formally prevent logarithmic collapse in the merge phase. The maximum w-diameter in our tests was 7 which is less than $2\log_2(|V|)$ in all cases. This highlights the importance of parallelising the merge phase and explains why it performs well in practice.

Finally the running time and correctness of the w-diameter algorithms is consistent with our theory. The worst case running time for Dynamic is quadratic, but as predicted it is not exhibited and in practice it is only around twice as slow as the Double BFS algorithm. The Dynamic and Naive BFS algorithms produce the same results, while the Double BFS algorithm produces a suboptimal w-path in the bunny, present, fuel and hydrogen atom data sets.

5 W-Structure Simplification

The topological complexity of a scalar field is largely governed by the number of critical points. Some methods for simplification remove critical points in pairs using an auxiliary topological data structure such as the Morse-Smale Complex, persistent homology or the Contour Tree. Once selected, pairs are ranked by persistence, or more advanced metric such as volume or surface area [9], and cancelled in that order.

It is well know that the sequence of simplifications do not always agree with persistence. For example in Morse-Smale complexes there are blocking structures known as strangulations [21] and in ϵ-simplification [18] compound function value changes can occur when following the persistence order. W-structures are a similar kind of blocking structures in the case of contour tree simplification.

What is less well known is the relationship between branch decomposition and persistent homology (first observed [29]). In the merge phase, the contour tree is build up of branches taken from the join and split tree. The critical pairs defined by those branches in turn correspond to the 0th persistence pairs of f and $-f$. In this chapter we will consider whether the branch decomposition of the contour tree also corresponds to those persistence pairs.

5.1 *Persistent Homology Overview*

The building blocks of persistent homology [17] are sequences of nested simplicial complexes called filtrations. In a filtration we start from the empty set and iteratively add simplices to obtain the full complex. Throughout this section will consider the ascending and descending filtrations on Fig. 6 of the simplicial mesh from Fig. 1. The ascending filtration of M is made up of the sub-level sets M_i which contain all vertices whose value is lower than or equal to i and all other simplices between them. The descending filtration of M is made up of the super-level sets M^i which contain all vertices whose value is bigger than or equal to i and all other simplices between them. In short, this is exactly the same as the join and split tree computation in the sweep and merge algorithm we discussed in Subsect. 2.2.

Persistent homology describes how the n-dimensional connectivity of the simplicial complexes in a filtration changes. The n-dimensional connectivity is described by an algebraic structure called the nth homology group. Since contour and merge trees only capture connected components and not higher dimensional connectivity such as holes and voids we will only need to consider the 0th homology group. When a connected component appears in the progression of the filtration we say that a 0th homology class is *born*. When two connected components merge together the 0th homology class that corresponds to the younger component *dies* and the 0th homology class that corresponds to the older one *persists*.

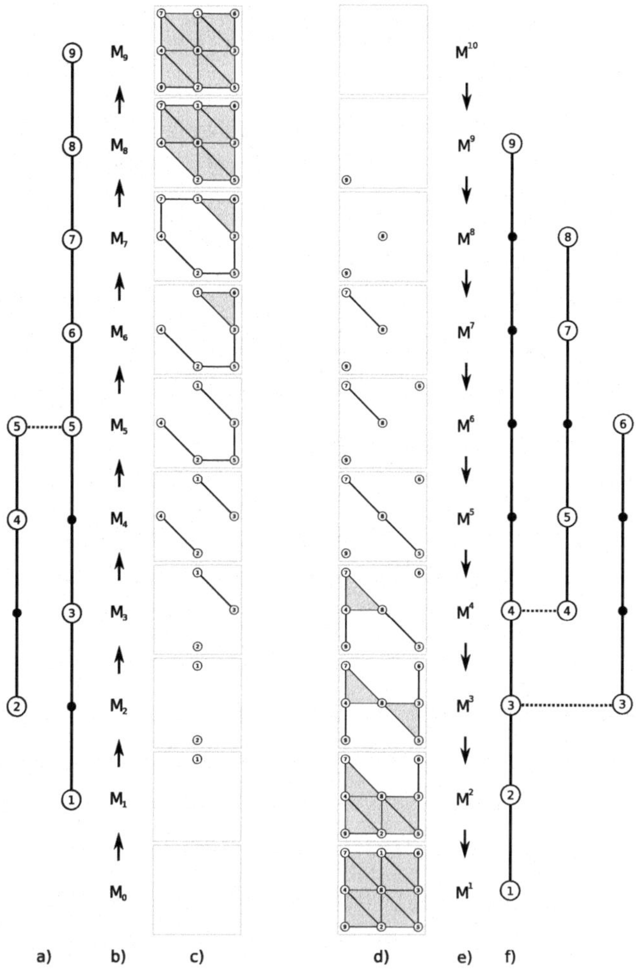

Fig. 6 Ascending **c** and Descending **d** filtration of Fig. 1 **a** Direction of travel of the two filtrations **b**and **d**. Branch decomposition of the split **a** and the join tree **f** with additional vertices corresponding to connected components

The output of persistent homology is the so called persistence pairs. A persistence pair is a record of the birth and death of a 0th homology class. The persistence pairs of a filtration give a basis for topological simplification.

5.2 Comparison of Critical Point Pairs

The proposition we aim to resolve is whether persistence pairs are equivalent to branch decomposition cancellation pairs. Since monotone paths in the contour tree correspond to monotone paths in the simplicial mesh [8] then branches obtained via branch decomposition correspond to valid topological cancellations in the simplicial mesh [29]. Therefore they are comparable to cancellations given by persistence pairs.

We start by computing the persistence pairs of the ascending filtration in Fig. 6 c). One connected component appears in the complex M_1, another one in the complex M_2 and they merge in M_5. When the two merge the older one born in M_1 persists and the younger one born in M_2 dies. We record this with the persistence pair $(2, 5)$. When the filtration is done we are left with a 0th homology class corresponding to the single connected component of the mesh. This 0th homology class is called essential and we give it an infinite persistence pair $(1, \infty)$.

In the descending filtration in Fig. 6 d) we can see that three 0th homology classes are born in the complexes M^9, M^8 and M^6. The one born in M^8 dies in M^4 when it merges with the one born in M^9 (because M^9 is older). The one born in M^6 dies in M^3 when it merges with the one born in M^9. The one born in M^9 does not die in the descending filtration because it represents the connected component of M itself. Therefore the persistence pairs are $(6, 3)$, $(8, 4)$ and $(9, \infty)$.

The two infinite persistence pairs can be resolved with extended persistence [13]. The idea behind extended persistence is that we take the essential classes from the ascending and descending pass and pair those which correspond to the same connected component. More generally we know that in a simple domain extended persistence always pairs the global minimum with the global maximum. In our example the extended persistence pair for the ascending filtration is $(1, 9)$ and the extended persistence pair for the descending filtration is $(9, 1)$.

Finally consider the branch decomposition of the contour tree in Fig. 2. While the first produced branch $(6, 3)$ is the same in the contour tree and in the join tree branch decomposition, the third branch of the join tree branch decomposition $(1, 9)$ does not occur as a pair in the contour tree. The same holds for the branch decomposition of the split tree and the persistent homology pairs of the ascending and descending filtration - they all pair the global minimum 1 with the global maximum 9. This cannot occur in the branch decomposition of the contour tree because branches represent monotone paths. There is no monotone path between the 1 and 9 in the mesh and therefore no monotone path in the contour tree. The result then follows.

6 Conclusion

In this paper we introduced the theory of a pathological case in contour tree parallel computation called a *W-structure*. We developed three algorithms to detect and extract W-structures and showed that they appear in real life data. We also showed that W-structures cause fundamental theoretical issues. They lead to an example that contour tree simplification is not equivalent to persistent homology.

Future work in this direction will focus on whether there is a form of persistent homology which matches the branch decomposition form of simplification, on algorithmic improvements for which W-structures do not pose a parallel bottleneck, and if need be, on further empirical studies to inform algorithmic development.

Acknowledgements This work was supported by the University of Leeds School of Computing, including a scholarship supporting the first author. We would also like to thank Ingrid Hotz, Talha Bin Masood, Filip Sadlo and Julien Tierny for organising the TopoInVis 2019 workshop.

References

1. Open SciVis Datasets. https://klacansky.com/open-scivis-datasets. Accessed 30 Jan 2020
2. Acharya, A., Natarajan, V.: A parallel and memory efficient algorithm for constructing the contour tree. In: 2015 IEEE Pacific Visualization Symposium (PacificVis), pp. 271–278 (2015). https://doi.org/10.1109/PACIFICVIS.2015.7156387
3. Biasotti, S., Giorgi, D., Spagnuolo, M., Falcidieno, B.: Reeb graphs for shape analysis and applications. Theor. Comput. Sci. **392**(1–3), 5–22 (2008)
4. Boyell, R.L., Ruston, H.: Hybrid techniques for real-time radar simulation. In: Proceedings of the 1963 Fall Joint Computer Conference, pp. 445–458. IEEE (1963)
5. Bremer, P.T., Hamann, B., Edelsbrunner, H., Pascucci, V.: A topological hierarchy for functions on triangulated surfaces. IEEE Trans. Visual. Comput. Graph. **10**(4), 385–396 (2004)
6. Carr, H., Geng, Z., Tierny, J., Chattopadhyay, A., Knoll, A.: Fiber surfaces: generalizing isosurfaces to bivariate data. Comput. Graph. Forum **34**, 241–250 (2015)
7. Carr, H., Snoeyink, J.: Representing interpolant topology for contour tree computation. In: H.C. Hege, K. Polthier, G. Scheuermann (eds.) Topology-Based Methods in Visualization II, Mathematics and Visualization, pp. 59–74. Springer, Berlin (2009)
8. Carr, H., Snoeyink, J., Axen, U.: Computing contour trees in all dimensions. Comput. Geom. **24**(2), 75–94 (2003)
9. Carr, H., Snoeyink, J., van de Panne, M.: Simplifying flexible isosurfaces using local geometric measures. In: Proceedings of the Conference on Visualization 2004 (VIS 2004), pp. 497–504. IEEE Computer Society, Washington, DC (2004). https://doi.org/10.1109/VISUAL.2004.96
10. Carr, H., Snoeyink, J., Van De Panne, M.: Flexible isosurfaces: simplifying and displaying scalar topology using the contour tree. Comput. Geom. **43**(1), 42–58 (2010)
11. Carr, H., Tierny, J., Weber, G.: Pathological and test cases for reeb analysis . In: Topology-Based Methods in Visualization 2017 (TopoInVis 2017), pp. 27–28. Tokyo, Japan, February 2017
12. Carr, H.A., Weber, G.H., Sewell, C.M., Ahrens, J.P.: Parallel peak pruning for scalable SMP contour tree computation. In: 2016 IEEE 6th Symposium on Large Data Analysis and Visualization (LDAV), pp. 75–84 (2016). https://doi.org/10.1109/LDAV.2016.7874312
13. Cohen-Steiner, D., Edelsbrunner, H., Harer, J.: Extending persistence using Poincaré and Lefschetz duality. Found. Comput. Math. **9**(1), 79–103 (2009)
14. Connolly, M.L.: Shape complementarity at the hemoglobin $\alpha 1 \beta 1$ subunit interface. Biopolymers **25**(7), 1229–1247 (1986)
15. Dewdney, A.K.: Computer recreations. Sci. Am. **17**, 18–30 (1985)
16. Edelsbrunner, H., Harer, J.: Computational Topology, An Introduction, 1 edn. Americal Methematical Society, Providence (2013)

17. Edelsbrunner, H., Letscher, D., Zomorodian, A.: Topological persistence and simplification. In: Proceedings 41st Annual Symposium on Foundations of Computer Science, pp. 454–463. IEEE (2000)
18. Edelsbrunner, H., Morozov, D., Pascucci, V.: Persistence-sensitive simplification functions on 2-manifolds. In: Proceedings of the Twenty-Second Annual Symposium on Computational Geometry (SCG 2006), pp. 127–134. ACM, New York (2006). https://doi.org/10.1145/1137856.1137878.
19. Gueunet, C., Fortin, P., Jomier, J., Tierny, J.: Contour forests: fast multi-threaded augmented contour trees. In: IEEE Symposium on Large Data Analysis and Visualization. Baltimore (2016). https://hal.archives-ouvertes.fr/hal-01355328
20. Gueunet, C., Fortin, P., Jomier, J., Tierny, J.: Task-based augmented merge trees with fibonacci heaps. In: 2017 IEEE 7th Symposium on Large Data Analysis and Visualization (LDAV), pp. 6–15 (2017). https://doi.org/10.1109/LDAV.2017.8231846
21. Gyulassy, A., Natarajan, V., Pascucci, V., timo Bremer, P., Member, B.H.: A topological approach to simplification of three-dimensional scalar functions. In: IEEE Transactions on Visualization and Computer Graphics, pp. 474–484 (2006)
22. Landge, A.G., et al.: In-situ feature extraction of large scale combustion simulations using segmented merge trees. In: Proceedings of the International Conference for High Performance Computing, Networking, Storage and Analysis (SC 2014), pp. 1020–1031 (2014). https://doi.org/10.1109/SC.2014.88
23. Li, C., Ovsjanikov, M., Chazal, F.: Persistence-based structural recognition. In: Proceedings of the IEEE Conference on Computer Vision and Pattern Recognition, pp. 1995–2002 (2014)
24. Maadasamy, S., Doraiswamy, H., Natarajan, V.: A hybrid parallel algorithm for computing and tracking level set topology. In: 2012 19th International Conference on High Performance Computing, pp. 1–10 (2012). https://doi.org/10.1109/HiPC.2012.6507496
25. Matsumoto, Y.: An Introductino to Morse Theory (Translation of Mathematical Monograms), vol. 208, 1st edn. American Mathematical Society, Providence (2002)
26. Morozov, D., Weber, G.: Distributed merge trees. In: Proceedings of the 18th ACM SIGPLAN Symposium on Principles and Practice of Parallel Programming, pp. 93–102 (2013). https://doi.org/10.1145/2517327.2442526
27. Morozov, D., Weber, G.H.: Distributed contour trees. In: Bremer, P.T., Hotz, I., Pascucci, V., Peikert, R. (eds.) Topological Methods in Data Analysis and Visualization III, pp. 89–102. Springer International Publishing, Cham (2014)
28. Pascucci, V., Cole-McLaughlin, K.: Parallel computation of the topology of level sets. Algorithmica **38**(1), 249–268 (2004). https://doi.org/10.1007/s00453-003-1052-3
29. Pascucci, V., Cole-McLaughlin, K., Scorzelli, G.: Multi-resolution computation and presentation of contour trees. In: Proceedings of the Conference on Visualization, Imaging, and Image Processing, pp. 452–290 (2004)
30. Pascucci, V., Tricoche, X., Hagen, H., Tierny, J.: Topological Methods in Data Analysis and Visualization: Theory, Algorithms, and Applications. Springer Science & Business Media, Berlin (2010)
31. Rosen, P., Tu, J., Piegl, L.A.: A hybrid solution to parallel calculation of augmented join trees of scalar fields in any dimension. Comput-Aided Des. Appl. **15**(4), 610–618 (2018). https://doi.org/10.1080/16864360.2017.1419648
32. Shi, Y., Li, J., Toga, A.W.: Persistent Reeb graph matching for fast brain search. In: Wu, G., Zhang, D., Zhou, L, (eds.) Machine Learning in Medical Imaging. MLMI 2014. Lecture Notes in Computer Science, vol. 8679. Springer, Cham (2014). https://doi.org/10.1007/978-3-319-10581-9_38
33. Tarjan, R.E.: Efficiency of a good but not linear set union algorithm. J. ACM **22**, 215–225 (1975)
34. Verovšek, S.K., Mashaghi, A.: Extended topological persistence and contact arrangements in folded linear molecules. Front. Appl. Math. Stat. **2**, 6 (2016)
35. Zomorodian, A.J.: Topology for Computing, 1 edn. Cambridge University Press, Cambridge (2009)

Mergemaps: Treemaps for Scientific Data

Adhitya Kamakshidasan and Vijay Natarajan

Abstract Topology driven methods for analysis of scalar fields often begin with an exploration of an abstract topological structure such as the merge tree. Such abstractions are hard to interpret and time-consuming, particularly for feature-rich data. Current visualization schemes often place less emphasis on enriching user experience, human perception, or interaction. In this work, we aim to bridge that gap by utilizing treemaps towards effective topological analysis. We present *mergemaps*, a treemap based interactive design, to better understand merge trees. To aid the perceptual understanding of large merge trees, we provide fusing and diffusing operations to reduce its hierarchical size while preserving topological features. We show multiple examples where our design leads to easy interpretations.

1 Introduction

Topological methods for data analysis have proven to be useful in multiple contexts ranging from exploring cosmic filaments [27] to extracting voids in proteins [29]. Contour trees, Reeb graphs, Morse-Smale complexes, persistence diagrams [5, 9, 11, 13, 26], to name a few, provide abstract representations that aid in topological analysis of scalar fields. Despite these abstractions being extremely powerful, they are still yet to gain widespread popularity because their interpretation requires background in algebraic topology and Morse theory [10]. To this extent, multiple attempts have been made to provide user interfaces that convey topological information in an

Electronic supplementary material The online version of this chapter (https://doi.org/10.1007/978-3-030-83500-2_2) contains supplementary material, which is available to authorized users.

A. Kamakshidasan
Inria Saclay – Île-de-France, Palaiseau, France

V. Natarajan (✉)
Indian Institute of Science, Bangalore, India
e-mail: vijayn@iisc.ac.in

I. Hotz et al. (eds.), *Topological Methods in Data Analysis and Visualization VI*, Mathematics and Visualization, https://doi.org/10.1007/978-3-030-83500-2_2

intuitive manner. In this work, we aim at improving the user experience of exploring a particular topological structure called the merge tree.

1.1 Related Work

A *merge tree* traces the connectivity evolution of sub/super-level sets in a scalar field, a *contour tree* contains the combined information of both sub- and super-level sets. From a data analysis perspective, merge trees have been used in various interesting applications. Bock et al. [3] used it for extracting fishes from micro-CT scans, Valsangkar et al. [35] track cyclones by understanding how their contours join and split, Doraiswamy et al. [8] identify congestion in New York roads.

Weber et al. [37] introduced the notion of topological landscapes, which uses a terrain metaphor for presenting the contour tree. Topological landscapes capture the nesting behavior of contours, with the intuition that humans better perceive topographic information. Their two-step algorithm involves placing branches by adaptively subdividing a mesh, and rebalancing the mesh to improve space utilization. While the resulting landscape allows users to grasp high-level features, the number of triangles increases sharply for deep hierarchies and the terrain can contain large empty spaces.

Building on the same paradigm, Harvey and Wang [12], observed the connection between topological landscapes and treemaps. Their proposed method, called Denali, computes a landscape corresponding to each edge interpreted as the root of the tree, and chooses the best landscape by defining a metric distance between each of them. While this method is not computationally expensive, it can result in skinny boundaries. Bekatayev et al. [1] proposed a solution that preserves the geometry proximity while constructing the topological landscapes by routing edges across a Voronoi diagram, but works only for small-sized trees. Demir et al. [7] layout branches as square boxes and render landscapes with a first-fit box packing scheme, followed by hierarchically triangulating the corresponding grid. This algorithm is an improvement over the original algorithm by Weber et al., but the resulting landscape looks artificial and lacks a spatial context. Topological landscapes have also been used to study higher dimensional point clouds [20]. Mergescapes [16] proposed force-directed landscape layouts, constructed directly from a merge tree as opposed to the branch decomposition. Other visual representations based on tree drawing or 1D profiles [14, 21] have also been proposed for contour trees and merge trees. In contrast, the nesting behavior is better perceived in a treemap based representation. In addition, the treemap based approach also supports various visual analysis tasks.

Existing techniques that present the contour tree using a landscape metaphor are affected by the limits of human perception and interaction. While a contour tree is well suited to be represented as a terrain, the interactive study of terrains is perceptually difficult. In particular, we believe that the simultaneous exploration of sub-level and super-level sets is difficult using topological landscapes. Therefore, instead of directly representing the contour tree, we study the merge tree, which captures either sub- or super-level set connectivity. Focusing on the merge tree enables the exploration of simple representations that utilize and highlight its hierarchical properties.

1.2 Summary of Results

We present *mergemaps*, a treemap based visual presentation of a merge tree.[1] We construct a mergemap, by processing the branch tree representation of the merge tree to compute an intermediate proxy called the *aggregate tree*, that stores the hierarchy of persistence pairs. We believe that a mergemap simplifies the process of comprehending and interacting with the merge tree. We build upon existing work on treemaps and landscape metaphors and extend it to visualize merge trees. We also provide fusing and diffusing operations to reduce hierarchical clutter and hence improve user experience. We demonstrate the utility of our designs for understanding scientific datasets. Our designs are simple to implement and we hope that they would be adopted by the community.

2 Background

In this section, we will provide a brief background about the two essential ingredients for constructing mergemaps: merge trees and treemaps.

2.1 Merge Tree

Consider a scalar function $f\colon D \to \mathbb{R}$ defined on a simply connected manifold domain D. A value c in the range of f is called an *isovalue*. Given an isovalue, an *isocontour* or *level set* is defined as the collection of all points $x \in D$ such that $f(x) = c$. A merge tree captures the connectivity of sub-level sets $f^{-1}(-\infty, c]$ (*join tree*) or super-level sets $f^{-1}[c, \infty)$ (*split tree*) of f. Both join and split trees are referred to as merge trees. For the sake of convenience, we use only the split tree for explanation. Figure 1 shows the merge trees and contour tree of the height function defined on a simply connected domain.

A split tree is constructed by sweeping the domain in decreasing order of function value. It records the points at which the number of connected components of the super-level set changes. Nodes of a split tree consist of maxima, split saddles, and the global minimum. In practice, a split tree can be conceptualized as a rooted binary tree, in which every interior node has at least two children. The root is defined by

[1] Video illustrating mergemaps at https://youtu.be/xuj9jG4E3lM.

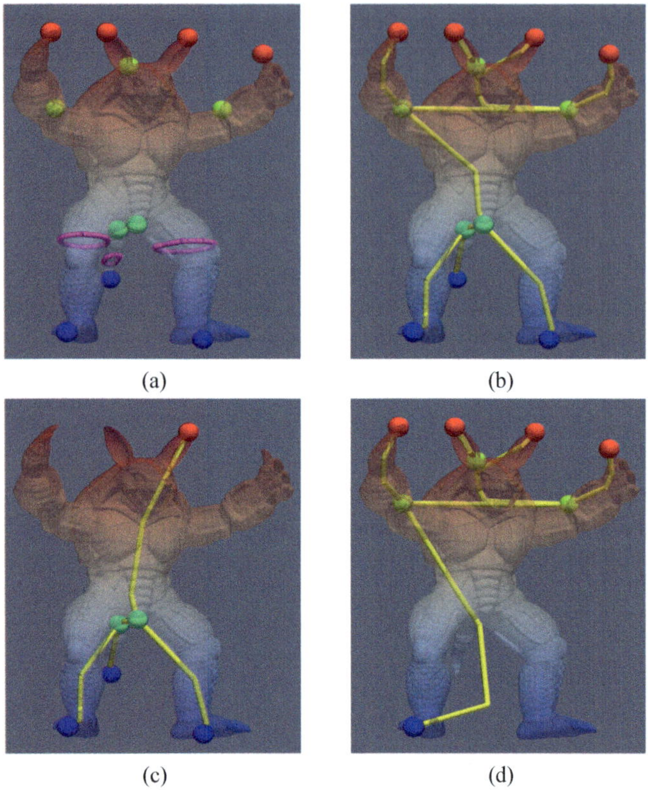

Fig. 1 Height field of Armadillo rendered using a blue-red colormap ( ). **a** critical points of the field and a level set, shown in pink. **b, c, d** show the contour tree, join tree, and split tree respectively

the global minimum, the leaves are maxima, and the interior nodes are saddles. All maxima can be paired with saddles based on the notion of topological persistence, except for the global maximum which is paired with the global minimum. Each such pair represents a topological feature and its *persistence* can be defined as the absolute difference between the two scalar function values.

Similarly, a join tree can be defined using minima, join saddles and the global maximum. The *contour tree* contains the combined information present in a split and a join tree, it captures the evolution of level set connectivity. Collectively, minima, maxima, join saddles and split saddles are called critical points.

A merge tree can be decomposed into a set of branches, such that each branch contains a persistence pair. This generates a nested hierarchy of branches, wherein each parent branch has a persistence greater than that of its children. This hierarchy of branches is called the *branch decomposition* representations of the merge tree.

2.2 Treemap

Trees are generally visualized as node-link diagrams. Such a visualization is inadequate while exploring large datasets, since navigating the structure is difficult and content information is often hidden within the individual nodes [36]. To counter this, Shneiderman [28] proposed treemaps as a technique to display tree structures using a two-dimensional nested space-filling approach similar to that of Venn diagrams. The original algorithm requires dividing the display into nested rectangles, each with an area corresponding to the weight of the associated node.

The utility of a treemap can be understood with an example, see Fig. 2. The file system hierarchy of a computer can be comprehended using a treemap. In this case, the tree to be visualized consists of files and folders. The leaves of the tree represent the files. All interior nodes, including the root, represent the folders. Each file is represented as a box with an area corresponding to its file size. All folders are represented as containers with an area equivalent to the sum of the file sizes of its children. Every node of the input tree serves as a container for its children.

By definition, treemaps take an input of n weights, a hierarchy upon these weights (the tree), and a shape (generally a rectangle). Since these weights correspond to the leaves of the tree, the size of a parent container (present as an interior node of the tree) should be equal to the sum of the sizes of its children.

Several treemap layouts have been presented in the literature – for improved presentation of the tree hierarchy, better display of values associated with nodes, enhanced aspect ratio of rectangles, and multiple other criteria. We refer the reader to the survey by Schulz et al. [25] for a more elaborate discussion on these variants. In this paper, we choose an appropriate existing variant based on the requirement. For our case studies in Sect. 4, we use squarified treemaps [4], zoomable treemaps [2], spatial treemaps [40], and cascaded treemaps [18].

3 Mergemap

A *mergemap* is a treemap based visual representation of a merge tree. In this section, we will describe an algorithm for constructing a mergemap, show how to interact with it, and use its hierarchical properties to improve perception.

3.1 Motivation

Our main goal is to be able to represent a merge tree using a treemap and allow for better perceptual and interactive analysis. However, treemaps cannot directly be used out-of-the-box for representing merge trees. We describe the reason using an example. Figure 2 shows the directory structure as an input tree, visually depicted as

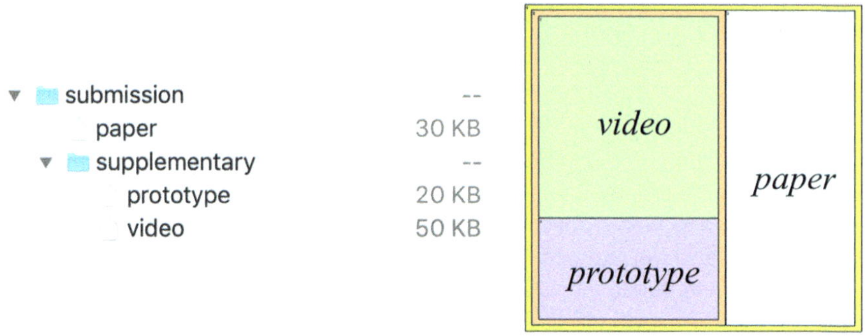

Fig. 2 An example file directory and its treemap visualization

a treemap. Each box in the treemap corresponds to a file, and each container corresponds to a folder. The area of a treemap box is proportional to the corresponding file's size. Note that the folder does not require additional space, therefore the area of the corresponding containers is equal to the sum of the areas of its children. In contrast, the internal nodes of a merge tree (the saddles), always have an associated weight (scalar function value) and cannot be shown in a treemap. Therefore, a merge tree should first be converted into an intermediate structure that preserves the topological abstraction, and whose visual representation is amenable to interaction, perception, and analysis.

3.2 *Algorithm*

There are three basic steps for generating a mergemap. First, we compute the branch decomposition. Second, using the branch decomposition, we construct an aggregate tree to introduce imposter nodes. Third, we visualize this aggregate tree using a treemap. Optionally, the hierarchy of the treemap is reduced using a sequence of fusing/diffusing operations and is spatially ordered. Figure 3 shows a mergemap and the output of the different steps.

We use the algorithm by Pascucci et al. [22], to convert a merge tree into a persistence based branch decomposition. The root branch is referred to as the trunk. The trunk has the global maximum and global minimum at each of its ends. Depending on which merge tree is used, all other branches have a minimum or maximum, and a saddle. Each branch can have an importance value like persistence, hypervolume or volume associated with it. For the purpose of our discussion, we assume that this branch decomposition has only one value associated with each branch, say persistence.

While the branch decomposition is often displayed as a collection of L-shaped branches, it can also be represented as a rooted tree, whose nodes represents individual branches. Such a representation is called a persistence hierarchy [23]. Understanding

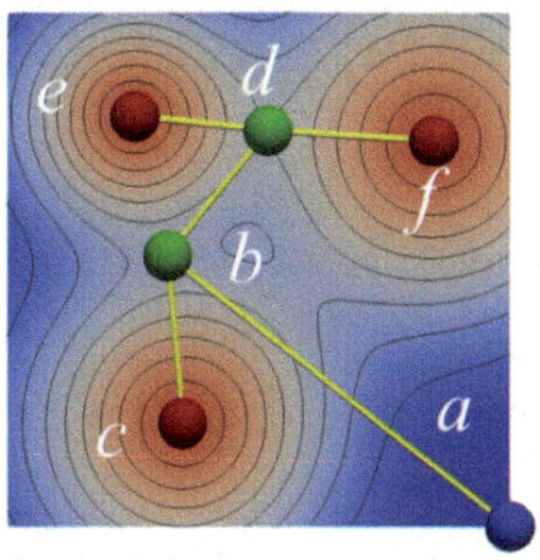

(a) 2D scalar field with Gaussians rendered using a blue-red colormap ( )

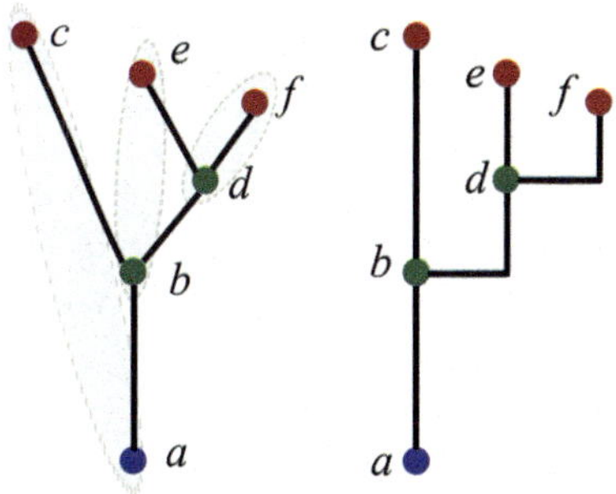

(b) Merge tree and its branch decomposition

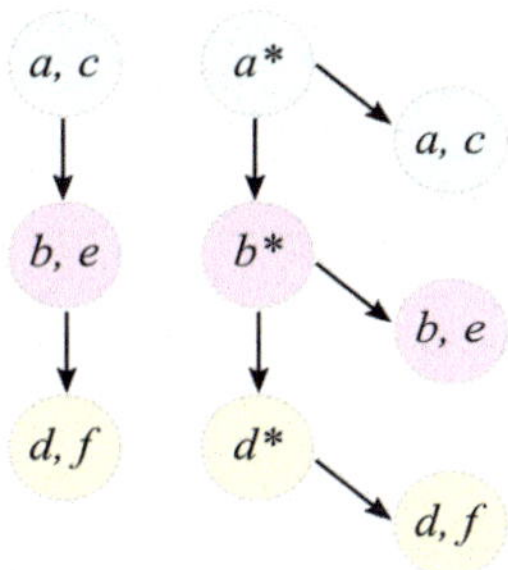

(c) Persistence hierarchy and aggregate Tree

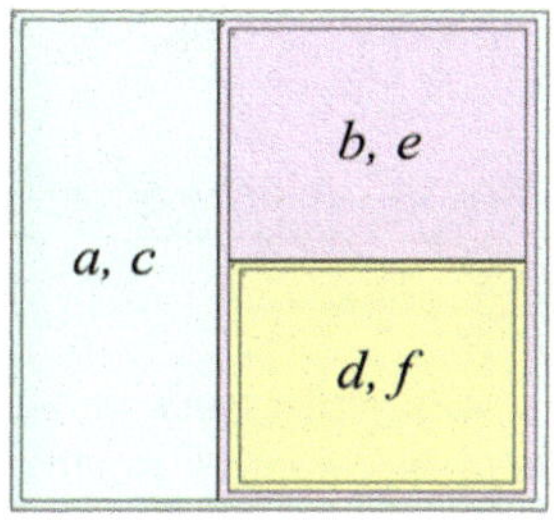

(d) Mergemap

Fig. 3 Constructing a mergemap for a scalar field

the branch decomposition in terms of persistence hierarchy, makes visualization easier.

The persistence hierarchy cannot be directly presented as a treemap. Each node of the persistence hierarchy has a value associated with it, whereas a treemap requires values to be associated only with leaf nodes. We construct a new tree, called the aggregate tree, whose leaves store values corresponding to all nodes of the persistence hierarchy tree. The aggregate tree can be presented as a treemap.

Every node in the persistence hierarchy is duplicated (without edges) and inserted as a child of the same node. The nodes are duplicated during a preorder traversal of the persistence hierarchy. After duplication, every non-leaf node is assigned a value equal to the sum of its children. This tree contains twice the number of nodes as the persistence hierarchy and is suitable for visualization using a treemap. Algorithm 1 shows this procedure.

Let us consider the example shown in Fig. 3 to understand the algorithm. The input scalar field has six critical points, *(a, b, c, d, e, f)*, associated with the split tree. The branch decomposition consists of 3 branches: *(a, c)*, *(b, e)*, *(d, f)*. These

Algorithm 1. Create Aggregate Tree

```
function MakeImposter(node)
    node ← Duplicate(node)
    for child ∈ node.children do
        if child not processed then
            imposter ← MakeImposter(child)
            node.value ← node.value + imposter.value
    Mark(node, processed)
    return node

function Duplicate(node)
    imposter ← Copy(node)
    imposter.value ← node.value
    imposter.parent ← node
    Delete(imposter.children)
    Insert(node.children, imposter)
    Mark(imposter, processed)
    return node

AggregateTree ← MakeImposter(node)
```

pairs have a persistence value and an implicit hierarchy defined upon them. Since the pairs cannot be directly visualized using a treemap, we create an aggregate tree. For each node in the persistence hierarchy, an *imposter* is created, by duplicating all properties of the original node, except for parent-child relationships. This is then inserted back as a child of the original node. The imposter can be considered as a symbolically perturbed saddle with function value lower than its parent.

This aggregate tree satisfies some desirable properties:

(i) the imposter will appear as a leaf node in the aggregate tree,
(ii) the branch will be represented both as an internal node as well as a leaf node in the aggregate tree, and
(iii) the internal node that represents the branch contains a copy that retains its original value.

When a node has imposters for itself and its initial children, it takes up a value equivalent to the sum of all its children. We call such a node as an *aggregate node* and denote it as $saddle^*$. In our example, d^* will be equal to the value of *(d, f)*, b^* will be equal to the sum of *(b, e)* and d^*, and a^* will be equal to the sum of *(a, c)* and b^*. Topologically, an aggregate node has a value equal to the total persistence of all branches beneath it.

Using these imposters and aggregate nodes, we can now show the branch decomposition using a treemap. The leaf nodes, *(a, c), (b, e), (d, f)* are shown as boxes, while their respective parents, a^*, b^* and d^* are shown as containers. Each branch has been represented twice, using a box and a container. The boxes show the exact value represented by the branch, while the containers express the hierarchical relationship amongst the branches.

3.3 *Interaction*

To explore a large dataset, a mergemap allows a user to navigate through multiple interesting features. Interaction with mergemaps is straightforward, and can be done in two ways: focus + context exploration, or using external widgets. In this section, we look at some best practices for interacting with mergemaps.

3.3.1 Focus + Context Exploration

A direct way to use a mergemap is by linking it with subvolumes of the domain [38]. A user can select individual boxes to interactively view the corresponding volumes. Meta-data associated with a branch is visible by hovering over its box in the mergemap and more specific information like branch identifiers and persistence values may be displayed as labels on top of a box. Even though containers and boxes represent the same branch, providing the same interactive capabilities is redundant. Containers can be used for size reduction operations (Sect. 3.4) and boxes for volume selection. A rectangle's area can also be selected for annotation using text or colour.

In order to focus on a single feature and its properties without getting distracted by the rest of the treemap, we use zoomable treemaps [2]. Zoomable treemaps provide capabilities that allow users to navigate up and down the hierarchy of the tree using animated rolling up and drilling down views. Zooming into a container causes a new treemap associated with its branches to be rendered into the original area. Zooming out causes the original context to be restored. A user may use this interaction capability to select the smallest features of the branch decomposition without the need for excessive simplification.

3.3.2 Linked Widgets

While interacting with a large branch decomposition, it can be cumbersome to repeatedly select features, one after another, in a hit-and-trial fashion. Hence, the designs presented in this paper may be used in conjunction with interaction widgets and tools. For instance, 'top/bottom k-persistent features' is a useful query to support. The traditional persistence diagram, where persistence pairs are plotted as vertical bars on the diagonal, is not best suited for selection. An alternative representation, called barcodes, shows the pairs as a sorted bar chart. This representation may be used in conjunction with mergemaps and standard brushing and linking by simply changing colors or shades of the corresponding boxes. Topological spines [6] is also an interesting candidate for a linked widget, since it preserves the underlying geometry of the scalar field.

3.4 Operations

User perception may be affected due to two different reasons. First, the mergemap shows a branch both as a container and a box, even when it has few or no children. This may result in clutter. Second, a branch decomposition often contains large nested hierarchies. To address this difficulty, we propose few operations that reduce the size of the branch decomposition. Our proposed operations are generic in nature. Other operations may be introduced to cater to dataset and application specific requirements. We first describe these operations and then describe how they may be used together with mergemaps.

3.4.1 Reduction Operations

We present three operations: *fuseSaddle*, *diffuseBranch*, and *fuseBranch*. These operations adapted from [31, 37], either fuse different branches into one or spreads branches or flatten a deep hierarchy. Figure 4 shows an illustration for each operation.

(i) ***fuseSaddle:*** If the critical value at the saddle of a branch is close to that of its parent, then both branches may be fused together so that they share the same saddle. This is a saddle stabilization operation.
(ii) ***diffuseBranch:*** If a branch has multiple nested children, the hierarchy below the branch may be flattened by spreading all nested descendant branches as immediate children. This is a hierarchy compression operation.
(iii) ***fuseBranch:*** If the minima of two branches in a join tree are in close proximity in the spatial domain, then both branches may be fused into a single branch. Children of both branches are hierarchically placed into the merged branch. The new branch is pushed up the hierarchy, its saddle value is set to the larger of the two saddles, and the value of the minimum is set to the smaller of the two minima. In a split tree, the saddle value is set to the smaller of the two saddles and the value of the maximum is set to the larger of the two maxima. This is a proximity-based simplification operation.

The branch decomposition represents a spatial containment relationship between parent and child branches. So, it may be possible to develop a method to compute the corresponding simplified scalar field via local changes [33]. For example, *fuseSaddle* may be realized via local changes to scalar values in the neighborhood of the preimage of the merged saddle point. We do not study this problem further since the intent here is to improve user perception.

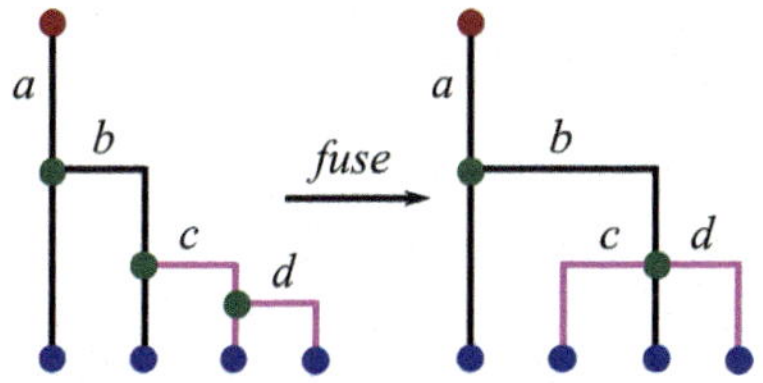

(a) fuseSaddle: Saddle of branch *c* is fused with saddle of branch *d*

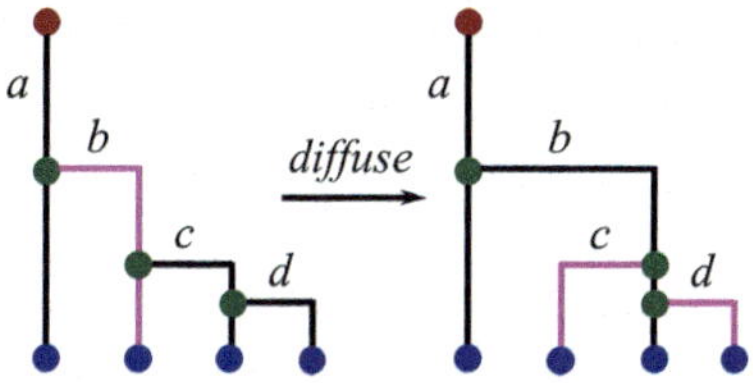

(b) diffuseBranch: Children of branch *b* are diffused to a common parent

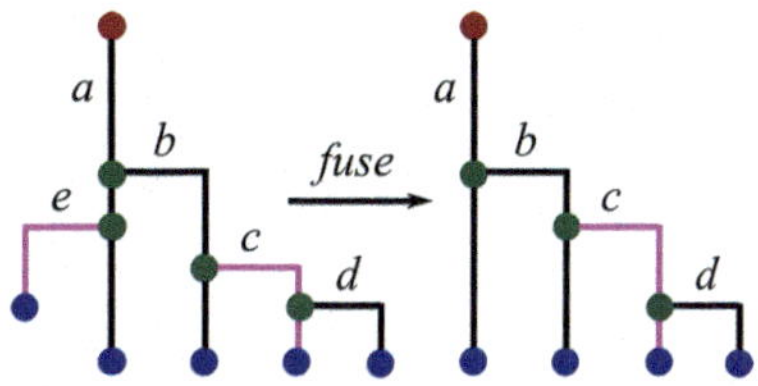

(c) fuseBranch: Branch *e* is fused with branch *c*

Fig. 4 Examples of reduction operations

3.4.2 Reducing Mergemaps

The above operations can be performed in two ways, directed by uniform thresholds or by a human-in-the-loop. We first describe the use of a uniform threshold to direct each operation.

The *fuseSaddle* operation traverses up the hierarchy from the leaves up to the trunk, merging branches whose saddle values are closer than a given saddle value proximity threshold. The *diffuseBranch* operation selects all branches at a chosen depth threshold and diffuses the descendants for each branch. The *fuseBranch* operation identifies groups of branches that are pairwise closer than a given spatial proximity threshold. For each group, a single branch is inserted into the hierarchy replacing all branches in the group.

Rectangles in a mergemap are easy to select. So, it makes sense to ask a user to manually identify the branches to be fused or diffused. To perform a fuse operation, the user selects all containers that are to be merged. Children of the selected branch are placed within a single container. In case of a diffuse operation, the user selects a single container and all descendants are hierarchically compressed.

From our experience, it is best to reduce a mergemap in the following order: perform persistence based simplification of the scalar field [34] and construct a branch decomposition, use uniform thresholds for the fuse/diffuse operations before the mergemap is rendered, finally apply the operations manually based on expert input.

3.5 Area Distortion

The size of containers in the mergemap is not exactly equal to the sum total size of its boxes. Each container has a constant padding that acts as a small cushion. This padding eats into the area of the children. So the area of the boxes in the illustrations is not exactly proportional to the value associated with it, but slightly smaller. The padding helps users perceive the hierarchy of the tree and hence select containers to perform the reduction operations. On a similar note, treemaps cannot directly show non-positive scalar values.

4 Case Studies

We describe four applications to demonstrate the utility of the mergemap. The datasets used in this section are available in the public domain [17, 39]. We use TTK [32] for computing the merge trees and persistence based simplification. Previous designs [12, 37] have also used the same datasets in Sects. 4.2 and 4.3, and therefore we compare mergemaps with their representation for those case studies.

4.1 Ethane-1,2-diol

The Ethane-1,2-diol dataset is a 3D electron density distribution over a small molecule. Higher density regions correspond to atom centers. This is a relatively small dataset, the merge tree contains only 20 critical points. We use a squarified treemap [4] for the layout. This layout creates approximate squares as opposed to elongated rectangles, for easy selection and comparison. After performing a few reduction operations, we get the mergemap shown in Fig. 5. Using brushable persistence barcodes, we assign a common color to similarly sized boxes. The similarly sized boxes directly correspond to similar bonding regions in this dataset. The orange boxes correspond to Carbon-Hydrogen bonds, brown boxes correspond to Oxygen-Hydrogen bonds, and the pink boxes correspond to the Carbon-Carbon bond.

4.2 Fuel

The dataset represents fuel density in a combustion chamber after fuel is injected. Understanding its structure is important for finding better combustion schemes. Past work [31, 37] have shown that the dataset exhibits radial symmetry. We attempt to replicate their results using mergemaps. Figure 6 shows our results. Initially, even after uniformly diffusing the mergemap, there were too many colors and nodes that

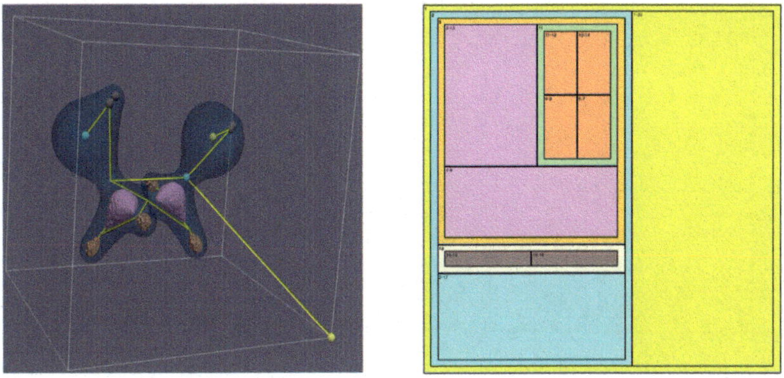

(a) (b)

Fig. 5 Bonds in Ethane-1,2-diol: **a** Isosurfaces and the merge tree. **b** Mergemap. Similar sized boxes are assigned a common color. We observe that this corresponds to similar bonding regions

affected the ability to locate the interesting features. However, since each container represents a topological feature, we were able to quickly locate the turbulent region. In order to remove clutter due to the presence of the other features, we zoomed into this turbulent container. Next, we used brushable persistence bar codes and found several boxes of the same size, which directly corresponded to symmetrical features.

We now compare the interface of mergemaps with that of Denali (proposed by Harvery and Wang [12]) and topological landscapes (proposed by Weber et al. [37]) for this dataset, see Fig. 6 (g)–(i) from [12]. One, mergemaps supports an interactive query-driven approach that helps locate the symmetrical structures. Denali and topological landscapes do not allow for such interaction. Two, Denali presents the (unrooted) contour tree. It considers the landscape corresponding to every possible root edge, and then chooses the best landscape by defining a metric distance between them. As a result, each such landscape can lead to a different interpretation and analysis. For instance, it is perceptually difficult to understand why the turbulent part of the fuel dataset (highlighted in yellow) appears to be below the stable part (highlighted in green) in Denali's landscape. Three, topological landscapes often have large spaces in the terrain, that diminishes simple comparative perceptual tasks.

4.3 Silicium

In the mergemaps that we have studied so far, if two containers are adjacent to each other, it implies that the corresponding nodes have the same depth in the persistence hierarchy tree. This notion of adjacency between containers can also be extended to show spatial relationships amongst them. We use a spatially ordered treemap [40] for this purpose.

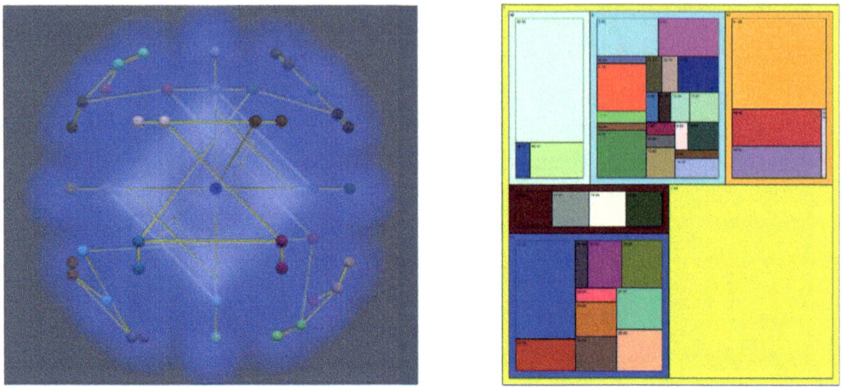

(a) Fuel and its uniformly diffused mergemap. The turbulent feature container is shown in light blue.

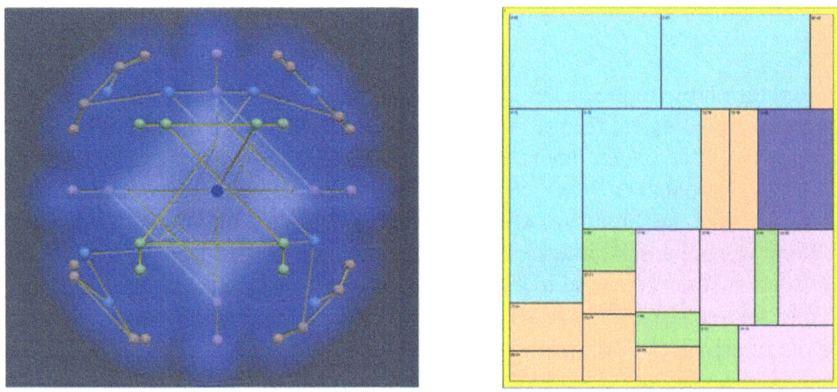

(b) Turbulent feature of fuel and its mergemap after zooming and annotation.

Fig. 6 Volume rendering of the density field in the Fuel dataset using a blue-red color map ( ) and the corresponding mergemaps

We illustrate this using a silicium grid, where the atoms appear as maxima. Such a grid is tightly packed, and the atoms are very close to each other at regular intervals. Scientists are generally interested in understanding impurities in such datasets and the atoms that are affected by them. Understanding spatial relationships is significant for this study. For this dataset, we suppress the hierarchy of the merge tree by fusing all saddles together, resulting in a bush where all maxima are connected to a single saddle.

There is one hurdle in depicting the spatial proximity. The coordinates of the maxima are in 3D and the treemap has a 2D layout. We perform dimensionality reduction on the coordinates of the maxima using Principal Component Analysis (PCA) to project them onto the plane. Using the reduced principal coordinates and

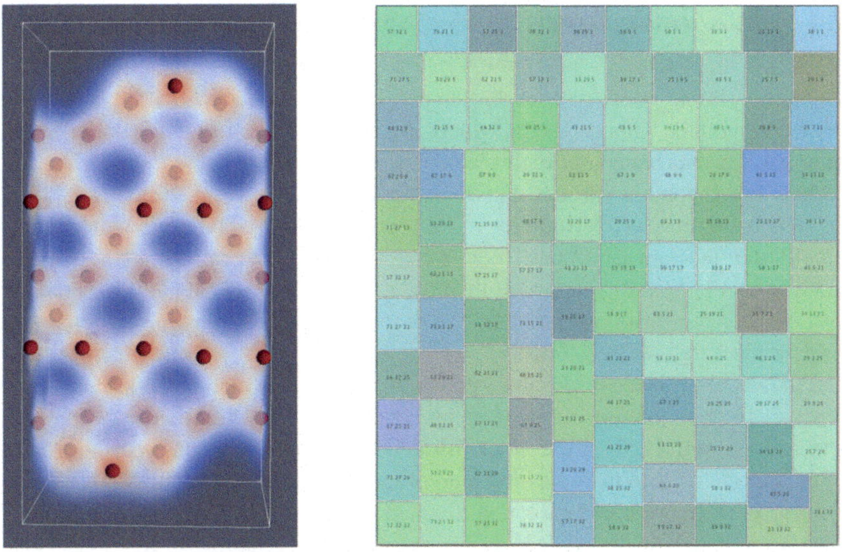

Fig. 7 Volume rendering of the Silicium dataset using a blue-red colormap ( ) and maxima of the density field (*left*). Spatially ordered treemap where the box size indicates persistence (*right*)

their persistence values as input to a spatially ordered treemap, we can visualize spatial relationships amongst the maxima, see Fig. 7.

All the boxes are of the same size. So, we infer that there are no impurities in the dataset. We expect impurities to have a very small or large persistence and hence appear disproportionately in the mergemap. If we identify such a box in a spatially ordered mergemap, we can define a radius and select all atoms that are affected by this impurity. This technique could potentially be used to understand higher dimensional datasets that are projected onto the plane using a topology preserving dimensionality reduction method [41].

We now compare the interface of mergemaps with that of topological landscapes for this dataset, see Fig. 12 from [37]. In both representations, it is difficult to distinguish individual maxima/minima because their sizes are similar. Mergemaps provides an additional guarantee that spatially proximal critical points appear close to each other in the visual representation. Further, we believe that a terrain representation is cumbersome to perceive contour trees, since it is difficult to analyze both maxima and minima at the same time without constantly rotating the 3-D view of the terrain. This is probably why a flipped version of the terrain is used to show minima. Mergemaps avoid this issue by focusing on merge trees.

4.4 von Kármán Street

The von Kármán street dataset is obtained from a simulation of 2D viscous flow behind a cylinder. This dataset is widely used as a demonstration for understanding time-varying data using topological analysis. Maxima of the velocity magnitude field directly correspond to vortices. Previous topological analysis of this dataset [15, 19, 24, 30] have reported periodicity in its vortex shedding. Here, we study a single time step of the data set. In particular, we are interested in answering two questions. (a) What is the spatial structure of the flow and the vortices? (b) How are individual vortices different from one another? We propose a minor variant of mergemaps that helps answer the two questions.

First, we introduce an additional constraint while constructing the branch decomposition from its merge tree that helps capture the spatial structure. A branch is attached as a child only if the index of its saddle end point in the preorder traversal of the merge tree lies in between the preorder indices of the end points of its parent branch. Further, we use a cascaded treemap [18] to display the aggregate tree. By design, cascaded treemaps use layering and offsetting rectangles to convey depth and hence showcase the structure of a tree.

Differences between vortices can be observed by studying maximum-saddle pairs in the merge tree. This requires the function values associated with the maximum and the saddle to be presented in the mergemap. So, instead of inserting a single imposter node to represent a branch, we insert two imposter nodes each one representing the two critical points. The size of a container now represents the sum total of function values of all critical points beneath it *i.e.*, the approximate hypervolume corresponding to the descendant branches. The size of a box represents the scalar function value associated with the critical point. The box of a saddle and its container are assigned the same color. One undesirable outcome of this variant is that a low persistent feature with high function value will be shown as a large container. For example, given a height function defined over a hand, low persistent features near the tip of the fingers, will be shown as large rectangles.

We now answer both questions about the dataset. Figure 8 shows how the mergemap captures the structure of the merge tree. It indicates that the domain can be split into three parts. The clockwise vortices on the top and counter clockwise vortices on the bottom of the vortex street correspond to the left and right parts of the mergemap, respectively. The bottom part of the mergemap corresponds to instabilities behind the cylindrical obstacle. We are able to answer the second question even though the padding results in area distortion. The lower region of the left and right part contains similar large boxes indicating that the vortices near the cylinder are extremely stable. However, as we move up, going deeper into the nested hierarchy, the size of the boxes reduce. This indicates that the vortices away from the cylinder obstacle have considerably lost speed. We also observe that both vortex streams "mirror" each other. To check whether our findings from a single time step are true for other time intervals, we computed and rendered the mergemap for each

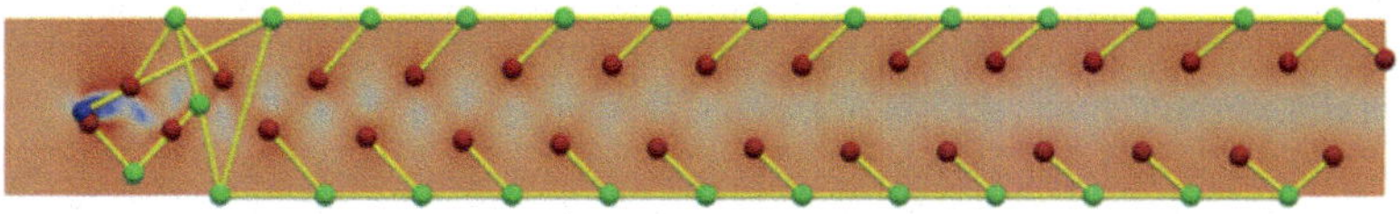

(a) First time step of the von Kármán street rendered using a blue-red colormap ( ) and merge tree of the magnitude field.

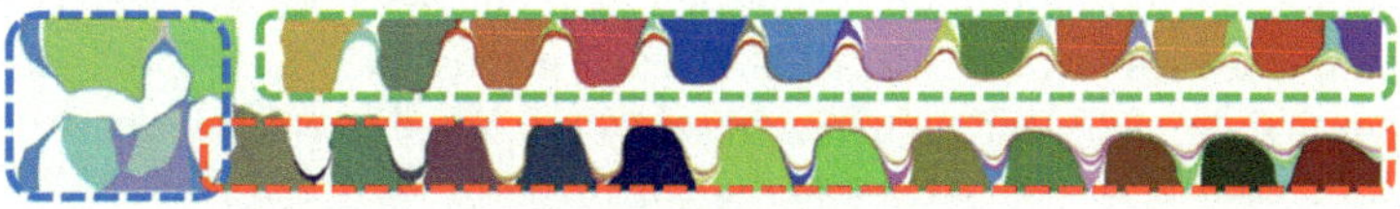

(b) Segmentation using a merge tree

(c) Cascaded mergemap for von Kármán street

Fig. 8 Cascaded mergemap captures the structure of the von Kármán street. Circled boxes in the mergemap correspond to the segments shown in (b) time step. The mergemap for all the time steps contains three parts but several small noisy features appear and disappear behind the cylindrical obstacle.

5 Conclusions

We have presented a treemap based design, that enables improved perception and interaction while exploring merge trees. We also discuss the best practices to interact with and analyze data using such a presentation. We demonstrate their utility on multiple datasets. They are simple to implement and lead to easy interpretations for better topological analysis.

Acknowledgments This work is partially supported by a Swarnajayanti Fellowship from the Department of Science and Technology, India (DST/SJF/ETA-02/2015-16), a Mindtree Chair research grant, and the Robert Bosch Centre for Cyber Physical Systems, Indian Institute of Science, Bangalore.

References

1. Beketayev, K., Weber, G.H., Morozov, D., Abzhanov, A., Hamann, B.: Geometry-preserving topological landscapes. In: Proceedings of the Workshop at SIGGRAPH Asia, WASA 2012, pp. 155–160. ACM, New York (2012)
2. Blanch, R., Lecolinet, E.: Browsing zoomable Treemaps: Structure-aware multi-scale navigation techniques. IEEE Trans. Visual Comput. Graphics **13**(6), 1248–1253 (2007)
3. Bock, A., Doraiswamy, H., Summers, A., Silva, C.T.: Topoangler: interactive topology-based extraction of fishes. IEEE Trans. Vis. Comput. Graph. **24**(1), 812–821 (2018)
4. Bruls, M., Huizing, K., van Wijk, J.: Squarified treemaps. In: Proceedings of the Joint Eurographics and IEEE TCVG Symposium on Visualization, pp. 33–42. Press (1999)
5. Carr, H., Snoeyink, J., van de Panne, M.: Simplifying flexible isosurfaces using local geometric measures. IEEE Vis. **2004**, 497–504 (2004)
6. Correa, C., Lindstrom, P., Bremer, P.-T.: Topological spines: a structure-preserving visual representation of scalar fields. IEEE Trans. Visual Comput. Graphics **17**(12), 1842–1851 (2011)
7. Demir, D., Beketayev, K., Weber, G.H., Bremer, P.-T., Pascucci, V., Hamann, B.: Topology exploration with hierarchical landscapes. In: Proceedings of the Workshop at SIGGRAPH Asia, WASA 2012, pp. 147–154. ACM, New York (2012)
8. Doraiswamy, H., Ferreira, N., Damoulas, T., Freire, J., Silva, C.T.: Using topological analysis to support event-guided exploration in urban data. IEEE Trans. Vis. Comput. Graph. **20**(12), 2634–2643 (2014)
9. Doraiswamy, H., Natarajan, V.: Computing reeb graphs as a union of contour trees. IEEE Trans. Vis. Comput. Graph. **19**(2), 249–262 (2013)
10. Edelsbrunner, H., Harer, J.L.: Computational Topology. An Introduction. American Mathematical Society, Providence (2010)
11. Gyulassy, A., Kotava, N., Kim, M., Hansen, C.D., Hagen, H., Pascucci, V.: Direct feature visualization using Morse-smale complexes. IEEE Trans. Visual Comput. Graphics **18**(9), 1549–1562 (2012)
12. Harvey, W., Wang, Y.: Topological landscape ensembles for visualization of scalar-valued functions. Comput. Graph. Forum **29**(3), 993–1002 (2010)

13. Heine, C., et al.: A survey of topology-based methods in visualization. Comput. Graph. Forum **35**(3), 643–667 (2016)
14. Heine, C., Schneider, D., Carr, H., Scheuermann, G.: Drawing contour trees in the plane. IEEE Trans. Visual Comput. Graphics **17**(11), 1599–1611 (2011)
15. Kamakshidasan, A., Natarajan, V.: Topological analysis of the 2D von kármán street. IEEE VIS Poster (2019)
16. Kamakshidasan, A., Natarajan, V.: Understanding merge trees with force-directed landscapes. EEE VIS Poster (2019)
17. Klacansky, P.: Open scivis datasets, April 2019. https://klacansky.com/open-scivis-datasets/
18. Lü, H., Fogarty, J.: Cascaded treemaps: examining the visibility and stability of structure in treemaps. In: Proceedings of Graphics Interface (2008)
19. Narayanan, V., Thomas, D.M., Natarajan., V.:Distance between extremum graphs. In: 2015 IEEE Pacific Visualization Symposium (PacificVis), pp. 263–270, April 2015
20. Oesterling, P., Heine, C., Janicke, H., Scheuermann, G., Heyer, G.: Visualization of high-dimensional point clouds using their density distribution's topology. IEEE Trans. Visual Comput. Graphics **17**(11), 1547–1559 (2011)
21. Oesterling, P., Heine, C., Weber, G.H., Scheuermann, G.: Visualizing nd point clouds as topological landscape profiles to guide local data analysis. IEEE Trans. Visual Comput. Graphics **19**(3), 514–526 (2013)
22. Pascucci, V., Cole-McLaughlin, K., Scorzelli, G.: Multi-resolution computation and presentation of contour trees. In: Proceedings of the IASTED Conference on Visualization, Imaging, and Image Processing (2005)
23. Rieck, B., Leitte, H., Sadlo, F.: Hierarchies and ranks for persistence pairs. In: Workshop on Topology-Based Methods in Visualization (TopoInVis), February 2017
24. Saikia, H., Seidel, H.-P., Weinkauf, T.: Extended branch decomposition graphs: structural comparison of scalar data. Comput. Graph. Forum **33**(3), 41–50 (2014)
25. Schulz, H.-J., Hadlak, S., Schumann, H.: The design space of implicit hierarchy visualization: a survey. IEEE Trans. Visual Comput. Graphics **17**(4), 393–411 (2011)
26. Shivashankar, N., Natarajan, V.: Parallel computation of 3d Morse-Smale complexes. Comput. Graph. Forum **31**(3), 965–974 (2012)
27. Shivashankar, N., Pranav, P., Natarajan, V., van de Weygaert, R., Bos, E.G.P., Rieder, S.: Felix: a topology based framework for visual exploration of cosmic filaments. IEEE Trans. Visual Comput. Graphics **22**(6), 1745–1759 (2016)
28. Shneiderman, B.: Tree visualization with tree-maps: 2-d space-filling approach. ACM Trans. Graph. **11**(1), 92–99 (1992)
29. Sridharamurthy, R., Masood, T.B., Doraiswamy, H., Patel, S., Varadarajan, R., Natarajan, V.: Extraction of robust voids and pockets in proteins. In: Linsen, L., Hamann, B., Hege, H. (eds.) Visualization in Medicine and Life Sciences III, Towards Making an Impact. Mathematics and Visualization, pp. 329–349. Springer (2016)
30. Sridharamurthy, R., Masood, T.B., Kamakshidasan, A., Natarajan, V.: Edit distance between merge trees. IEEE Trans. Visual Comput. Graphics **26**(3), 1518–1531 (2020)
31. Thomas, D.M., Natarajan, V.: Symmetry in scalar field topology. IEEE Trans. Visual Comput. Graphics **17**(12), 2035–2044 (2011)
32. Tierny, J., Favelier, G., Levine, J.A., Gueunet, C., Michaux, M.: The topology toolkit. IEEE Trans. Vis. Comput. Graph. (2017). https://topology-tool-kit.github.io/
33. Tierny, J., Pascucci, V.: Generalized topological simplification of scalar fields on surfaces. IEEE Trans. Visual Comput. Graphics **18**(12), 2005–2013 (2012)
34. Tierny, J., Pascucci, V.: Generalized topological simplification of scalar fields on surfaces. IEEE Trans. Visual Comput. Graphics **18**(12), 2005–2013 (2012)
35. Valsangkar, A.A., Monteiro, J.M., Narayanan, V., Hotz, I., Natarajan, V.: An exploratory framework for cyclone identification and tracking. IEEE Trans. Vis. Comput. Graph. **25**(3), 1460–1473 (2019)
36. Vicente, K.J., Hayes, B.C., Williges, R.C.: Assaying and isolating individual differences in searching a hierarchical file system. Hum. Factors **29**(3), 349–359 (1987). PMID: 3623569

37. Weber, G., Bremer, P., Pascucci, V.: Topological landscapes: a terrain metaphor for scientific data. IEEE Trans. Visual Comput. Graphics **13**(6), 1416–1423 (2007)
38. Weber, G.H., Dillard, S.E., Carr, H., Pascucci, V., Hamann, B.: Topology-controlled volume rendering. IEEE Trans. Visual Comput. Graphics **13**(2), 330–341 (2007)
39. Weinkauf, T., Theisel, H.: Streak lines as tangent curves of a derived vector field. IEEE Trans. Vis. Comput. Graph. **16**(6):1225–1234 (2010)
40. Wood, J., Dykes, J.: Spatially ordered treemaps. IEEE Trans. Visual Comput. Graphics **14**(6), 1348–1355 (2008)
41. Yan, L., Zhao, Y., Rosen, P., Scheidegger, C., Wang, B.: Homology-preserving dimensionality reduction via manifold landmarking and tearing. CoRR, abs/1806.08460 (2018)

Notes on Percolation Analysis of Sampled Scalar Fields

Wiebke Köpp, Anke Friederici, Marco Atzori, Ricardo Vinuesa, Philipp Schlatter, and Tino Weinkauf

Abstract Percolation analysis is used to explore the connectivity of randomly connected infinite graphs. In the finite case, a closely related percolation function captures the relative volume of the largest connected component in a scalar field's superlevel set. While prior work has shown that random scalar fields with little spatial correlation yield a sharp transition in this function, little is known about its behavior on real data. In this work, we explore how different characteristics of a scalar field—such as its histogram or degree of structure—influence the shape of the percolation function. We estimate the critical value and transition width of the percolation function, and propose a corresponding normalization scheme that relates these values to known

W. Köpp and A. Friederici contributed equally to this work.

Electronic supplementary material The online version of this chapter (https://doi.org/10.1007/978-3-030-83500-2_3) contains supplementary material, which is available to authorized users.

W. Köpp (✉) · A. Friederici · T. Weinkauf
Division of Computational Science and Technology, KTH Royal Institute of Technology, Stockholm, Sweden
e-mail: wiebkek@kth.se

A. Friederici
e-mail: ankef@kth.se

T. Weinkauf
e-mail: weinkauf@kth.se

M. Atzori · R. Vinuesa · P. Schlatter
Linné FLOW Centre, KTH Mechanics, Royal Institute of Technology, Stockholm, Sweden
e-mail: atzori@mech.kth.se

R. Vinuesa
e-mail: rvinuesa@mech.kth.se

P. Schlatter
e-mail: pschlatt@mech.kth.se

I. Hotz et al. (eds.), *Topological Methods in Data Analysis and Visualization VI*, Mathematics and Visualization, https://doi.org/10.1007/978-3-030-83500-2_3

results on infinite graphs. In our experiments, we find that percolation analysis can be used to analyze the degree of structure in Gaussian random fields. On a simulated turbulent duct flow data set we observe that the critical values are stable and consistent across time. Our normalization scheme indeed aids comparison between data sets and relation to infinite graphs.

1 Introduction

Percolation theory has been initiated in 1957 by Broadbent and Hammersley [3]. It is widely used today to characterize and model complex random systems by studying random connectivity in a graph or lattice using statistics. Examples can be found in many domains such as fluid dynamics, cosmology, geology, material science, epidemiology, and others [15].

Percolation is largely a theoretical tool to understand the topology of infinite graphs. One of its central components is the *percolation probability*, which describes the likelihood of the existence of an infinite connected subgraph under certain threshold criteria on the vertices or edges of the original graph. Only few previous works apply percolation theory to real data such as fluid flow simulations [11].

This paper explores the computational aspects and potential pitfalls when computing and analyzing a *percolation function*, closely related to the percolation probability, for real data defined on finite lattices. We give the following contributions:

- We discuss the consequences of computing percolation functions on sampled data. This includes how the percolation function is influenced by the dimensions of the grid. We also describe a normalization to the input data that is crucial for comparing percolation functions between different data sets and to the theory (Sect. 3).
- We propose a method for analyzing the percolation function and its features, combined with a comprehensive visualization for parameter- and time-dependent data sets (Sect. 3.2).
- We research the sensitivity of the percolation function to the amount of structure in data by designing a family of Gaussian random fields with varying degree of structure (Sect. 5.1).
- Finally, we apply our framework to fully developed turbulent flows (Sect. 5.2).

2 Related Work and Background

Consider an infinite graph $\mathbf{L}$. We define for each of its vertices to be *open* with probability $0 \leq p \leq 1$, and *closed* otherwise. Based on this, we define the *open* subgraph $\mathbf{L}'$ using the *open* vertices and their adjacent edges only. Percolation theory studies the structure of $\mathbf{L}'$ depending on the value of p. To do so, we observe the connected components of $\mathbf{L}'$: for small values of p, this subgraph consists of many small connected

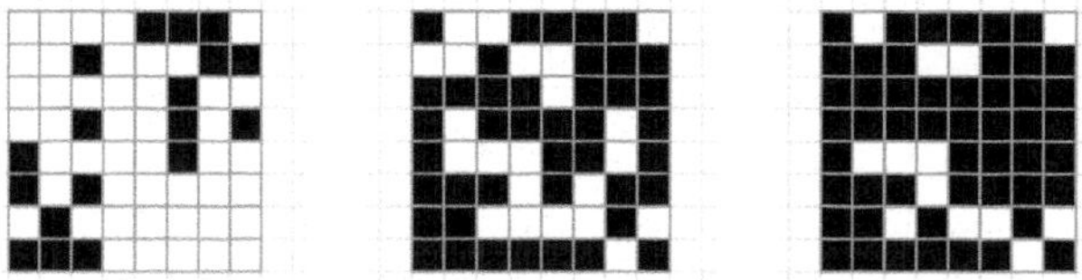

Fig. 1 Examples for site percolation on a 2D lattice with 4-connectivity for $p = 0.2, 0.6$ and 0.8. The percolating cluster of open sites (black) forms approximately at p_c^{2D} (middle)

components which are finite in size (remember that $\mathbf{L}$ and $\mathbf{L}'$ themselves are infinite). For a *critical value* p_c, large-scale structures form and the open subgraph contains an infinite connected component pervading the entire domain: the *percolating cluster*, see Fig. 1. Intriguingly, the transition from finite components to an infinite connected component is sharp: for $p < p_c$, the open subgraph $\mathbf{L}'$ contains finite connected components only, whereas the picture immediately changes for $p > p_c$, for which we see an infinite percolating cluster in $\mathbf{L}'$. The critical value p_c is often referred to as *percolation threshold*. In random media such as porous rocks, the percolation threshold denotes the point where global physical properties of the medium change qualitatively. For example, a porous rock is impermeable before p_c, but lets liquids through after p_c.

The percolation threshold p_c depends *solely* on the connectivity in the infinite graph $\mathbf{L}$. Different topologies have been researched in the mathematics community [21] such as 2D and 3D uniform lattices, triangle meshes, or bow-tie lattices. One also distinguishes between *site* and *bond* percolation, which refers to considering the vertices or edges of the $\mathbf{L}$ as open/closed, respectively. We are concerned with site percolation in this paper, but our work transfers to bond percolation straightforwardly.

The 2D lattice with 4-connectivity (infinite structured grid in 2D) has a site percolation threshold $p_c \approx 0.5927$ [5]. The 3D lattice with 6-connectivity (infinite structured grid in 3D) has a site percolation threshold $p_c \approx 0.3116$ [5]. These values are defined by considering the open subgraph and have been estimated through simulations on finite grids, see Fig. 2a.

Level-set percolation [1, 14], which we are concerned with here, applies percolation theory to real data by considering (seemingly) random scalar data values at the vertices of a finite lattice. In this setup, we are looking at the superlevel set[1] of the scalar field $f(\mathbf{x})$ defined as the set of voxels fulfilling $f(\mathbf{x}) \geq p$. The superlevel set can be equated with the open subgraph $\mathbf{L}'$ from before. Again, we are interested in the threshold value p_c where the connected components of the superlevel set pervade the entire domain of the scalar field.

To determine the existence of the percolating cluster and the percolation threshold p_c in this scenario, we have to define a *percolation function*. Similar to Moisy and Jiménez [11], we choose a function based on the volumes of the connected components:

[1] Level-set percolation traditionally refers to the study of the superlevel set, but all the analysis steps presented in this paper can be applied just as well to the sublevel set.

$$P_{\max}(p) = \frac{V_{\max}}{V_{\text{total}}} \tag{1}$$

where V_{total} denotes the total volume of the superlevel set for a given threshold p, and $V_{\max}$ is the volume of its largest connected component. Figure 2b plots $P_{\max}(p)$ for 2D and 3D random noise data sets.

The percolation function can be computed efficiently using an iterative Union-Find algorithm whose non-iterative version has first been suggested in the context of percolation by Hoshen and Kopelman [8]. The algorithm uses similar ingredients as a merge tree computation [4], has been adapted to work in a distributed setting for large-scale simulation data [7], and works as follows: We traverse all sample points $\mathbf{x}$ in decreasing order of their value $f(\mathbf{x})$. We set $p = f(\mathbf{x})$. Let $\mathcal{N}$ denote the set of components assigned to the neighbors of $\mathbf{x}$. We distinguish three cases:

- $\mathcal{N} = \emptyset$: We *create* a new component containing only $\mathbf{x}$.
- $|\mathcal{N}| = 1$: We *extend* the single component $\mathcal{N}$ with $\mathbf{x}$.
- $|\mathcal{N}| > 1$: We *merge* all of the components in $\mathcal{N}$ and add $\mathbf{x}$ to the result.

V_{total} and $V_{\max}$ are recorded in equidistant intervals in order to graph $P_{\max}(p)$. We will show later in Sect. 3 how to normalize p with respect to the histogram of the data, which serves several purposes around the comparison of percolation functions. The computation of $P_{\max}$ essentially gathers statistics on the connected components of the superlevel set. Alternative statistics such as the number of components are not as expressive, as they are rather featureless. For further exploration, see [7].

It is of interest to automatically analyze the percolation function in order to obtain the critical value p_c and other characteristics such as the width Δ of the percolation transition. Stauffer and Aharony [16] suggest to set p_c simply where a percolation function ranging between 0 and 1 first assumes value 0.5. Similarly, they propose to define the transition's width Δ as the interval where the function ranges between either 0.1 and 0.9, or 0.2 and 0.8, both of which they found to yield empirically suitable estimations. Ziff [20] proposed more involved estimation methods. In contrast to our approach, the goal behind these estimations is to derive p_c for an infinite lattice from a number of simulations on finite lattices. We estimate p_c and Δ by fitting a suitable function parametrized with these values to the percolation function, see Sect. 3.2.

It is the purpose of this paper to provide methods and guidance for the computation, analysis, and visualization of $P_{\max}$, and to discuss the shape of this curve under varying conditions.

3 From Infinite to Finite

Percolation theory was conceived in the context of noise functions defined on infinite domains. When applying it to measured or simulated data, we have finite domains and not necessarily random data. These aspects affect the percolation function and some

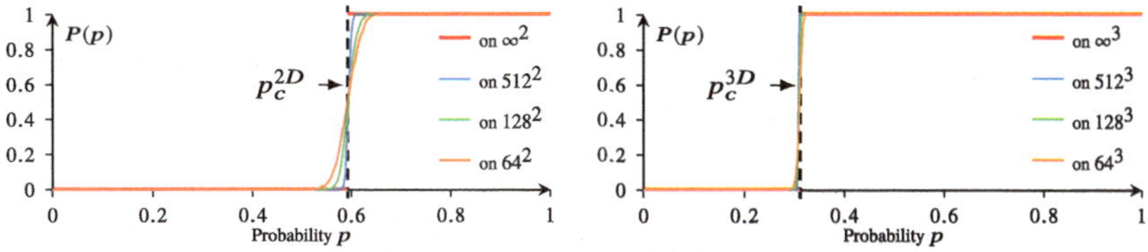

(a) Percolation probability $P(p)$ for 2D and 3D structured grids of different sizes. All values are estimated with 1000 random uniform noise samples of the respective size.

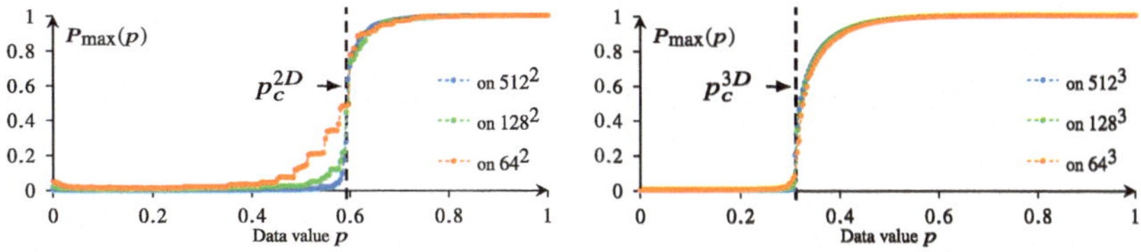

(b) Percolation function $P_{\max}$ for a single sample per 2D and 3D structured grid size.

Fig. 2 Percolation probability $P(p)$ and percolation function $P_{\max}$ for uniform random noise in 2D and 3D structured grids. The dashed lines denoted with p_c^{2D} and p_c^{3D} mark the theoretical critical values on infinite lattices. We can see that $P_{\max}$ features a sharp transition around these values even for finite examples

care has to be taken when computing it. As described by Newman and Ziff [12], an approximation for the percolation threshold p_c on an infinite lattice can be obtained through determining the value at which the percolating cluster forms when sampling uniform random values on the vertices of a sufficiently large finite lattice. Repeating this procedure a large number of times and averaging over the function that is 1 where the percolating cluster exists and 0 elsewhere yields an estimation of the percolation probability $P(p)$, as shown in Fig. 2a.

Based on these graphs, we can conclude that the size of the domain affects the transition width. The location of the percolation threshold on the other hand depends on the grid's dimensionality. Similar characteristics can be observed in Fig. 2b, where the percolation function for a single sample per grid size is shown. This is analogous to analyzing simulated or measured scalar data, as only a fixed set of data values is available then.

Note that in structured data, it is possible that the data intrinsically has different dimensionality than the underlying lattice. Consider, for example, a data set where the same 2D slice repeats over the third dimension. For uniform random data, we would expect p_c to be close to the theoretical two-dimensional value in this case.

As discussed next, the shape of the percolation function is further affected by value distribution and structure in the data.

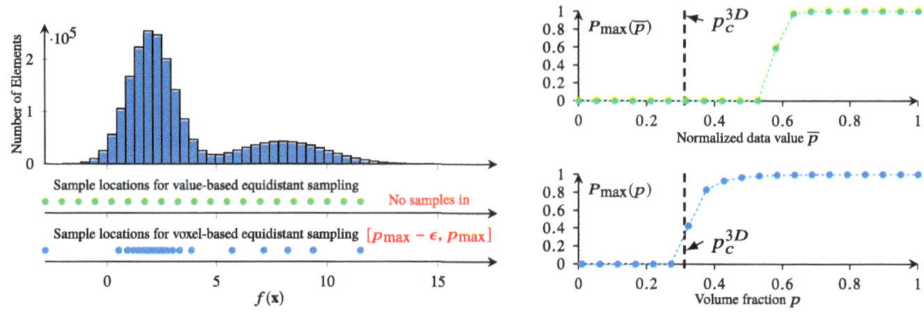

(a) Histogram of the bimodal normal distribution $f(\mathbf{x})$ and corresponding sample locations.

(b) Value-based and voxel-based percolation functions for $f(\mathbf{x})$.

Fig. 3 Histogram, sampling schemes and percolation function for a scalar field $f(\mathbf{x})$ sampled randomly from a mixture of two Gaussians on a grid of size 128^3. Both value-based and voxel-based sampling yield a sharp transition in the percolation function. However, in case of value-based sampling, the transition is only sharp due to a large fraction of all voxels being processed around the transition. It is also not located near the theoretical threshold p_c^{3D}. This hinders the comparison between data sets and to known results on infinite graphs. Comparability can be achieved by computing the percolation function over the percentage of voxels in the superlevel set

3.1 *The Extremes of the Value Range*

We compute $P_{\max}(p)$ in the interval $[p_{\min}, p_{\max}]$ which corresponds to the data value range in the scalar field $f(\mathbf{x})$. Only few voxels will be part of the superlevel set around $p_{\max}$. The volume computations for Eq. (1) can be rather erratic in this range, since they depend on the connectivity between a rather small number of data samples. For example, consider the very first voxel in our algorithm (global maximum): it is easy to see that $P_{\max}(p_{\max}) = 1$ in this case, since this single voxel is the only connected component in the superlevel set. Yet, the existence of an infinite cluster in an infinite domain would have probability 0.

Reliable values are obtained only after having iterated over a sufficiently large number of data points. We thus only start recording the percolation function after having reached a value $p_{\max} - \epsilon$. In this work, we set ϵ such $[p_{\max} - \epsilon, p_{\max}]$ amounts to the highest 1% of all data points.

3.2 *Histogram Distribution*

Percolation thresholds in infinite domains are known for different random distributions such as uniform and Gaussian noise. See for example Fig. 2. However, a measured or simulated scalar field will feature an arbitrary value distribution. This leads to different percolation functions shapes. Is it possible to align these cases such that we can utilize the theoretical knowledge? For example, it would be interesting to compare percolation thresholds as a way to judge the amount of randomness in data.

To this end, we propose a simple normalization scheme. Remember from Sect. 2 that the parameter p refers to the percentage of open sites/vertices in a lattice. This relates directly to the number of vertices in the superlevel set while computing the percolation function. Hence, instead of setting $p = f(\mathbf{x})$ as it is done in previous work such as by Moisy and Jiminéz [11], we set p in relation to the number of voxels in the superlevel set, which also corresponds to V_{total}. More precisely, from now on p will denote the percentage of voxels in the superlevel set. Computation-wise, this corresponds to the position of a given data value in the sorted data list. Meanwhile, the symbol $\overline{p}$ will refer to the scalar data value $f(\mathbf{x})$. We also normalize this value to the range [0, 1] to ease comparisons between different percolation functions. Essentially, this procedure shifts the data values such that the histogram matches a uniform distribution. This allows us to relate the level of structure in a given data set to uniform random noise via their percolation functions. An example for how this affects sampling locations and percolation function is given in Fig. 3.

4 Analysis and Visualization of Percolation Curve Ensembles

4.1 Analysis of a Single Percolation Curve

We are interested in analyzing the percolation function in Eq. (1) to determine the critical value p_c and the width Δ of the percolation transition.

The percolation function in purely random data follows an S-shape. Suitable candidates for approximating curves of this shape come from the family of sigmoidal functions, which include the logistic function, the hyperbolic tangent, and the error function. Due to its prevalence in the analysis of percolation curves from Monte-Carlo simulations [13, 19], we use an adapted version of the *error function*, defined as

$$\operatorname{erf}(x) = \frac{2}{\sqrt{\pi}} \int_0^x e^{-t^2} \mathrm{d}t. \tag{2}$$

Note that the error function is related to the normal cumulative distribution function for mean μ and variance σ^2

$$\Phi(x, \mu, \sigma) = \frac{1}{2}\left(1 + \operatorname{erf}\left(\frac{x - \mu}{\sigma\sqrt{2}}\right)\right). \tag{3}$$

As such, it ranges between –1 and 1, is monotonically increasing, symmetric with respect to the y-axis and has its point of maximal slope at $x = 0$. By inserting our two parameters p_c and Δ we get:

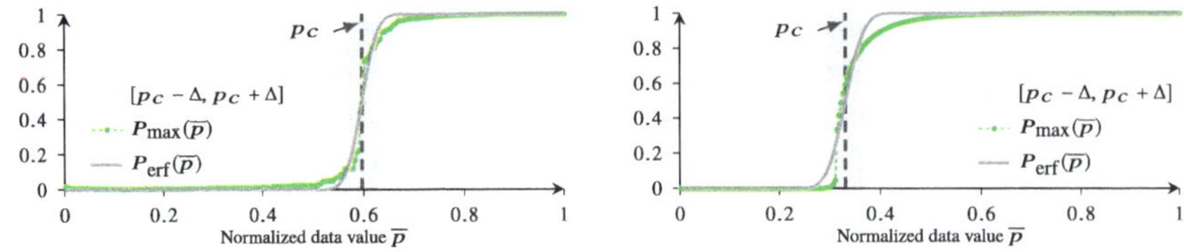

(a) 2D structured grid, 128^2 sites. (b) 3D structured grid, 128^3 sites.

Fig. 4 A fit of function P_{erf} (gray) to the percolation functions $P_{\max}$ (green) for uniform random noise in 2D and 3D structured grids estimates percolation threshold p_c and transition width Δ

$$P_{\text{erf}}(p, p_c, \Delta) = \frac{1}{2}\left(1 + \text{erf}\left(\frac{p - p_c}{\Delta}\right)\right) \quad \text{with } \Delta > 0. \tag{4}$$

By design, P_{erf} has an inflection point and maximal absolute slope at p_c.

Furthermore, the values obtained at p_c, $p_c - \Delta$ and $p_c + \Delta$ are close to the ones suggested by Stauffer and Aharony [16] for the analysis of percolation functions:

$$P_{\text{erf}}(p_c, p_c, \Delta) = \frac{1}{2}(1 + \text{erf}(0)) = \frac{1}{2} \tag{5}$$

$$P_{\text{erf}}(p_c - \Delta, p_c, \Delta) = \frac{1}{2}\left(1 + \text{erf}\left(\frac{-\Delta}{\Delta}\right)\right) \approx 0.07865 \tag{6}$$

$$P_{\text{erf}}(p_c + \Delta, p_c, \Delta) = \frac{1}{2}\left(1 + \text{erf}\left(\frac{\Delta}{\Delta}\right)\right) \approx 0.92135 \tag{7}$$

In order to fit the function P_{erf}, we use a non-linear least squares fitting algorithm available in SciPy [9]. We obtain an initial guess for p_c and Δ by estimating p_c as the point of maximal slope for a fitted polynomial and Δ as the half of the interval where that polynomial ranges between 0.1 and 0.9. While a polynomial of any degree is not able to capture the asymptotic behavior of the percolation function, estimates based on polynomials serve well as initialization to the actual curve fitting. Figure 4 shows the fitted function P_{erf} for the 2D and 3D examples from Fig. 2. In both cases, the fitted curve nicely resembles the shape of the sampled percolation function. Note that for the 3D example, the estimated $p_c = 0.33533$ does not quite reach $p_c^{3D} \approx 0.3116$. We attribute this to the approximate nature of the percolation function in Eq. (1). To confirm this, we conducted another experiment in this data set testing for each threshold whether there exists an actual percolating cluster defined as a connected component of the superlevel set spanning the entire domain in any dimension. Indeed, we find the percolating cluster at $p = 0.311294$, which is much closer to the theoretical $p_c^{3D} \approx 0.3116$. We further observe that the percolation function is more asymmetric in the 3D case. While the current estimation scheme cannot capture the asymmetric nature of the curve, it yields sufficiently indicative values. An investigation of alternative estimation schemes is left for future work.

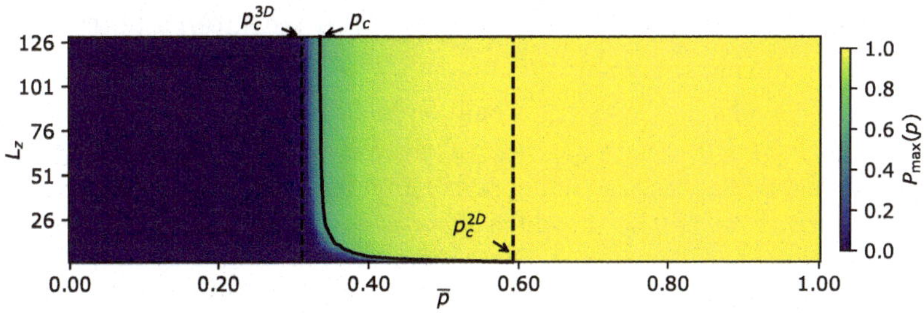

Fig. 5 Percolation function for random uniform noise of dimension $128 \times 128 \times L_z$. The estimated value p_c for $L_z = 1$, i.e. the 2D case, is close to the site percolation threshold for the infinite square lattice $p_c^{2D} \approx 0.5927$. As we increase the dimensionality, p_c moves quickly towards the threshold for the infinite cubic lattice $p_c^{3D} \approx 0.3116$

4.2 Analysis of Percolation Curve Ensembles

For the analysis of multiple percolation curves for a data set varying over a parameter t, such as time or dimension, the procedure in Sect. 4.1 is simply repeated for every sample of t. To then easily assess the development of p_c and Δ over parameter t while still being able to see the connection to the sampled curves, we visualize all three together in a heatmap. The heatmap consists of texels, one for each sample (p, t) with its color encoding the function value $P_{\max}(p)$. The x-axis of the heatmap corresponds to threshold p, and the y-axis varies over parameter t. Figure 5 shows an example: Random uniform noise is sampled on a structured grid of size $128 \times 128 \times L_z$. The row of colored squares for $L_z = 1$ at the very bottom of the plot corresponds to the curve in Fig. 4a, whereas the top row with $L_z = 128$ is shown in Fig. 4b. The estimated values for p_c and Δ are graphed on top of the heatmap.

5 Experiments

In the following, we discuss the shape of the percolation curve under varying conditions. We examine a family of synthetic data sets with varying degree of randomness in Sect. 5.1 to understand how the interplay of structure and randomness caries over to the shape of the percolation function. A simulated flow data set is analyzed in Sect. 5.2, where we showcase the utility of percolation analysis and observe the effects of our algorithmic choices such as histogram normalization.

5.1 Randomness and Structure: Gaussian Random Fields

The original percolation theory [3] is built on homogeneous and isotropic randomness without correlation between data values. We want to explore the impact of structure in a data set on the resulting percolation function. To this end, we employ *Gaussian random fields* (GRFs). which allow us to construct a family of fields with varying degrees of randomness. A GRF is a stochastic process defined by a mean and a covariance function and can be understood as a probability distribution over functions, just like its one-dimensional version, the *Gaussian process*. The unique property of a GRF is that the values in every finite subset of sample locations have multivariate Gaussian distribution. There are many ways to sample from GRFs. We choose a rather efficient method using a fast Fourier transform, which computes a GRF in $\mathcal{O}(n \log n)$ for a grid with n points. Gaussian white noise has a constant power spectrum. To introduce more structure into the field, lower frequencies need to be more pronounced. This is achieved by multiplying the power spectrum $P(k) = k^{-\alpha}$ for frequencies k with the power spectrum of generated Gaussian noise. Using an inverse Fourier transform, we get our desired scalar field with a level of structure depending on α. We analyze the percolation for varying levels of structure α in Fig. 6a, where the percolation function is shown over the threshold p for increasing structure parametrized by α. We focus on the percolation threshold p_c and the width of the percolation transition Δ, which both change over α. The percolation function and a rendering of the scalar field are shown for three values of α in Fig. 6b to g.

First, while all curves end with $P_{\max}(p) = 1$, not all of them start with the minimal value 0. When looking at the curve and scalar field at perfect white noise, $\alpha = 0$, we have several small connected components in the data for a low value of p. When increasing p, more of them will form and slowly grow larger, keeping the highest relative volume (cf. Eq. (1)) close to 0. Only once a certain point is reached, we observe a sharp increase in $P_{\max}(p)$ as more and more of them merge, rapidly forming a percolating connected component at approximately p_c. For $\alpha > 0$, the non-zero value at $p = 0$ stems from the existence of at least one large connected component for the superlevel set of the largest percent of scalar values. This indicates a high level of structure. Indeed, in Fig. 6a we can observe that the percolation function begins below 5% for all $\alpha \leq 3$, but not at any α above that. At approximately that point, no classic sharp percolation transition is visible anymore. Another interesting change is observed at around $\alpha \approx 3$: where the function was very smooth before, large jumps can be seen to appear in the individual percolation functions at higher levels of structure. They begin to form as several small discontinuities as can be seen in Fig. 6e at $\alpha = 3.5$ and develop into large disconnected segments for $\alpha = 10$ in Fig. 6g. Each such visible discontinuity marks the merging of the largest connected components with another large one. After a merge, the volume of that component decreases in relation to other structures growing, before merging again.

In the heatmap, we can see that the size of these segments grows with increasing α, as fewer and larger connected components form early on. At that point, with no values around zero and considerable jumps in the function, the percolation value p_c

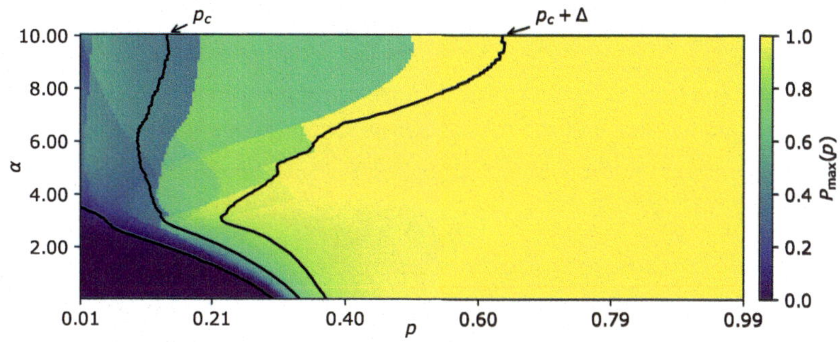

(a) Percolation function $P_{max}(p)$ for multiple $\alpha_i \in [0, 10]$ with $\alpha_{i+1} - \alpha_i = 0.05$. Individual functions along with the respective fields for $\alpha = 0, 3.5$ and 10 are shown below.

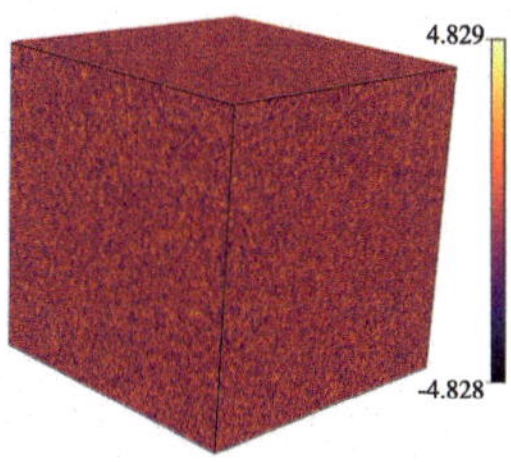

(b) GRF for $\alpha = 0.0$.

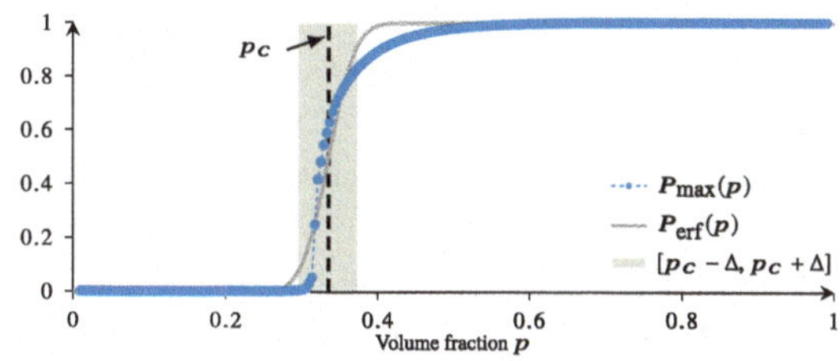

(c) Percolation function for the field in Figure 6b.

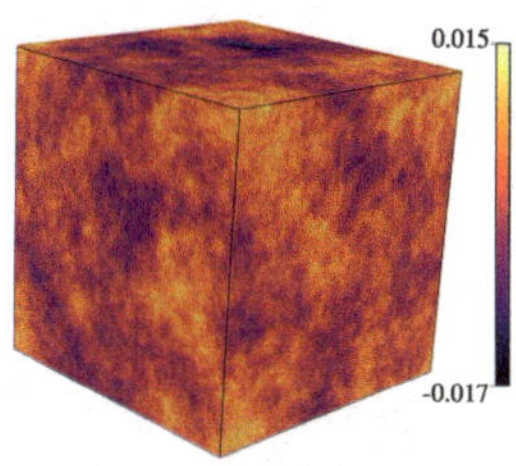

(d) GRF for $\alpha = 3.5$.

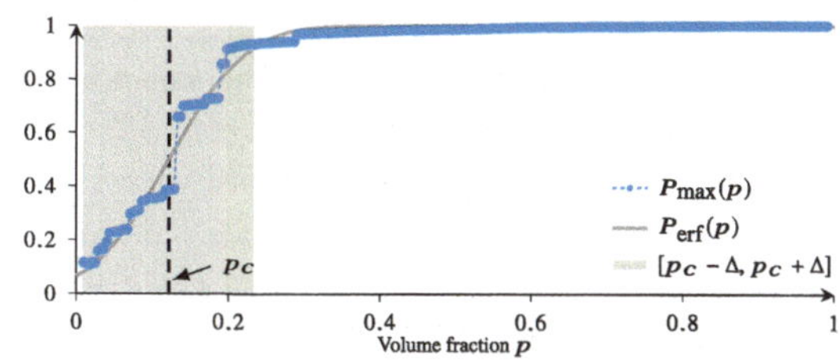

(e) Percolation function for the field in Figure 6d.

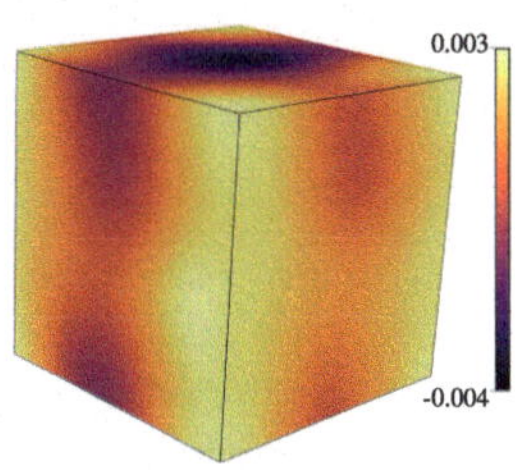

(f) GRF for $\alpha = 10.0$.

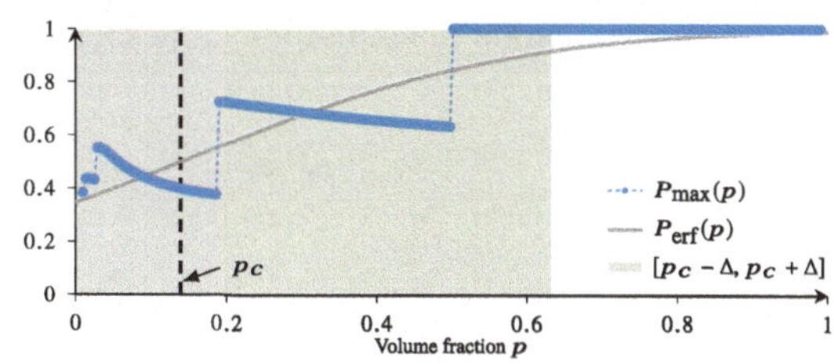

(g) Percolation function for the field in Figure 6f.

Fig. 6 Percolation function for generated GRFs with power spectrum $P(k) = k^{-\alpha}$ of dimension 128^3. Three samples of $\alpha \in \{0, 3.5, 10\}$ are shown separately. We observe the transition until about $\alpha \approx 3$, at which point $P_{max}(0)$ does not start from zero anymore, large discontinuities form and the width Δ of the fitted function increases considerably loses its meaning as the "point of steepest decrease" and the width Δ of the fitted function grows.

All these observations indicate that there is a strong correlation between the level of structure and the shape of the percolation function. While it is up to definition to pinpoint the exact value of α where no percolation transition occurs, several indicators point to the region of $\alpha \approx 3$.

5.2 *Turbulent Flow: Duct Data Set*

One common application for percolation analysis is the study of turbulence in fluid dynamics. It has, among other applications, been employed to study the impact of the Reynolds number on the transition from laminar to turbulent flow [10] and to find optimal thresholds to separate highly turbulent structures from the surrounding flow [11].

We analyze a duct flow simulated to investigate intense Reynolds-stress events in fully-developed turbulent flow [17]. It is sampled over $193 \times 194 \times 1000$ data points at an approximately square cross-section and is periodic in the stream-wise direction. As an indicator of Reynolds stress, the scalar combination $\overline{uv}$ has been employed [2]. Here, $\overline{u} = \frac{u - u_{avg}}{u_{rms}}$ denotes the normalized stream-wise directional component of the flow, perpendicular to the normalized cross-stream component $\overline{v}$. This average u_{avg} and the root mean squared deviation u_{rms} are accumulated over a large number of stream-wise slices and time samples. The data points close to the wall are disregarded, as turbulence does not show in that region. Percolation analysis aids in finding the exact wall distance in which to disregard values, see [6].

5.2.1 Stability of the Percolation Function

To find a sensible threshold for intense Reynolds stress events, it is common to compute the percolation function for n_t time slices and average all function values

$$P_{\max}(p) = \frac{1}{n_t} \sum_{t=1}^{n_t} P_{\max}(p, t). \tag{8}$$

The final analysis is applied to this rather smooth averaged curve. However, we are not aware of any work assessing the temporal evolution of the percolation function for real simulated data, which we will do in the following.

We compute the percolation function for a number of $n_t = 2000$ time slices and visualize them in Fig. 7a. All time slices have highly similar statistics and are normalized with the same average and root square error. Note that the heatmap plot is rotated by 90 degrees as time t is plotted horizontally. This full range visualization

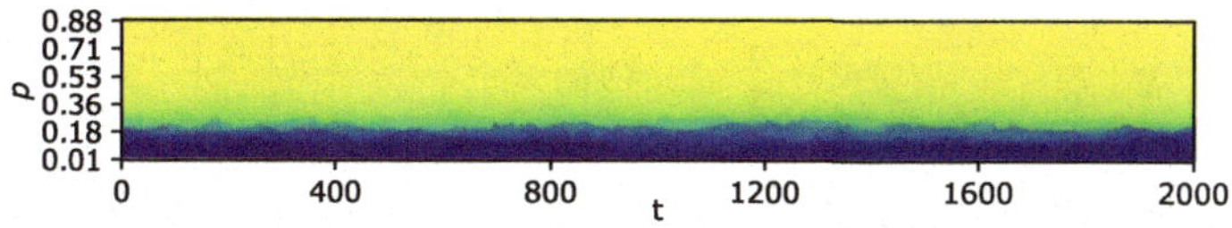

(a) Percolation function $P_{\max}(p, t)$ for all 2000 time slices of the duct data set.

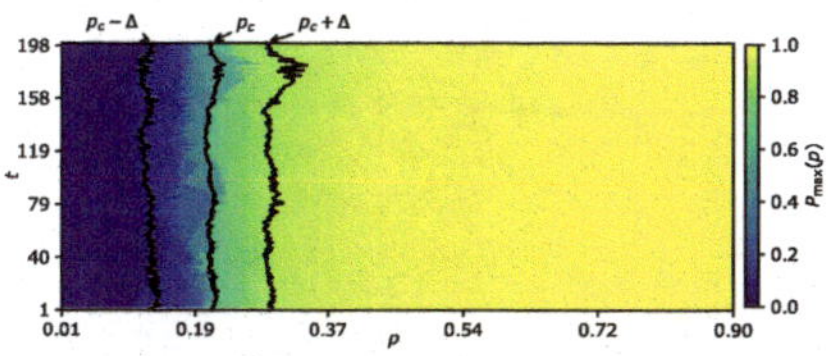

(b) $P_{\max}(p, t)$ for $t \in [1, 200]$.

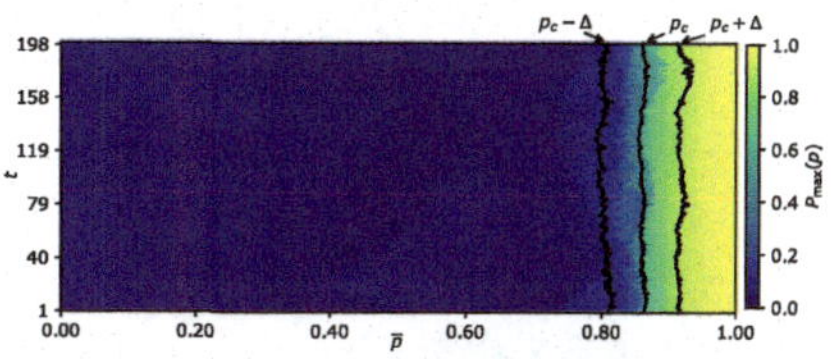

(c) $P_{\max}(\overline{p}, t)$ for $t \in [1, 200]$.

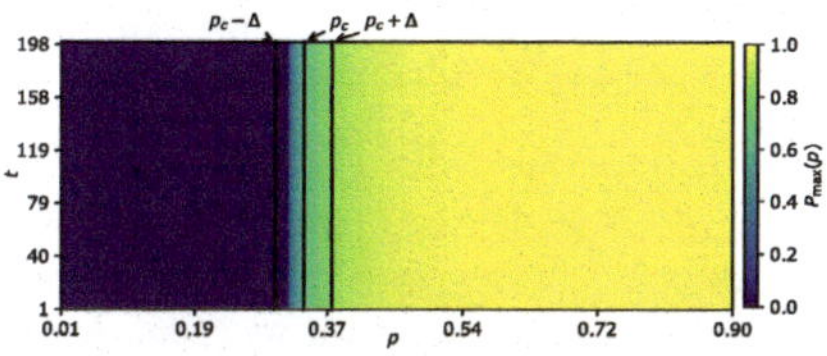

(d) $P_{\max}(p, t)$ after permutation

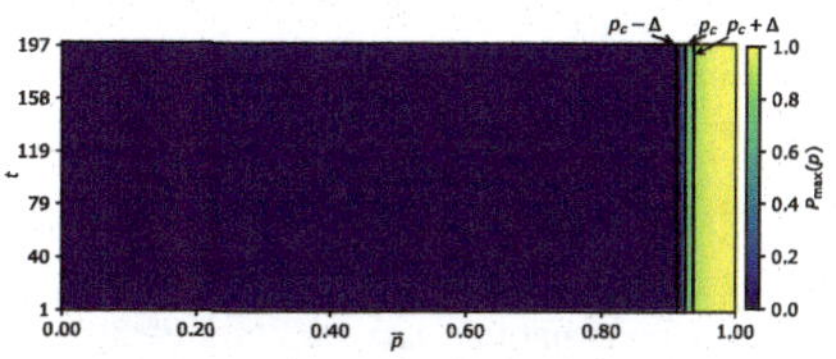

(e) $P_{\max}(\overline{p}, t)$ after permutation.

Fig. 7 Percolation function for the Reynolds-stress field of the duct data set The first visualization gives an overview of all 2000 time slices, while the following plots are depicting the first 200 times. In the second row, the percolation function is displayed for both the volume-normalized (left) and the original scalar values (right). Below them, the same function is plotted after permuting the scalar field. We observe that the normalization of the original scalar values does have an impact on the percolation function, but mostly keeps the shape intact. Permuting, on the other hand, removes all variation between the individual time slices

gives an overview for a long time span, which shows us that the percolation function is rather stable.

For a more detailed view, Fig. 7b shows the first $n_t = 200$ time slices. The approximated percolation threshold p_c and the transition width Δ are shown as well. The visualization reveals only minor variations in the percolation function, showing that the analysis is stable for temporal development.

5.2.2 Effect of Histogram Normalization

As discussed in Sect. 3, all visualizations of the duct shown so far are plotted for a normalized histogram of p, which is in practice the volume of the sublevel set. We compare this to working on the values of the scalar field itself. Especially in the case where percolation is used as an indicator for an optimal threshold, it becomes necessary to analyze the function by value. On the other hand, in order to connect these results to percolation theory, the comparison should always be voxel-based.

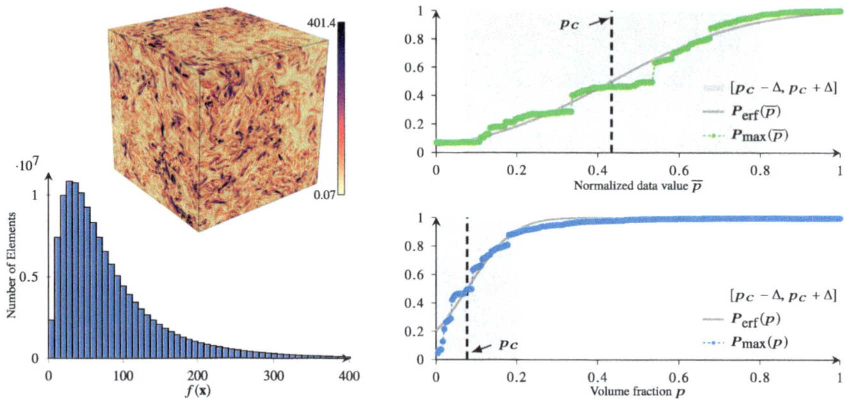

(a) Histogram and volume rendering of $f(\mathbf{x})$ in the sampled range

(b) Value-based and voxel-based percolation function for $f(\mathbf{x})$

Fig. 8 Percolation functions, rendering and histogram of the vorticity magnitude of an isotropic homogeneous turbulence data set provided by Yeung et al. [18]. Comparing this with Fig. 2b, we can conclude that this data structurally differs from uniform noise, but is still random enough to not show any larger discontinuities

Figure 7c shows the visualization for $P_{\max}(\overline{p})$ with $t \in [1, 200]$. Figure 7b and 7c are overall very similar. Note that also in the value-based plot, the percolation transition p_c is very stable. However, the respective values of p_c and the width Δ are rather different. This shows that a percolation transition can appear in an irregularly distributed scalar field, but that normalization of these values has a huge impact on the actual function parameters. We explore this further using an isotropic turbulent flow kindly provided by Yeung et al. [18]. This is a 512^3 sub-volume of a 4096^3 vorticity magnitude field. A visualization is provided in Fig. 8a. The histogram reveals the non-uniform data value distribution in this data set. The effect on the value-based percolation function $P_{\max}(\overline{p})$ is drastic: contrary to previous examples, the typical sharp transition between 0 and 1 is lost (Fig. 8b, top). However, we can recover the sharp transition with the voxel-based percolation function $P_{\max}(p)$ (Fig. 8b, bottom). When comparing it to uniform random noise as shown in Fig. 2b, we can see that the transition happens at much larger values of p. This indicates that this data is not random and has indeed some structure.

5.2.3 Effect of De-correlation

To look deeper into the role that structure plays on the shape of the percolation function as opposed to the value distribution, we remove all structure from the field. By randomly permuting the input scalar field values, we keep the histogram intact, but remove any correlation.

The results for the duct data set are shown for both the volume-based sampling scheme in Fig. 7d and the original data values in Fig. 7e, which were permuted via the same random seed. As a histogram normalization for permuted values is the same as sampling a uniform random distribution for each point, the results are very close to the theoretical 3D percolation function shape shown in Fig. 2a. Also, the variation between time slices is virtually zero. When permuting the original scalar values $\overline{p}$, the function is again stable over time slices, but the percolation transition p_c is shifted notably compared to the theoretical values. The latter confirms that the histogram normalization is effective in enabling the comparison of real data with the theoretical results.

6 Conclusions and Future Work

In this work, we presented a framework for analyzing the structure of existing scalar fields with the help of percolation theory. It was established that a normalization of the scalar values is needed in order to compare the translation and spread of the percolation function to the theoretical values of uniform white noise or other scalar fields. These parameters are obtained by fitting a Gauss error function. To visualize the results, we have shown curve plots of 1D percolation functions and a colormap-based representation of the evolution of percolation across a smoothly changing field.

The analysis was applied to two main data series, a Gaussian random field of increasing correlation and the time series of a fully turbulent duct flow. We could show that for an increase in structure and correlation, three main features point to the existence of a percolation transition: the percolation function assuming a value of almost zero for small p, the absence of large jumps in the curve and a low transition width Δ. For a time series of a statistically stable scalar field on the other hand, we observe a very consistent percolation function.

This work has made a first step in gaining insight into the underlying structure of scalar fields by means of their percolation function. However, more experiments need to be made to determine an arithmetic correlation between different kinds of structure and the parameters of the resulting percolation curve.

Acknowledgments This work was supported through grants from the Swedish Foundation for Strategic Research (SSF, Project BD15-0082) and the Swedish e-Science Research Centre (SeRC). The presented concepts have been implemented in the Inviwo framework.

References

1. Alexander, K.S., Molchanov, S.A.: Percolation of level sets for two-dimensional random fields with lattice symmetry. J. Stat. Phys. **77**(3), 627–643 (1994). https://doi.org/10.1007/BF02179453

2. Atzori, M., Vinuesa, R., Lozano-Durán, A., Schlatter, P.: Characterization of turbulent coherent structures in square duct flow. J. Phys. Conf. Ser. **1001**(1), 012008 (2018). https://doi.org/10.1088/1742-6596/1001/1/012008
3. Broadbent, S.R., Hammersley, J.M.: Percolation processes: I. Crystals and mazes. Math. Proc. Camb. Philos. Soc. **53**(3), 629–641 (1957). https://doi.org/10.1017/S0305004100032680
4. Carr, H.: Topological manipulation of isosurfaces. Ph.D. thesis, The University of British Columbia (2004)
5. Deng, Y., Blöte, H.W.J.: Monte Carlo study of the site-percolation model in two and three dimensions. Phys. Rev. E **72**(1), 016126 (2005)
6. Friederici, A., Atzori, M., Vinuesa, R., Schlatter, P., Weinkauf, T.: An efficient algorithm for percolation analysis and its application to turbulent duct flow. In: Euromech Colloquium 598: Coherent structures in Wall-bounded Turbulence (2018)
7. Friederici, A., Köpp, W., Atzori, M., Vinuesa, R., Schlatter, P., Weinkauf, T.: Distributed percolation analysis for turbulent flows. In: 2019 IEEE 9th Symposium on Large Data Analysis and Visualization (LDAV) (2019)
8. Hoshen, J., Kopelman, R.: Percolation and cluster distribution. I. Cluster multiple labeling technique and critical concentration algorithm. Phys. Rev. B **14**, 3438–3445 (1976). https://doi.org/10.1103/PhysRevB.14.3438
9. Jones, E., Oliphant, T., Peterson, P., et al.: SciPy: open source scientific tools for Python (2001). http://www.scipy.org/. [Online]
10. Lemoult, G., Shi, L., Avila, K., Jalikop, S.V., Avila, M., Hof, B.: Directed percolation phase transition to sustained turbulence in Couette flow. Nat. Phys. **12**, 254–258 (2016). https://doi.org/10.1038/nphys3675
11. Moisy, F., Jiménez, J.: Geometry and clustering of intense structures in isotropic turbulence. J. Fluid Mech. **513**, 111–133 (2004). https://doi.org/10.1017/S0022112004009802
12. Newman, M.E.J., Ziff, R.M.: Efficient monte carlo algorithm and high-precision results for percolation. Phys. Rev. Lett. **85**, 4104–4107 (2000). https://doi.org/10.1103/PhysRevLett.85.4104
13. Rintoul, M.D., Torquato, S.: Precise determination of the critical threshold and exponents in a three-dimensional continuum percolation model. J. Phys. A Math. Gen. **30**(16), L585–L592 (1997). https://doi.org/10.1088/0305-4470/30/16/005
14. Rodriguez, P.F., Sznitman, A.S.: Phase transition and level-set percolation for the Gaussian free field. Commun. Math. Phys. **320**(2), 571–601 (2013). https://doi.org/10.1007/s00220-012-1649-y
15. Sahini, M., Sahimi, M.: Applications of Percolation Theory. CRC Press, London (2014)
16. Stauffer, D., Aharony, A.: Introduction To Percolation Theory. Taylor & Francis, London (1994)
17. Vinuesa, R., Schlatter, P., Nagib, H.M.: Secondary flow in turbulent ducts with increasing aspect ratio. Phys. Rev. Fluids **3**, 054606 (2018). https://doi.org/10.1103/PhysRevFluids.3.054606
18. Yeung, P.K., Donzis, D.A., Sreenivasan, K.R.: Dissipation, enstrophy and pressure statistics in turbulence simulations at high reynolds numbers. J. Fluid Mech. **700**, 5–15 (2012). https://doi.org/10.1017/jfm.2012.5
19. Yonezawa, F., Sakamoto, S., Hori, M.: Percolation in two-dimensional lattices. I. A technique for the estimation of thresholds. Phys. Rev. B **40**, 636–649 (1989). https://doi.org/10.1103/PhysRevB.40.636
20. Ziff, R.M.: Results for a critical threshold, the correction-to-scaling exponent and susceptibility amplitude ratio for 2d percolation. Physics Procedia **15**, 106 –112 (2011). https://doi.org/10.1016/j.phpro.2011.06.009. http://www.sciencedirect.com/science/article/pii/S1875389211003403. Proceedings of the 24th Workshop on Computer Simulation Studies in Condensed Matter Physics (CSP2011)
21. Ziff, R.M., Scullard, C.R.: Exact bond percolation thresholds in two dimensions. J. Phys. A Math. Gen. **39**(49), 15083–15090 (2006). https://doi.org/10.1088/0305-4470/39/49/003

Distributed Task-Parallel Topology-Controlled Volume Rendering

Jan-Tobias Sohns, Gunther H. Weber, and Christoph Garth

Abstract Topology-controlled volume rendering has proven to be a useful tool for exploration of volumetric data by highlighting the global, high-level structure of data sets. However, topological analysis is difficult to parallelize on distributed memory systems – and thus to utilize for in situ visualization – due to the global nature of topological descriptors.

This chapter presents and evaluates a task-parallel formulation of topology-controlled volume rendering applicable to visualization of large scalar field data. It evaluates previous efforts towards parallel topology extraction and introduces a distributed computation schema for augmented contour trees. Through data partitioning into rectilinear blocks, the algorithm is designed to be in-situ suitable. The use of a task-parallel framework aims at latency hiding and dataflow-specific scheduling. It thereby also allows for combining contour tree computation and subsequent volume rendering. The technique divides the scalar field with separate transfer functions according to the branch decomposition of the full data set while each local block only has to keep track of its own vertex augmentation. Beyond describing the approach and its implementation in the task-parallel framework HPX, initial experiments on scaling behaviour are presented.

1 Introduction

Computer simulation techniques have become ubiquitous in the investigation of both scientific and engineering problems. Owing to increased computational ability,

J.-T. Sohns (✉) · C. Garth
TU Kaiserslautern, Kaiserslautern, Germany
e-mail: j_sohns12@cs.uni-kl.de

C. Garth
e-mail: garth@cs.uni-kl.de

G. H. Weber
Lawrence Berkeley National Laboratory, Berkeley, USA
e-mail: ghweber@lbl.gov

I. Hotz et al. (eds.), *Topological Methods in Data Analysis and Visualization VI*, Mathematics and Visualization, https://doi.org/10.1007/978-3-030-83500-2_4

Fig. 1 Distributed topology-controlled volume rendering of the jet data set. Branch pruning minimal persistence is set to 3 (max 16.63)

such simulations have outgrown the ability to retain the data produced by them for post hoc analysis, and thus a re-thinking of established post-processing visualization work flows is necessary. In situ techniques that derive analysis artifacts from the data while it is produced have shown promise in this context. In their most extreme form, visualization images are generated in situ and examined post hoc. Choosing an appropriate set of images allows sufficient flexibility in exploring simulation results [1].

Concurrently, the ever-increasing complexity of data describing real-world problems has necessitated the development of efficient abstraction and reduction techniques. Among these, topological feature extraction relies on a solid mathematical foundation to achieve these goals. In particular, topology-controlled volume rendering [2] (Fig. 1) can be employed to automatically define transfer functions for volume rendering directly on the structure of the data, as represented by the contour tree [3]. Topology-controlled volume rendering was previously restricted to small data sets, due to the global nature of contour trees that make them hard to compute in a distributed manner.

In this book chapter, we report on a proof of concept distributed pipeline for topology-controlled volume rendering. It is aimed at making bigger, possibly distributed data sets accessible to the technique. To leverage latency-hiding and dataflow-based asynchronous scheduling, our implementation is based on a task-parallel formulation of topological analysis, transfer function generation, and volume rendering. We present early findings on the performance and scalability of our implementation. The experiments are designed to get a first impression of the applicability of topology-controlled volume rendering for bigger data sets. Further investigation in more extensive settings has to be done in the future to get a complete understanding of the practical capability.

The presented work relies heavily on topological descriptors such as the contour tree [4], merge tree [3] and branch decomposition [5]. The reader is referred towards the proposing papers for in-depth information on these concepts. Further on, the augmented version of the trees is used as in [6]. The augmentation is a segmentation of data in the spatial domain, where each data point is affiliated with an arc in the corresponding tree.

The manuscript is structured as follows: After briefly reviewing related work in Sect. 2, we present our approach in Sect. 3 and implementation in Sect. 4. Results and benchmarks on typical datasets are presented in Sect. 5, and we conclude and reflect on further improvements in Sect. 6.

2 Related Work

Topology-controlled volume rendering is first presented by Weber et al. [2]. They use the vertex affiliation with branches in the augmented branch decomposition of the contour tree to specify individual user-designed transfer functions for each branch. Weber et al. also realize that a simplification of the topological descriptor is unavoidable to prevent a cluttering of the scene. The visual results are satisfying and promise a powerful analysis technique for 3D scalar data. However, the drawback of their algorithm is the sequential nature of the contour tree computation, which dominates the runtime. It is calculated separately with the algorithm introduced by Carr et al. [3].

Parallel contour tree computation was examined before the practical applications arose. Pascucci et al. [7] describe a parallel algorithm that can also be used in a distributed setting. Based on a divide-and-conquer formulation, it lists a merge routine for merge trees of two adjacent data subsets. This approach still forms the basis for most subsequent algorithms as well as ours. Although vital for many use cases, the initial method does not keep track of the vertex augmentation of the contour tree.

Gueunet et al. [6] present a parallel algorithm that creates an augmented contour tree by dividing the data in range space. Dividing data sets in range space is uncommon in distributed settings, since simulations and measurements are usually allocated in domain space. This makes their algorithm suitable if the data can be divided in range space without too much effort, yet is not fitted for domain space distribution. Nonetheless, it provides a fast parallel computation in a shared-memory system and shows the need of topological simplification again.

Another efficient approach was taken by Carr et al. [8], who demonstrate a highly thread-parallel algorithm to compute contour trees whilst retaining correct augmentation. More fast and efficient parallel algorithms [9] have been proposed recently that successfully focus on shared-memory computation, which this work tries to overcome.

In the wake of the shift towards using graphical processing units for scientific computations, Rosen et al. [10] published an augmented merge tree construction algorithm using OpenCL on a GPU. It outperforms the CPU version by a magnitude in their benchmarks on 2D scalar fields. Unfortunately, it is mentioned in the article that the computational benefits are not necessarily carried over to an extension into 3D.

Morozov et al. [11, 12] specifically target their algorithm on distributed systems and achieved competitive results. They allow for a distributed computation and representation of merge and contour tree, which can handle specific topological requests. These requests do not suffice for the pursued volume rendering and the algorithm does not produce the desired augmented contour tree.

Another important addition to this field was made by Bremer et al. [13], who uses a streaming fashion to gather the merge tree and its augmentation. Their sequential algorithm serves as inspiration for updating the augmentation through a search process and performing on-the-fly simplification on the topological complexity.

Landge et al. [14] focus on the augmentation of very high or low function values and compute only the locally relevant parts of a merge tree on distributed blocks. Their analysis is fast, can be computed in situ and easily highlights smaller features. The focus on local features of a distributed merge tree is advantageous, since it eases the distributed computation and is sufficiently accurate for most analysis purposes.

Recent advances in parallel programming have shown that the algorithm of Landge et al. [14] can be implemented efficiently in task-parallel frameworks. Petruzza et al. [15] show that the *Legion* [16] framework is faster for low core counts but does not exhibit good scalability, while using *Charm++* [17] to schedule the tasks runs and scales just as well as the original data-parallel implementation. Therefore, local merge tree analysis can be done in distributed task-parallel settings.

Some techniques like topology-controlled volume rendering [2] rely on the contour tree and a branch decomposition thereof to assign transfer functions consistently to branches. However, it is not immediately obvious to construct a complete contour tree from distributed merge trees. Hence, both full merge trees, namely join and split tree, are required to cover the whole value range with a branch decomposition. The work in this chapter aims to overcome previous distributed restriction on local merge tree analysis. The full topology-controlled volume rendering pipeline is presented for distributed settings, facilitating the approach for larger data.

3 System Design

As other works on this topic already recognized, the size of currently produced data sets warrants splitting them up into smaller blocks. In some cases data is already split up by its production process. The global manner of contour trees leads to high communication effort between these distributed data blocks during their computation. We aim to overcome this challenge by using a task-parallel setup that allows for defining complex dependency structures and arbitrary block granularity to suit most simulation outputs.

Due to noise or necessary symbolic perturbation the topology of data sets can be too complex to read useful information out of a volume rendering. In many cases

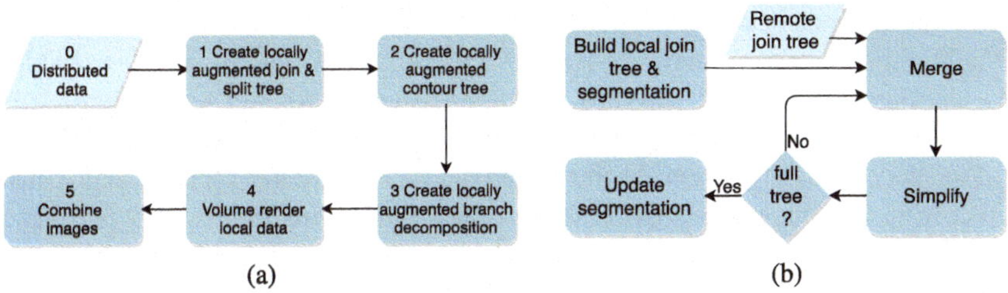

Fig. 2 Work flow per data block. **a** complete pipeline, references to Sect. 3 **b** locally-augmented join tree assembly, explained in Sects. 4–4.3

expensive computation and communication costs can be cut down by simplifying the tracked topological descriptors in early stages of their calculation without losing essential features. In the previous chapter current methods are described, which struggle either with data size, topological complexity or missing augmentation. Our approach works on arbitrarily distributed blocks of regular data according to the schema shown in Fig. 2a.

The key idea is to generate consistent global topological information on all blocks while keeping track of the local augmentation only and reducing the topological complexity early. The data is assumed to already be distributed to blocks (0). On each block the locally-augmented, simplified merge trees are constructed first (1). The merge trees are then combined on each block following the established scheme of Carr et al. [3] to create a locally-augmented contour tree (2). Subsequently, the contour tree is transformed into a locally-augmented branch decomposition covering the topology of the whole scalar field (3). This contour tree is identical over all blocks, whereas each block keeps track of a version with their local augmentation. It is now possible to define global transfer functions that are identical for all blocks, thereby guaranteeing a coherent color mapping.

Assuming the transfer functions are assigned automatically, the blocks can be rendered individually as soon as their precomputation finishes (4). In line with the task-parallel principle, rendering time can be hidden within preprocessing time. To conclude, the resulting images are composited regarding their depth level yielding the topology-controlled volume rendering of a scalar field (5). In the following the algorithm of creating a locally-augmented simplified merge tree will be elaborated in detail.

4 Implementation

Initially, all blocks are assumed to be filled with a coherent fragment of the scalar field with a single layer of ghost cells. For each block, the locally-augmented, simplified merge trees are assembled by first computing the augmented merge trees of the local data. Any sequential algorithm can be used. The established contour tree construction algorithm [3], executed in a later phase, requires coinciding nodes in both merge trees. Intending to reduce communication, nodes are exchanged mutually inside each block on a local stage. Then, local trees are progressively combined with unaugmented trees of tree-neighboring blocks following a reduction schema extended from [14].

The procedure for join trees is illustrated in Fig. 2b, split trees follow analogously. A recurring sequence of merging two trees and simplifying the result is performed. The methods will be called *Merge* and *Simplify* and are repeated on all blocks until the local trees cover the topology of the full data set. After that, the augmentation is corrected accordingly with the *Update* method. The functions as well as the remaining pipeline steps will be explained in chronological order hereinafter.

4.1 Merge

Whenever the merge procedure is called, the currently present unaugmented merge tree is merged with a remote one via the algorithm Landge et al. [14] call *Join Routine*. Their code follows the first proclaimed algorithm for the joining of merge trees [7] closely, yet is more intuitively formulated. The combined tree is iteratively formed by traversing the individual trees in value order starting from the lowest valued leaf.

4.2 Simplify

All nodes of the new tree are checked to decide whether the tree can be simplified by removing them. A node is obsolete when it has become regular through the merge process, recognizable through a degree of two. This was already proposed in the first parallel algorithm [7]. Regular nodes are eliminated without losing topological information.

As mentioned before, we want to reduce the branch decomposition to a distinguishable number of branches for rendering. The *Merge* function's runtime depends linearly on the size of the individual trees. Therefore, the usual chronology of simplifying the topology at the final stage, e.g. the branch decomposition, induces linear overhead in the magnitude of global topology per processor for each merge. However, reducing the tracked topological complexity as early as possible should lower the workload to a small linear overhead of simplified topology per merge. Branch pruning [2, 18] is employed here as a simple and well-established simplification tool that works on both topological structures. To keep the merge tree simplification comprehensible, it has to conducted as if branches were pruned in the branch decomposition.

The idea of branch pruning is removing less relevant branches from the tree. Persistence [19] is chosen as the relevance measure, though other measures work just as well as long as they can be tracked throughout the merging process. Consistent with the branch decomposition the more persistent component is considered to survive if two components merge, while the less persistent component is annexed.

Based on the previous line of reasoning, leaves with a persistence value lower than a predefined threshold are removed from the tree if they are adjacent to a saddle. Exceptions are the most persistent ones in comparison to their siblings, which represent branches that extend above the saddle node. Removing them would prune a branch partly, yet only complete branches are supposed to be eliminated. Further on, boundary nodes regarding the new tree and nodes relevant in opposing join/split tree are retained for stitching in later phases. The conditions hold for regular vertices as well.

To eliminate chains of removable nodes, the simplification routine is looped until no more nodes can be removed. In our benchmarks the single digit loop iterations accrued minuscule overhead.

4.3 Update

After the addition and removal of nodes through joining and simplification, the mapping of local vertices onto arcs can become incorrect. An update method is deployed to achieve the correct augmentation according to the new join tree without recomputing the labels from scratch.

When removing nodes, a new label candidate is saved for every removed node. For regular nodes, the candidate is the predecessor of the node in leaf direction. For leaf nodes, the governing saddle is chosen as the candidate, which corresponds to the parent branch in the branch decomposition. If the candidate is removed as well, the recursive candidates are traversed until an existing node is reached. Path compression is implemented to shorten candidate chains.

4.4 Branch Decomposition Assembly

The contour tree and consequently the branch decomposition can be sequentially assembled for each block individually with any sequential algorithm. With the presented algorithm design both block size and tree sizes can be reduced to low complexity through parameters. Thus, the construction algorithm runs on small data sizes per block and all blocks can run in parallel. Hence, a short runtime is expected for the transformation performed by the *libtourtre* [20] library. The augmentation with local vertices of the merge trees is implicitly carried over to the branch decomposition.

4.5 Transfer Function Assignment

The augmentation of the data points onto branches of the branch decomposition allows assigning a specific transfer function to each branch. The original topology-controlled volume rendering algorithm [2] stopped after computing the topology to allow the user to define custom transfer functions on each branch. Thereby it is possible to create customized functions and yield pleasing visual results. Since fine-tuning the transfer functions requires deep domain knowledge and many iterations, it breaks the algorithm into two separate steps that have to be started individually.

It would be beneficial to the presented method, if these parts could run continuously through forgoing the interactive break. Therefore, a basic automatic transfer function assignment is chosen for exemplary results which imitates colored isosurfaces. A distinguishable [21] color is used per branch with an opacity peak at the saddle value. Overall opacity is increased with branch depth to allow visible inclusion relationships. The main branch is colored in light grey with little opacity to ease perceiving branch location in the volume.

Since we are aware that automatic transfer function assignment is a complex topic with numerous sophisticated solutions, this part leaves room for further work. Exemplary, the implementation also allows user-defined transfer functions as well as the choice of commonly used predefined ones.

4.6 Rendering

Whenever the transfer functions are defined for a block, it can be rendered. To stay in line with the distribution of data blocks, the rendering implementation is drawn upon a task-parallel volume renderer [22] augmented with a branch lookup per sample [2]. The approach divides work in both screen and object space. Each block is rendered individually with the resulting images being composited in depth order afterwards. Additionally, the image is split up into rectilinear tiles allowing further steering of task-size. These tiles are concatenated to create the full resolution image.

5 Results

The following chapter will shine a light on questions arising from this approach. The algorithmic idea will be validated first on shared memory. Then benchmarks are run that examine performance for changing input size as well as resources in distributed settings. In consideration of ever-growing data size, the focus is set on scaling properties.

Fig. 3 Distributed topology-controlled volume rendering of the 'CT bones' data set of TTK [23]. Branch pruning minimal persistence is set to 171 (max 255)

5.1 Experimental Design

The implementation is tested on the Cori supercomputing system of NERSC. Per node two Intel Xeon E5-2698 v3 Haswell processors are provided with a combined total of 32 cores at 2.3 GHz and 128 GB of memory. For the test cases, a combustion simulation of a jet turbine with native $256 \times 512 \times 256$ data points is used. To facilitate increasing data size, the jet data set is super-sampled for higher resolution runs. Resampling a data set is suboptimal for scaling experiments as it theoretically will keep topological complexity near constant, since the number and relative position of features remain roughly the same. However, symbolic perturbation can influence the precise contour tree structure to varying degree and it can be initially examined if topology-controlled volume rendering can be applied for bigger data, even if further experiments are necessary to determine practical applicability.

Before starting with the analysis, a few preliminary considerations are taken first. This algorithm is most useful when analyzing huge distributed data sets such as the results of a distributed simulation. Therefore, it is assumed that the preceding application split up the data in ready to use blocks distributed over the hardware. This is mirrored by assuming that blocks are already loaded into the memory partitions. For now, data is read from disk, though this set-up is a precursor for the in situ case.

Dependencies in the task-parallel framework can be set in a way that allows intermingling of precomputation and rendering phase. However, the speedup through overlapping both phases has shown to be insignificant in comparison to the total runtime and the usual fluctuations of supercomputing environments. Hence, timings are taken separately with global barriers before and after each phase to ensure they don't slow each other down.

5.2 General Observations

As mentioned before, the biggest novelty of this approach lies in the creation of an augmented branch decomposition from distributed data blocks. Hence, the presented benchmarks address the topology computation. Further analysis towards the distributed rendering can be found in the proposing paper [22]. The implementation differences as covered in Sect. 4.6 induce a near constant overhead through a branch identification per sample. These changes are not expected to lead to changes in scaling behavior.

The implementation is based on assumptions that come up through observation of previous use cases. Figure 1 and Fig. 3 convey a clear visualization of topological features while features are still distinguishable. The simplification assumption seems to hold for two contemporary data sets.

5.3 Algorithm Validation

Before thinking about distributed settings, it needs to be determined if the algorithmic idea of subdivision into blocks combined with early topological simplification does behave as expected in the first place. To examine the algorithm's properties, the implementation is tested in a shared memory setting where constant data is divided into an increasing number of blocks (see Fig. 4). The used jet data set spans a value range from 0 to 16.63 and branches with less than 3 persistence are pruned. The final branch decomposition contains 25 branches as rendered in Fig. 1.

The examined computation, after which all blocks contain a simplified branch decomposition augmented with their local vertices as well as common transfer functions, will be called DAB (Distributed Augmented Branch decomposition) in the following analysis. Only summary timings are provided here, since the task-based paradigm relies on interleaved execution with latency hiding.

The timings shown in Fig. 4 reveal that subdivision into blocks amounts for a significant speedup. This can be attributed to a combination of better work distribution among processors for smaller tasks and the on-the-fly removal of short branches early in the construction process. Consequential, fewer arcs have to be processed in the remainder of the algorithm. Considering that computational resources are kept constant, this means less overall operations have to be done even with more merging steps. On the other hand, after a certain point the additional merging overhead outweighs the speedup from early arc removal and the execution time rises.

It has to be noted that the fastest achieved computation time is on par with other shared memory augmented contour tree algorithms [8] which do not rely on simplification. Therefore, these are preferable for shared memory settings. However, the novel point of this approach is that it allows a subdivision without a shared memory and thus can be applied to much bigger data sets, which will be examined in the following sections.

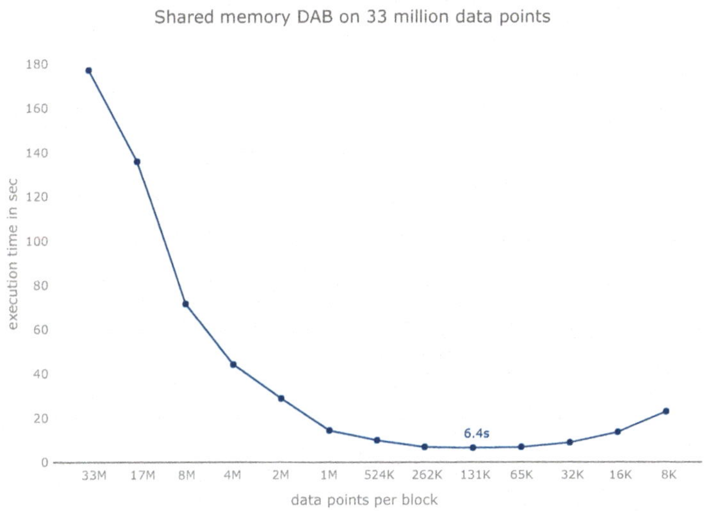

Fig. 4 Computing the DAB on a single Cori Haswell node for the native jet data set with 33M vertices while dividing into different block sizes

5.4 Strong Scaling

Strong scaling examines the change in execution time that can be achieved for a fixed problem size by increasing the number of processing units. Multiple measures can be deployed to indicate strong scaling behavior. A test case that inspects strong scaling of the DAB computation was constructed via re-sampling the jet data set to $1024 \times 1024 \times 1024$ vertices and benchmarking it on 1 to 128 nodes. With increasing processor count the data set is divided accordingly into more blocks, starting with 128 blocks on a single node.

The tests depicted in Fig. 5 showed that increasing hardware availability accelerates the DAB computation in accordance with Amdahl's Law of 95% parallel and 5% sequential parts. If one remembers that the conversion from merge tree to branch decomposition is still done sequentially per block, these results are in expected bounds.

Further on, for a fixed problem size the algorithm does not speed up infinitely with additional resources, but slows down after a certain node count. The overhead through additional merges dominates the cost as in the shared memory test. The slowdown through additional blocks and resources is even more extreme here, since network communication is required for each merge of two blocks on different nodes.

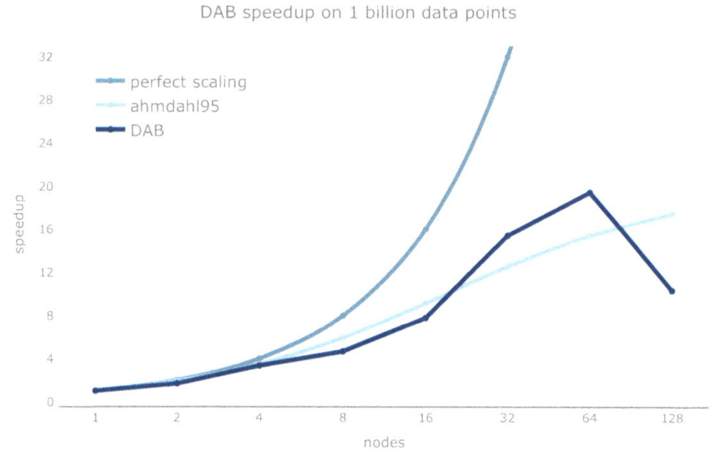

Fig. 5 Speedup for DAB computation on resampled jet data set with 1024 × 1024 × 1024 vertices using the single node case as the baseline. Perfect scaling and scaling according to Ahmdahl's Law with 95% parallel program parts is shown in lighter blue

5.5 *Weak Scaling*

One refers to weak scaling when examining the execution time of an algorithm with proportional increases in resources and problem size. Since the algorithm is designed to be applicable to future data sets of increasing size, weak scaling benchmarks are conducted through resampling the jet data set and computing the DAB on increasingly powerful hardware configurations. The baseline is chosen to be a resolution of 64 × 128 × 128 data points on a single processor. Processor count and number of data points are doubled in each iteration so that the ratio of approximately 1 million data points per processor remains constant. The last iteration handles 4 billion data points on a 2048 × 2048 × 1024 grid. Per processor 8 blocks are created in the single to 16 processors case, 4 blocks for 32 to 256 processors and 2 blocks from 512 processors on. The successive reduction is chosen to minimize communication while still performing local early simplification.

From inspecting Fig. 6, it is evident that the execution time increases overall with problem size. The bumps on 32 and 512 processors coincide with stagnant block counts on more resources. Comparing the edge cases, increasing data size 4096-fold while keeping relative resources constant triples the execution time. Communication accounts for significant overhead on rising problem size and resources, yet for the examined cases the overhead is sufficiently small to suggest that the algorithm can stay feasible under the weak scaling paradigm.

In all experiments the communication imposed a high runtime penalty. As a general guideline, results were best with block counts per processor of 2 or 4 and block sizes of ~1 million data points. This should be taken as a starting point for further experiments on new data sets.

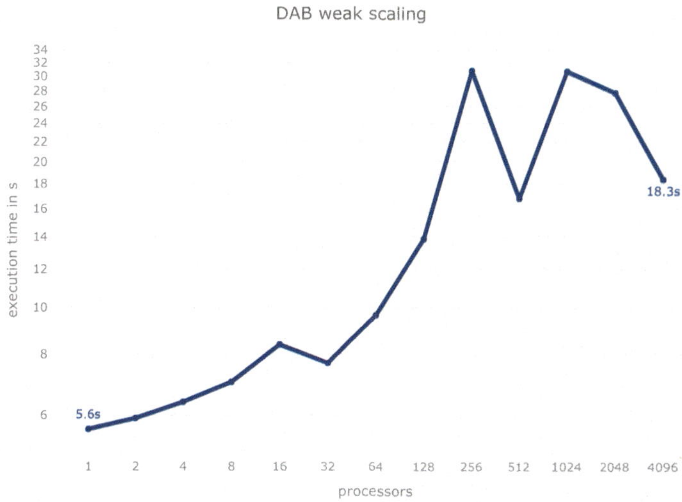

Fig. 6 Weak scaling for DAB computation keeping relative workload per processor constant through resampling the jet data set. Each processor handles ~1 million data points

An aspect not mentioned before is that the studies in the scope of this chapter revealed that contour tree algorithms generally require a substantial amount of memory during the computation. Most papers did not comment on the memory requirements and memory efficient algorithms exist already [13, 24]. However, computing the contour tree or a directly related topological descriptor for a 3D data set of 4 billion data points is challenging without running out of memory for algorithms not specifically designed for it. This chapter presents such an algorithm allowing analysis of huge data sets through on-the-fly topological simplification and minimal data duplication. If the rising communication overhead and the reduction on a small number of topological features is still practical on future data sets is up for future work.

6 Conclusion

The size of contemporary data sets is rising rapidly with no end in sight. Extracting topological features on them get increasingly time-consuming. Simultaneously, the processing power available in high performance compute clusters is growing annually. The idea suggests itself to leverage the distributed systems for preprocessing and rendering of 3D data sets. In this chapter, a task-parallel distribution is chosen to present a possible pipeline.

To put the presented work into perspective, the advantages and drawbacks of current parallel contour tree algorithms are compared first. The analysis revealed that while a plethora of shared memory solutions exist, distributed algorithms are scarce and come with restrictions.

Thereupon, an algorithm is proposed that allows distributed topology-controlled volume rendering. It relies on eliminating small topological features on-the-fly in a distributed construction process. It continues by merging trees so that the full contour tree is known globally while the augmentation is only locally present. Completing the pipeline, a possible distributed rendering technique is provided.

The algorithm was implemented in a task-parallel framework to provide portability and flexible parallelism granularity. To examine the effects of granularity as well as scaling performance, extensive benchmarks were run and analyzed.

The results suggest existing yet limited strong scaling ability. Benchmarks where data size and processor count was raised accordingly revealed that the algorithm can handle larger data sets with small performance decreases. The algorithm's main speedup as well as its limitation stems from the assumption that one is interested in the simplified topology only.

All things considered, this chapter presented a distributed augmented contour tree algorithm that exhibits limited strong and promising weak scaling capabilities. Whether this solution is sufficient for practical use cases or the simplification process can be circumvented in distributed settings is ground for future work.

Acknowledgements This material is based upon work supported by the U.S. Department of Energy, Office of Science, Office of Advanced Scientific Computing Research under Award Number DE-AC02-05CH11231 and used resources of the National Energy Research Scientific Computing Center (NERSC), which is a DOE Office of Science User Facility. We thank the Deutsche Forschungsgemeinschaft (DFG, German Research Foundation) - 252408385 - IRTG 2057 for providing accommodation during the research stay at Lawrence Berkeley National Lab (LBNL).

References

1. Ahrens, J.P., Jourdain, S., O'Leary, P., Patchett, J., Rogers, D.H., Petersen, M.: An image-based approach to extreme scale in situ visualization and analysis. In: Proceedings of the International Conference for High Performance Computing, Networking, Storage and Analysis, pp. 424–434. IEEE Press (2014)
2. Weber, G.H., Dillard, S.E., Carr, H.A., Pascucci, V., Hamann, B.: Topology-controlled volume rendering. Trans. Vis. Comput. Graphics **13**, 330–341 (2007)
3. Carr, H.A., Snoeyink, J., Axen, U.: Computing contour trees in all dimensions. In: Proceedings of the Eleventh Annual ACM-SIAM Symposium on Discrete Algorithms, pp. 918–926. Society for Industrial and Applied Mathematics (2000)
4. Oostrum, R., Kreveld, V., Bajaj, C., Pascucci, V., Schikore, D.: Contour trees and small seed sets for isosurface traversal. In: 13th ACM Symposium on Computational Geometry (1999)
5. Scorzelli, G., Pascucci, V., Cole-McLaughlin, K.: Multi-resolution computation and presentation of contour trees. In: IASTED Conference on Visualization, Imaging and Image Processing (2004)
6. Gueunet, C., Fortin, P., Jomier, J.: Contour forests: fast multi-threaded augmented contour trees. In: 6th IEEE Symposium on Large Data Analysis and Visualization, pp. 85–92 (2016)
7. Pascucci, V., Cole-McLaughlin, K.: Parallel computation of the topology of level sets. Algorithmica **38**, 249–268 (2003)

8. Carr, H.A., Weber, G.H., Sewell, C.M., Ahrens, J.P.: Parallel peak pruning for scalable smp contour tree computation. In: 6th IEEE Symposium on Large Data Analysis and Visualization, pp. 75–84 (2016)
9. Gueunet, C., Fortin, P., Jomier, J., Tierny, J.: Task-based augmented merge trees with fibonacci heaps. In: 7th IEEE Symposium on Large Data Analysis and Visualization, pp. 6–15 (2017)
10. Rosen, P., Tu, J., Piegl, L.A.: A hybrid solution to parallel calculation of augmented join trees of scalar fields in any dimension. Computer Aided Design Appl. **15**, 610–618 (2018)
11. Morozov, D., Weber, G.H.: Distributed merge trees. In: Proceedings of the 18th ACM SIGPLAN Symp. on Principles and Practice of Parallel Programming, pp. 93–102. ACM (2013)
12. Morozov, D., Weber, G.H.: Distributed contour trees. In: Topological Methods in Data Analysis and Visualization (2014)
13. Bremer, P.-T., Weber, G.H., Tierny, J., Pascucci, V., Day, M., Bell, J.: Interactive exploration and analysis of large-scale simulations using topology-based data segmentation. IEEE Trans. Vis. Comput. Graphics **17**, 1307–1324 (2011)
14. Landge, A.G., Pascucci, V., Gyulassy, A., Bennett, J., Kolla, H., Chen, J., Bremer, P.-T.: In-situ feature extraction of large scale combustion simulations using segmented merge trees. In: Proceedings of the International Conference on High Performance Compilation, Networking, Storage and Analysis, pp. 1020–1031 (2014)
15. Petruzza, S., Treichler, S., Pascucci, V., Bremer, P.: BabelFlow: an embedded domain specific language for parallel analysis and visualization. In: 2018 IEEE International Parallel and Distributed Processing Symposium (IPDPS), pp. 463–473 (2018)
16. Bauer, M., Treichler, S., Slaughter, E., Aiken, A.: Legion: expressing locality and independence with logical regions. In: SC 2012: Proceedings of the International Conference on High Performance Computing, Networking, Storage and Analysis, pp. 1–11 (2012)
17. Kale, L.V., Krishnan, S.: CHARM++: a portable concurrent object oriented system based on C++. In: Proceedings of the Eighth Annual Conference on Object-Oriented Programming Systems, Languages, and Applications, OOPSLA 1993, pp. 91–108, Association for Computing Machinery, New York (1993)
18. Carr, H.A., Snoeyink, J., van de Panne, M.: Simplifying flexible isosurfaces using local geometric measures. In: Proceedings of the Conference on Visualization 2004, pp. 497–504. IEEE Computer Society (2004)
19. Edelsbrunner, H., Letscher, D., Zomorodian, A.: Topological persistence and simplification. Discrete Comput. Geom. **28**(4), 511–533 (2002)
20. Dillard, S.: libtourtre: a contour tree library (2008). http://graphics.cs.ucdavis.edu/~sdillard/libtourtre/doc/html/. Accessed 11 June 2018
21. Green-Armytage, P.: A colour alphabet and the limits of colour coding. Color Design Creativity **5**, 1–23 (2010)
22. Biedert, T., Werner, K., Hentschel, B., Garth, C.: A task-based parallel rendering component for large-scale visualization applications. In: Eurographics Symposium on Parallel Graphics and Visualization, The Eurographics Association (2017)
23. Tierny, J., Favelier, G., Levine, J.A., Gueunet, C., Michaux, M.: The topology toolkit. IEEE Trans. Vis. Comput. Graphics (2017). https://topology-tool-kit.github.io/
24. Acharya, A., Natarajan, V.: A parallel and memory efficient algorithm for constructing the contour tree. In: IEEE Pacific Visualization Symposium, pp. 271–278 (2015)

Topology-Based Feature Design and Tracking for Multi-center Cyclones

Wito Engelke, Talha Bin Masood, Jakob Beran, Rodrigo Caballero, and Ingrid Hotz

Abstract In this paper, we propose a concept to design, track, and compare application-specific feature definitions expressed as sets of critical points. Our work has been inspired by the observation that in many applications a large variety of different feature definitions for the same concept are used. Often, these definitions compete with each other and it is unclear which definition should be used in which context. A prominent example is the definition of cyclones in climate research. Despite the differences, frequently these feature definitions can be related to topological concepts.

In our approach, we provide a cyclone tracking framework that supports interactive feature definition and comparison based on a precomputed tracking graph that stores all extremal points as well as their temporal correspondents. The framework combines a set of independent building blocks: critical point extraction, critical point tracking, feature definition, and track exploration. One of the major advantages of such an approach is the flexibility it provides, that is, each block is exchangeable. Moreover, it also enables us to perform the most expensive analysis, the construction of a full tracking graph, as a prepossessing step, while keeping the feature definition interactive. Different feature definitions can be explored and compared interactively based on this tracking graph. Features are specified by rules for grouping critical points, while feature tracking corresponds to filtering and querying the full tracking graph by specific requests. We demonstrate this method for cyclone identification and tracking in the context of climate research.

W. Engelke (✉) · T. B. Masood · I. Hotz
Department of Science and Technology, Linköping University, Norrköping, Sweden
e-mail: wito.engelke@liu.se

T. B. Masood
e-mail: talha.bin.masood@liu.se

I. Hotz
e-mail: ingrid.hotz@liu.se

J. Beran · R. Caballero
Department of Meteorology (MISU), Stockholm University, Stockholm, Sweden
e-mail: jakob.beran@misu.su.se

R. Caballero
e-mail: rodrigo.caballero@misu.su.se

I. Hotz et al. (eds.), *Topological Methods in Data Analysis and Visualization VI*, Mathematics and Visualization, https://doi.org/10.1007/978-3-030-83500-2_5

1 Introduction

Dynamic numerical simulations are prevalent and play a substantial role in understanding physical phenomena. Usually, such simulations result in feature-rich time-varying multi-fields and appropriate analysis and visualization methods are essential to exploit their full potential. This entails formal definitions of meaningful features, their algorithmic extraction and tracking, and finally a contextual visualization. There is a large body of work dealing with these aspects, however, efficient and robust tracking of semantic features at multiple scales is still challenging.

Inspecting existing work, many methods are proposing generic tracking algorithms of topological features as critical points [14], or contours [7, 8]. While these methods are valuable, they often cannot directly be applied to solve an application-specific tracking problem. On the other hand, there are methods inspired by applications proposing very specific algorithms for a fixed feature definition, for example for dissipation elements [18] or tracking of cyclones [22]. Even though these approaches work well in one setting, they often miss the necessary flexibility to be useful in a larger context. This is partially because physical phenomena can be vague in their description and no commonly accepted mathematical feature definition exists. Additionally, it is often quite unclear which descriptors work best for which tasks. A prominent example for such a phenomenon is a cyclone, where new tracking methods are continuously published [9], but still no efficient robust method that is satisfactory for general cyclone tracking exists. Often, methods have been designed for a specific event, are not generic, and depend on many parameters [12].

In this paper, we propose a framework that supports the design and comparison of features defined as sets of critical points, based on robust topological concepts i.e., *merge tree* and *Morse complex*. As an underlying principle, our framework disconnects the extraction and tracking of topological entities from the specific design of features for an application-specific task. The topological tracking and extraction of critical points results in a large directed graph containing all extrema and their temporal correspondence of the selected scalar fields, as well as a merge tree and its branch decomposition per time-step. These calculations are performed in a preprocessing step. Different feature descriptors can then be interactively explored. Tracking of features is formulated as tracking of groups of extrema and realized as queries to the tracking graph. As a concrete example, we focus on climate simulation data and the identification as well as tracking of cyclonic systems.

We demonstrate the framework in a meteorological context where understanding the dynamics of weather phenomena is essential to generate reliable predictions of intensity and frequency of extreme weather events in the future. In Europe, such hazards are mostly associated with extreme extra-tropical cyclones (ETCs), which are the focus of this paper. We show how our framework can be applied for efficient identification and tracking of multi-centered systems. This idea has been successfully applied in [10] for visualization of cyclonic regions. It formalizes a cyclone identification and tracking method that relies on a few clear principles but is still flexible enough to fulfill the domain scientists' demands.

The backbone of our method is characterized by a combination of different topological structures, the merge tree [2] and the Morse Complex [4, 16], facilitating the advantages of both. The merge tree is well suited for a hierarchical feature definition also supporting multi-centered cyclones, which can be defined as a set of extremal points based on user-specified criteria. Our tracking algorithm follows a method similar to Reininghaus et al. [14, 15], which is based on the Morse complex. Parallel implementations are used for both, the merge tree and Morse complex computation [1, 20]. The tracking itself is inherently local and fast.

Method overview

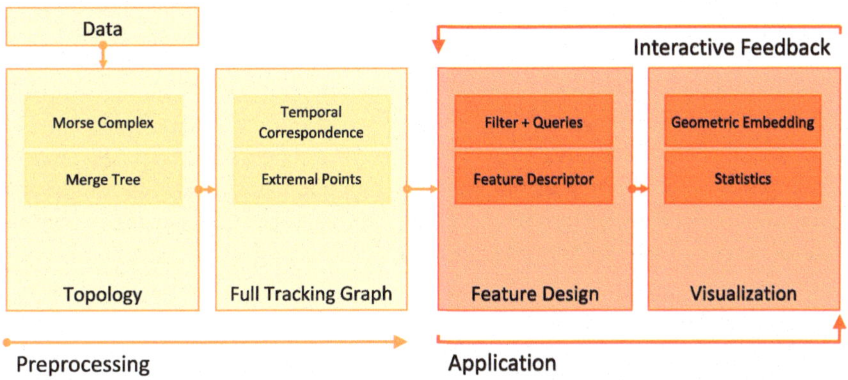

Fig. 1 **Overview:** Different building blocks of our approach. Our method consists of a preprocessing phase, which includes the topological data analysis. The result of this stage is the raw tracking graph containing the temporal correspondence between extrema and the extracted merge tree including its branch decomposition per time-step. In the second building block domain specific knowledge can be used to design features. This step includes the feature descriptor itself, visualizations and computed statistics. This information can also be used to redesign features and reissue tracking graph queries

In summary, the key aspect of our work is the separation feature definition and tracking. For both aspects robust methods from the field of topological data analysis are used. Figure 1 describes our pipeline in detail. The tracking approach relies on using the Morse complex to map every extrema in each time-step in both forward and backward directions to construct the raw tracking graph (see Fig. 4). The second building block is an interface for flexible and interchangeable feature definition as a set of critical points obtained from the merge tree including its branch decomposition computed per time-step. This information is enriched with domain specific criteria and heuristics to form possible feature definitions.

Our main contributions are:

- An extrema tracking method based on the Morse complex for time-varying scalar fields.
- A concept for flexible feature definitions based on sets of critical points obtained from a merge tree and its branch decomposition.
- Application of the above as an example to provide robust definition and tracking of cyclonic systems, possibly containing multiple cyclone centers.

2 Background

Our feature extraction and tracking is based on two different topological concepts, the merge tree (join or split tree) and the Morse complex where we use the descending/ascending manifolds. Both concepts are briefly described in the following.

Merge Trees: Features are groups of critical points that are defined by some rules acting on the merge tree of a selected scalar field $f : M \to \mathbb{R}$ defined on a smooth manifold M. Intuitively, a join tree keeps track of topological changes of sublevel sets (or superlevel sets in the case of split tree) when changing the level value a. The sublevel sets of f are defined as $M_a := f^{-1}(-\infty, a]$ for some $a \in \mathbb{R}$. Respectively, superlevel sets of f are defined as $M^a := f^{-1}[a, -\infty)$. Two points $x, y \in M$ are considered *equivalent*, $x \sim y$ if they have the same function value and they belong to the same connected component of the sublevel set M_a, respectively the superlevel set M^a. A *merge tree* is defined as the quotient space $M/\sim$; it results from identifying points specified by the equivalence relation $\sim$. The merge tree (resp. split tree) is a graph $G = (V, E)$ rooted at the absolute maximum (resp. absolute minimum), of the field. Its node-set V consists of local minima and saddle points where the sublevel-sets grow together. Its edges E represent the equivalent classes.

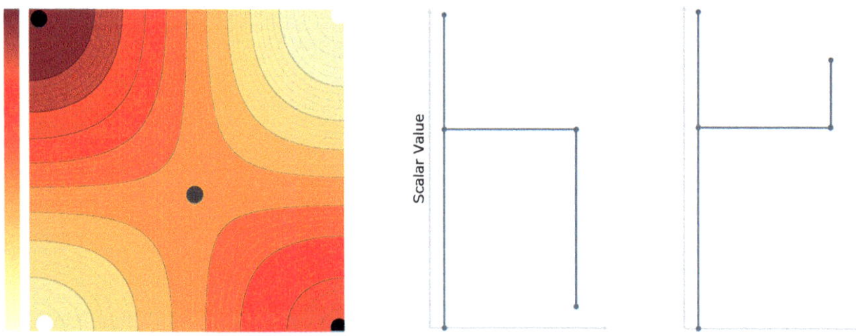

(a) (b) (c)

Fig. 2 Simple Example: A scalar field **a** with marked minima (white), maxima (black) and the saddle (gray) in **a** and the merge trees in form of a join **b** and split tree **c**

Figure 2 illustrates a simple example of join and split trees. It is constructed by tracking the evolution of the components of M_a as the parameter a is increased (resp. decreased). Specifically, leaves represent the creation of a component at local extrema, internal nodes represent the merging of components, and the root represents the entire space as a single component. A merge tree can be embedded in the domain M by visualizing its edges as straight lines that connect spatially-embedded critical points or also be visualized abstractly as a tree (Fig. 2(b) and (c)).

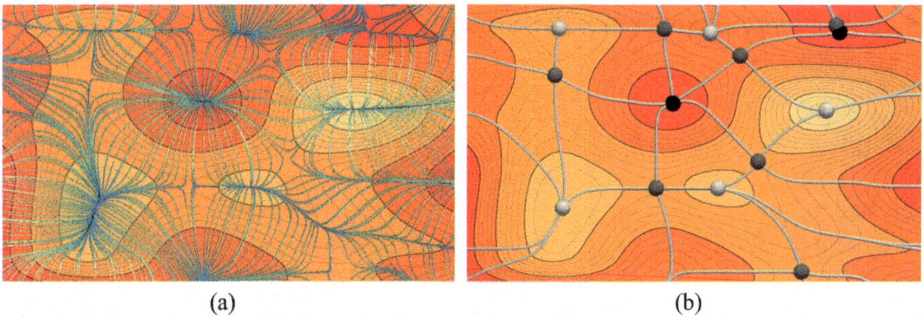

Fig. 3 Morse-Smale Complex: Example of the Morse decomposition of a domain. (a) shows a visualization of the gradient field derived from a given Morse function. In (b) the critical points of the function are visualized. White circles indicate locations of minima, black circles maxima, while the saddle points are depicted in gray. Critical points are connected via separatrices. Both images show the same sub-domain

We use a *branch decomposition* of the merge tree that decomposes the tree in a hierarchical structure. The root branch connects the two global extrema of the scalar field. All other branches connect a leaf with an interior node corresponding to a saddle point in a hierarchical way such that a persistence-based tree simplification corresponds to removing branches. For more details, we refer to the work by Pascucci et al. [11].

Morse Decomposition: A domain decomposition in ascending (or descending) manifolds forms the basis of our tracking algorithm. These are concepts related to the *Morse-Smale complex* of a function f given over a smooth manifold M, in our case of dimension two. A critical point of this function in the domain is a point where all spatial derivatives are equal to zero. In the following, we assume that the Hessian (second order derivatives) at the location of the critical points is non-singular and all critical points have pairwise different function values. Such functions are also called Morse functions. Based on the function f one can define a decomposition of M that carries geometric and topological information. For this, the gradient of the function f, a vector field on M, is considered. The critical points of f are the zeros in the gradient field (see Fig. 3(b)). The Morse-Smale complex then defines a decomposition into regions with uniform gradient flow behavior. This means all gradient lines have the same asymptotic behaviour, emerging from the same minimum and approaching the same maximum. See Fig. 3(a) for an illustration. These cells can also be interpreted as the intersections of the ascending and descending manifold of critical points. Ascending (resp. descending) manifolds of the critical points are defined as the set of points that flow towards (resp. emerge from) the same critical point. We are especially interested in the ascending manifolds of maxima and the descending manifolds of minima, which for the two dimensional case are topologically equivalent to open disks. We use discrete Morse theory for the computation of the cells [5, 20].

3 Full Tracking Graph Computation

During the preprocessing phase the full tracking graph of a selected scalar field of interest, $f : M \rightarrow \mathbb{R}$, is calculated. This graph provides tracks of all critical points (Fig. 4). It will later serve as a basis for the identification of feature tracks, merging, and splitting events, compare Sect. 4. Many algorithms that track critical points have been proposed. In general, the tracking is a two-step process, where critical points are extracted for all time-steps and then a critical point correspondence is established based on some heuristics. Examples are criteria using distances between critical points or contour overlap [7]. Sohn et al. [21] proposed a method to track the entire contour tree using volume overlap. Treating the time as an additional spatial dimension and assuming linear interpolation between the time-steps, the critical point tracks can be derived using the Reeb graph [3, 23]. In principle, all critical point tracking methods can be used in our framework. In the current implementation, we decided to follow the concept of combinatorial feature flow fields introduced by Reininghaus et al. [14]. After the segmentation of the domain into descending or ascending manifolds, the tracking is a local approach and only requires the evaluation of the next segment a critical point falls into.

Based on the Morse complex as a topologically meaningful partition of the domain, the tracking graph connects all extremal points in the forward and backward temporal directions across consecutive time-steps. Specifically, we use the descending manifolds for the case of minima tracking and ascending manifolds for maxima tracking. The tracking method for minima is described in more detail in the following. Maxima tracking works analogously.

Given an index set $I_t \subset \mathbb{N}$ specifying the minima, let $\{m_i^t, i \in I_t\}$ be the collection of minima of the scalar field f^t in the time-step t. Additionally, let $DM(m_i^t)$ denote the descending manifold of the minimum m_i^t. We say m_i^t is *forward mapped* to m_j^{t+1} if $m_i^t \in DM(m_j^{t+1})$. Note that m_i^t and the points in $DM(m_j^{t+1})$ belong to the same spatial domain, thus checking if m_i^t belongs to $DM(m_j^{t+1})$ is a valid operation. Furthermore, in discrete Morse theory, which we use in our implementation, the partition of the mesh vertices into descending manifolds of minima is complete. Combined, the above two conditions ensure that the forward mapping operation is well-defined for all minima m_i^t in time-step t, and each minimum m_i^t is forward mapped to a unique minimum m_j^{t+1} in the time-step $t+1$. Similarly, m_i^t is *backward mapped* to m_k^{t-1} if $m_i^t \in DM(m_k^{t-1})$. In other words, we check into which descending manifold a minimum falls, to determine the next position of the minimum forward (and backward) in time. In this way we define a forward and backward map for time-step t as sets of corresponding minima pairs in forward and backward temporal direction.

$$FM_t = \{(m_i^t, m_j^{t+1}) | i \in I_t, j \in I_{t+1} \text{ and } m_i^t \in DM(m_j^{t+1})\}$$

$$BM_t = \{(m_i^t, m_j^{t-1}) | i \in I_t, j \in I_{t-1} \text{ and } m_i^t \in DM(m_j^{t-1})\}$$

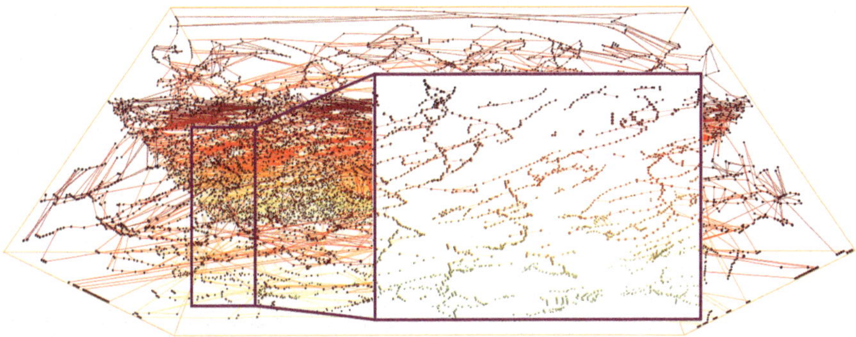

Fig. 4 Raw Tracking Graph: Geometric embedding of the complete tracking graph for a two-dimensional time-dependent pressure field given over the globe. The x- and y-axis represent longitude and latitude, whereas the z-direction corresponds to time. The tracking graph contains all minima as vertices and all elements from the forward and backward map as directed edges. This graph itself is independent of any feature definition or filter criterion. The inset shows a filtered subset of the graph. As filter criterion, a combination of edge length and spatial location was used

Please note that every minimum in time-step t is a member of exactly one pair in FM_t and BM_t. Combining these maps over all time-steps results in the sets $FM = \bigcup_t FM_t$ and $BM = \bigcup_t BM_t$. The two maps together form a directed graph $\mathcal{G} = (\mathcal{E}, \mathcal{V})$. Here, $\mathcal{V}$ contains all minima m_i^t from the scalar field across all available time-steps. Additionally, each element of the forward and backward map are represented by one directed edge in $\mathcal{E}$. For two consecutive time-steps the map corresponds to a $n : m$ mapping between the minima in the respective time-steps.

Graph Filtering and Queries: For the construction of the tracking graph all extremal points of the Morse complex are considered, except zero-persistence points. This avoids problems arising from the unstable geometric embedding of high-persistence critical points over time, compare Fig. 8. This complete tracking graph builds the foundation for further interaction and semantic feature design. For this purpose, we allow additional node- and edge-based property-vectors. With this, additional context information, for example, geometric length of an edge and other scalar values can further support the filtering and querying mechanism. This is important, since the domain partition the graph is based on is complete, and connections violating domain specific requirements may exist. Filtering can be based on spatial and temporal conditions or rely on edge and node properties. During graph queries, connected sub-components are extracted, which describe paths of critical points in the spatio-temporal domain. Here, queries can be formulated such that specific minima or minima groups are used as request input and current filter criteria are respected. This gives a powerful tool for exploration and analysis.

4 Feature Definition and Tracking of Cyclonic Systems

Feature Descriptor: As second building block of our method, the feature design plays an integral role. The feature descriptor is a set of rules that defines the grouping of the extremal points of one or several scalar fields. Currently, the merge tree is used as an interface to setup this set of rules. The specific descriptor is application- and task-dependent and typically also depends on a set of thresholds. Often, the feature descriptor itself is an active research topic and an interactive interface for designing such descriptor is of great benefit.

In this section, we describe the application of this concept to the topic of cyclone tracking. Relevant scalar fields in this context are pressure and vorticity fields. There is no generally accepted feature definition or ground truth for how these fields can be used in a feature descriptor. In fact, the definition of a cyclone is an open scientific question in the field of climate research. One common aspect, however, is that features are related to extremal points. Therefore, our method allows for a generic description of features as long as it is based on sets of extremal points. During the development of our method, where we closely collaborated with the domain scientists, we focused on cyclone tracking in pressure fields characterized by deep minima. During the course of this collaboration we experimented with different feature definitions. Using a global pressure threshold can be seen as a first example of a naïve cyclone definition grouping minima falling into the same component. Due to the seasonal and latitudinal variations in the pressure field, global thresholds are not the most promising concept but could still allow fast inspections of the extracted feature tracks and are therefore of interest. A combination of rules based on the merge tree's branch decomposition and the notion of persistence provides a more meaningful cyclone descriptor [24].

At first, one can build on the classic persistence-based filtering approach. A global persistence threshold can be interpreted as a single rule that groups all critical points of a sub-branch of the merge tree with the depth of the persistence threshold. However, on its own, this criterion is too rigid to account for the local variations of the pressure field. An alternative is to define locally varying persistence thresholds which can partially resolve this disadvantage. All of the aforementioned approaches and their combinations are possible and have their justification. There are also cases where it is more difficult to formalize a specific intuition. Here, a manual selection of regions of interest can be part of the setup.

Local Offset Threshold: In the following, we describe one approach that is of specific interest to our collaborators in more detail. This approach consists of two rules. The first rule defines a minimum m_i that qualifies as feature carrier for a feature $\mathscr{F}_i$, which is similar to the pure persistence-based approach, see Fig. 5(a). Therefore, a local persistence threshold δ is defined using the background pressure field. This threshold can be understood as the 'depth of a pressure minimum' to qualify as a cyclone and can depend on the location on earth. Each branch $br_i = br(b_i, d_i)$ where $b_i = f(m_i)$ is the birth and d_i is the death of this branch with persistence $(d_i - b_i)$ greater than δ creates a cyclonic feature. Clearly, the absolute value of the pressure minimum does not matter and instead, the relative depth of the minimum is captured.

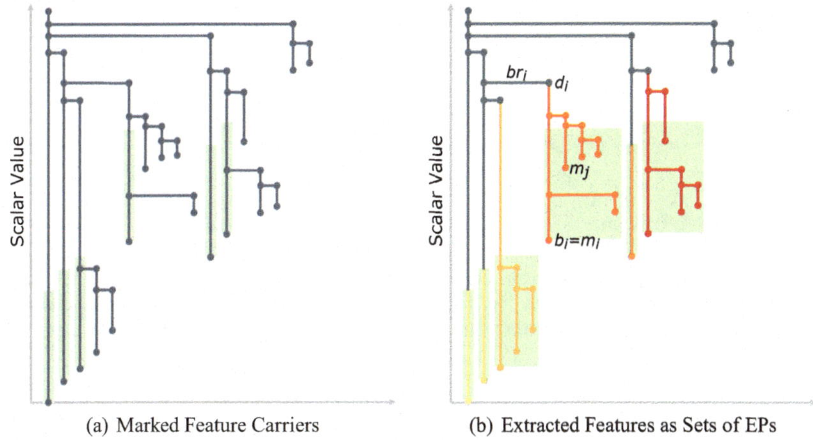

Fig. 5 Branch decomposition: A possible feature definition embedded in the branch decomposition. (a) Definition of feature carriers: The persistence threshold $\delta(x) = \delta$ (marked in green) is used to define a cyclonic feature $\mathscr{F}_i$. Only branches with persistence higher than δ are considered. (b) Definition of the full feature as set of extremal points. If a extremal point m_j merges into a given feature such that $s(m_j) \leq f(m_i) + \delta$, $(d_j - b_j) < \delta$ and br_j merges into br_i, it will be considered as a part of the feature $\mathscr{F}_i$. For each feature, the branch with the highest persistence can be used as the representative. Alternatively, the minimum with the lowest value can be used. For geometric representation, an iso-contour can be used

The second rule uses three criteria (i)-(iii) to define the set of minima attached to this feature. This results in a complete description of a possible feature definition; where each feature is an extremal point with additional extremal points attached to it, e.g., a set of extremal points, see Fig. 5(b).

$$\mathscr{F}_i = \{m_i, m_j | m_j \text{fulfilling criteria (i)-(iii)}\}.$$

with

$$\begin{array}{rl} (i) & f(m_j) \leq f(m_i) + \delta \\ (ii) & (d_j - b_j) < \delta \\ (iii) & br_j \text{ merges into } br_i \end{array}$$

Criterion (i) ensures that the scalar value of m_j is in the interval of the feature defining extremal point itself and δ. Criterion (ii) ensures that the attached extremal point does not span a feature by itself, and criterion (iii) ensures that m_j is part of a child branch of br_i. Each feature $\mathscr{F}_i$ is represented by its master branch br_i, which is the branch with the highest persistence within the feature (see Fig. 5(a)) or alternatively the lowest minimum m_i in the set. A possible geometric representation for this feature definition are the components of the iso-contour for the iso-value $s(m_i) + \delta$ which contains at least one minimum of the set $\mathscr{F}_i$ (see Fig. 6).

The influence of changing the parameter δ is demonstrated in Fig. 6. It can be seen how the size and the number of extracted features varies. In general, smaller values

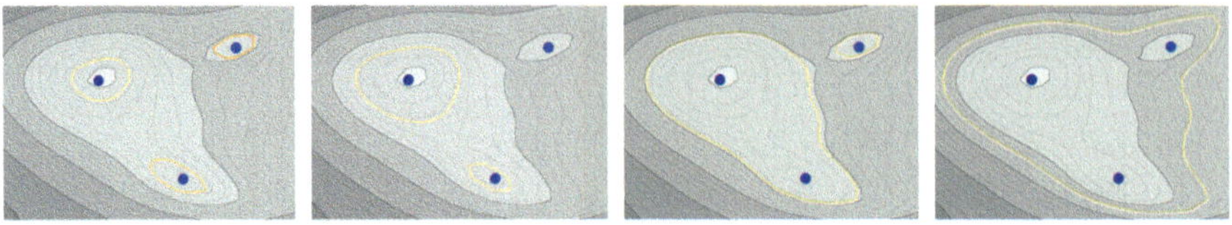

(a) $\delta = 2\%$ (b) $\delta = 5\%$ (c) $\delta = 10\%$ (d) $\delta = 15\%$

Fig. 6 Parameter Influence: The influence of the parameter δ in the feature definition. In (a) $\delta = 2\%$ of the scalar field range. Each of the three minima spans an individual feature. In (b) the value of delta is increased. As a result, the minimum in the top right is not fulfilling the criterion $(d_i - b_i) > \delta$ anymore and therefore does not carry a feature. Further increasing δ to 10% and 15% leads to a feature consisting of 3 minima and 2 separate regions (c) and one region (d) respectively

lead to smaller features, with respect to the number of extremal points contained in them. At the same time the total number of features is decreased.

Feature Tracking: The feature tracking is based on the full tracking graph. Considering two features $\mathscr{F}_i^t$ and $\mathscr{F}_j^{t+1}$ in consecutive time-steps, we define the forward tracking score between these features as:

$$score(\mathscr{F}_i^t, \mathscr{F}_j^{t+1}) = \sum_{m_i^t \in \mathscr{F}_i^t} \sum_{m_j^{t+1} \in \mathscr{F}_j^{t+1}} w(m_i^t, m_j^{t+1}) \cdot fm(m_i^t, m_j^{t+1})$$

with

$$fm(m_i^t, m_j^{t+1}) = \begin{cases} 1 \text{ if } (m_i^t, m_j^{t+1}) \in FM \\ 0 \text{ else} \end{cases}$$

where w determines the importance of the match between two extremal points in consecutive time-steps. The weight w can be proportional to the persistence of extremal points, overlap between the descending manifolds or sublevel sets or other reasonable criteria. Each weight criterion has its advantages and disadvantages. We leave the decision regarding which criteria and weights to use up to the users. In our implementation, by default, we use $w(m_i^t, m_j^{t+1}) = persistence(m_i^t)$ as the weight function.

Now, for a given feature $\mathscr{F}_i^t$ in the time-step t, $score(\mathscr{F}_i^t, \mathscr{F}_j^{t+1})$ can be computed for all features in the time-step $t+1$. Feature $\mathscr{F}_i^t$ is then mapped to the feature $\mathscr{F}_j^{t+1}$ which has the highest *score* value. In addition, we also compute the weight of extremal points in $\mathscr{F}_i^t$ which could not be matched to any feature in time-step $t+1$. If that weight is greater than the maximum matching score, then the feature $\mathscr{F}_i^t$ is not matched to any feature in the next time-step and it dies.

Tracking Events: As described in Sect. 3, the raw tracking graph can be filtered and queried. Recall that this graph only contains mappings between minima and that the used feature description is independent of these mappings. A minimum can (a) spawn a feature, (b) be part of a feature spawned by a different minimum, or (c) exist without any relation to a feature. Since this can also change over time, features can be born

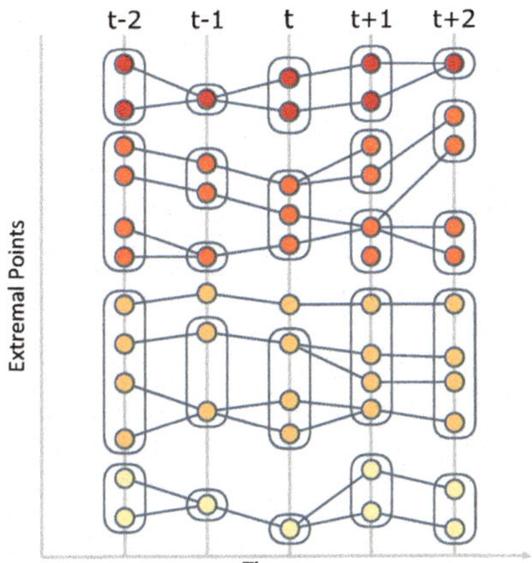

(a) Corespondence of Extremal Points

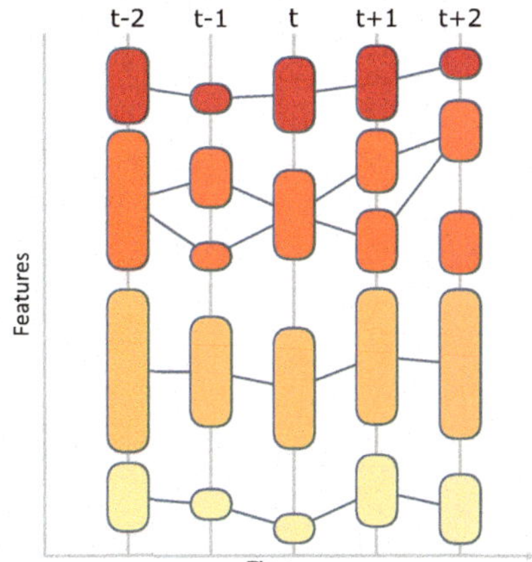

(b) Feature Tracks

Fig. 7 **Tracking Graph:** Abstract representation of the tracking graph. In both figures each time step is represented as a collection of extremal points (EP) aligned on a vertical axis overlaid by feature grouping these critical points. In (a) the connections between the critical points illustrate the point-to-point pairs from the sets *FM* and *BM*. The feature tracks live on top of the raw tracking graph for the extremal points, and are highlighted in (b). In the feature tracks, unmatched features can exist, supporting the detection of feature birth, death, merge as well as splitting events. However, note that in the raw tracking graph there are no unmatched EPs

(birth event) or vanish (death event). Furthermore, a strongly expressed minimum in one time step spawning its own feature can flatten out and be not pronounced enough to carry its own feature. If it is close to another strong minimum of similar scalar value the associated features can now merge. This process and the abstraction between features and minima is illustrated in Fig. 7. With this, features can not only spawn and vanish but also merge and split.

5 Implementation Details

Our un-optimized prototype implements all algorithmic stages, constructs and stores the raw tracking graph and supports feature descriptor design. Additionally, various visualization methods for scalar fields, points and edges as needed for the graph and two dimensional plots are implemented within the same framework. Our implementation uses asynchronous execution and parallel processing where applicable. Currently, the computation of the Morse complex is the most demanding part and takes about 720 ms for a 320×160 grid. During pre-computation of the grid we calculate neighborhood information for the dimension 0 and 1 cells [5]. Note that we are not handling dimension 2 cells since we are only interested in descending manifolds. Computation of the merge tree takes about 170 ms for the same grid size.

Here, we facilitate a union-find data structure as underlying concept. The given timings are measured on a single CPU workstation running a Xeon E5-1650 v4 with 3.6 GHz. Rendering is done using the OpenGL API and a combination of a G-Buffer for deferred shading as well as an A-Buffer for order independent transparency. The current implementation is a proof of concept and we are currently working on the integration of our approach into the *topology toolkit (TTK)* where it also would become open source and freely available to the community.

6 Case Study

In this section we present our case study. Here, we applied our method to the problem of tracking multi-center cyclones in pressure fields.

Data Set: The data set is given as a time-dependent scalar field containing the mean-sea-level (MSL) pressure of an atmospheric simulation. The data sets resolution is 320×160 in longitude and latitude, while the temporal resolution is 120 at 6 h intervals. Figure 8 shows three consecutive time-steps (i.e., time-steps 97, 98, 99) of a subdomain on the northern hemisphere. The pressure field is rendered by a mapping to a yellow to red color-scale. Additionally, the visualization is enriched with iso-contours in fixed intervals.

Feature Descriptor and Tracking Results: As feature descriptor, we applied a persistence-based metric as described in Sect. 4. As mentioned, Fig. 8 shows three consecutive time-steps for a smaller portion of the data-set. The red iso-line indicate the feature area, whereas the blue circles indicate the location of the local pressure minima. In addition, the circles are scaled according to the persistence value of these minima. It can be observed that the geometric embedding of a specific minimum is not stable, meaning the dominant minimum (i.e., the minimum with highest persistence) "jumps" from one minimum to another within a single time-step. Furthermore, a low persistence minimum can appear or vanish within one time-step (see Fig. 8(c)). The used feature descriptor handles both cases as it does not rely on a single critical point. Therefore, the geometric embedding of the feature itself is stable, even if the geometric embedding of individual minima is not stable.

Feature tracking is realized as filtering and querying of our raw tracking graph (see Fig. 4). With this, the forward and backward correspondence of features based on minima correspondence can be calculated. Figure 9 shows a visualization of all features across all time-steps for a specific value of δ. Furthermore, Fig. 9 (b)–(d) show the merge and split of two features over time.

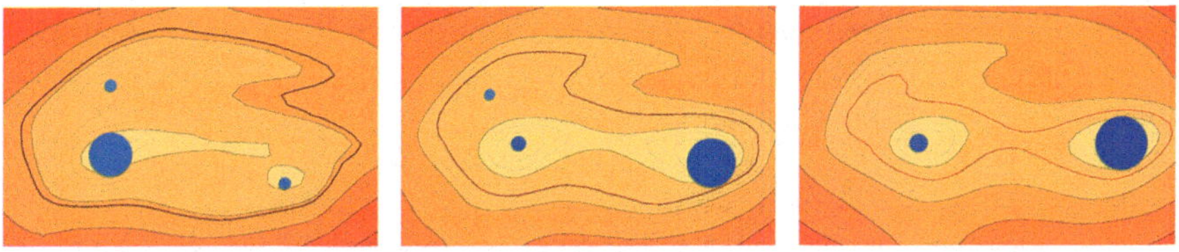

(a) $t = 97$ (b) $t = 98$ (c) $t = 99$

Fig. 8 Feature Descriptor: Three consecutive time-steps of the mean-sea-level pressure field. The spheres indicate the local minima of the field and are scaled by their persistence values. It can be seen that the geometric embedding of the point with maximum persistence is not stable. However, the resulting feature (red iso-line), defined as a combination of the extremal points, is stable

(a) Time Series of Feature Contours

(b) $t = 49$ (c) $t = 50$ (d) $t = 51$

Fig. 9 Feature Tracking: In (a) the iso-contours of all features across all time-steps are shown. The image sequence on the bottom shows the merge of two separate features. In time-step (b) the features are distinct and merge later in time-step (c). Our tracking graph also contains information about the backward direction which enables us to also detect splitting events. The feature in (c) splits into two separate features after (d)

7 Conclusion and Discussion

We presented a flexible concept to define and track features in scalar data-fields based on topological data analysis as a pre-processing step. The goal of this approach is not to define yet another cyclone tracking method but to give the application scientists a tool at hand that supports an easy and interactive investigation of changes in their feature definition. This concept has already been proven to be of value over the course of time when we collaborated with our partners in climate sciences who are experimenting with variations in the definition of cyclones. The core feature of the method that makes varying cyclone definitions possible is a separation of the topological analysis and the feature configuration. The preprocessing step consists of pure topological analysis of the data without any assumptions while following clear rules. Thus, it provides a well-defined foundation for an application-specific feature configuration.

A further advantage of this approach is that the computationally expensive steps are separated into a pre-processing step leaving enough headroom to allow for interactive exploration. Additionally, our current concept can be extended further and adapted to an in situ pipeline. Currently, we are using a critical point tracking method that is based on descending or ascending manifolds. While this tracking gives stable results when following extrema, its accuracy depends on the extraction method of the descending manifolds and can suffer from sub-optimal geometric embedding when using discrete methods. This is a well-known problem and solutions have been proposed in the literature, e.g. [13] or [6], which we plan to integrate in the future. Due to the modular design of our pipeline, the critical point tracking method can be easily exchanged.

In this paper, we have exemplified our concept for the application of cyclone extraction and tracking in pressure fields, but our approach is far more powerful and can be used for other applications as well. There are many possibilities for the extension of the work. One direction that we want to pursue is to extend the pre-analysis step towards a multi-field analysis. Another interesting idea would be to employ user guided classification of critical points into relevant and non-relevant features in a few selected time steps, and using learning algorithms to automatically extract relevant features in the complete time series. To improve the feature configuration and feature tracks, there are a lot of ideas in the literature, for example, to consider global optimization criteria for the tracks similar as proposed by Saikia et al. [17] or Schnorr et al. [19]. For this, a better interface for designing rules to be used as feature descriptors is required. Additionally, we want to extend our prototype in terms of context embedded visualizations and the extraction of statistics, allowing for comparison between different feature descriptors.

Acknowledgements This work was supported through a grant from the Swedish Foundation for Strategic Research (SSF, BD15-0082), the SeRC (Swedish e-Science Research Center) and the ELLIIT environment for strategic research in Sweden. We would also thank Jochen Jankowai for his support and many interesting discussions.

References

1. Acharya, A., Natarajan, V.: A parallel and memory efficient algorithm for constructing the contour tree. In: Proceedings of IEEE PacificVis 2015, pp. 271–278. IEEE (2015)

2. Carr, H., Snoeyink, J., Axen, U.: Computing contour trees in all dimensions. Comput. Geometr. **24**(2), 75–94 (2003)
3. Edelsbrunner, H., Harer, J., Mascarenhas, A., Snoeyink, J., Pascucci, V.: Time-varying Reeb graphs for continuous space-time data. Comput. Geometr. Theory Appl. **41**(3), 149–166 (2008)
4. Günther, D., Reininghaus, J., Wagner, H., Hotz, I.: Efficient computation of 3D Morse-Smale complexes and persistent homology using discrete Morse theory. Vis. Comput. **28**(10), 959–969 (2012)
5. Gyulassy, A., Bremer, P.T., Hamann, B., Pascucci, V.: A practical approach to Morse-Smale complex computation: scalability and generality. IEEE Trans. Vis. Comput. Graph. **14**(6), 1619–1626 (2008)
6. Gyulassy, A., Günther, D., Levine, J.A., Tierny, J., Pascucci, V.: Conforming Morse-Smale complexes. IEEE Trans. Vis. Comput. Graph. **20**(12), 2595–2603 (2014). https://doi.org/10.1109/TVCG.2014.2346434
7. Lukasczyk, J., Weber, G., Maciejewski, R., Garth, C., Leitte, H.: Nested tracking graphs. Comput. Graph. Forum **36**(3), 13–22 (2017)
8. Maadasamy, S., Doraiswamy, H., Natarajan, V.: A hybrid parallel algorithm for computing and tracking level set topology. In: 20th Annual International Conference on High Performance Computing, pp. 1–10 (2012)
9. Neu, U., .: IMILAST: a community effort to intercompare extratropical cyclone detection and tracking algorithms. Am. Meterorol. Soc. (AMS) J. **94**, 529–549 (2013)
10. Nilsson, E., Engelke, W., Friederici, A., Hotz, I.: Tracking and visualizing multi-center cyclones. In: LEVIA 2020: Leipzig Symposium on Visualization in Applications 2020 (2020)
11. Pascucci, V., Cole-McLaughlin, K., Scorzelli, G.: Multi-resolution computation and presentation of contour trees. In: Proceedings of the IASTED Conference on Visualization, Imaging, and Image Processing (VIIP 2004), pp. 452–290 (2004)
12. Peixoto, J.P., Oort, A.H.: Physics of Climate. American Institute of Physics, New York (1992)
13. Reininghaus, J., Günther, D., Hotz, I., Weinkauf, T., Seidel, H.P.: Combinatorial Gradient Fields for 2D Images with Empirically Convergent Separatrices. arXiv:1208.6523v1 (2012)
14. Reininghaus, J., Kasten, J., Weinkauf, T., Hotz, I.: Efficient computation of combinatorial feature flow fields. IEEE Trans. Vis. Comput. Graph. **18**(9), 1563–1573 (2012)
15. Reininghaus, J., Kotava, N., Günther, D., Kasten, J., Hagen, H., Hotz, I.: A scale space based persistence measure for critical points in 2D scalar fields. IEEE Trans. Vis. Comput. Graph. **17**(12), 2045–2052 (2011)
16. Robins, V., Wood, P., Sheppard, A.: Theory and algorithms for constructing discrete Morse complexes from grayscale digital images. IEEE Trans. Pattern Anal. Mach. Intell. **33**(8), 1646–1658 (2011)
17. Saikia, H., Weinkauf, T.: Global feature tracking and similarity estimation in time-dependentscalar fields. Comput. Graph. Forum **36**(3), 1–11 (2017)
18. Schnorr, A., Helmrich, D.N., Denker, D., Kuhlen, T.W., Hentschel, B.: Feature tracking by two-step optimization. IEEE Trans. Vis. Comput. Graph. **26**, 2219–2233 (2018)
19. Schnorr, A., Helmrich, D.N., Denker, D., Kuhlen, T.W., Hentschel, B.: Feature tracking by two-step optimization. IEEE Trans. Vis. Comput. Graph. (TVCG) **26**(6), 2219–2233 (2020)
20. Shivashankar, N., Senthilnathan, M., Natarajan, V.: Parallel computation of 2D Morse-Smale complexes. IEEE Trans. Vis. Comput. Graph. **18**(10), 1757–1770 (2012)
21. Sohn, B.S., Bajaj, C.: Time-varying contour topology. IEEE Trans. Vis. Comput. Graph. **12**(1), 14–25 (2006)
22. Valsangkar, A.A., Monteiro, J.M., Narayanan, V., Hotz, I., Natarajan, V.: An exploratory framework for cyclone identification and tracking. IEEE Trans. Vis. Comput. Graph. **25**(3), 1460–1473 (2018)
23. Weber, G., Bremer, P.T., Day, M.S., Bell, J.B., Pascucci, V.: Feature tracking using Reeb graphs. In: Topological Methods in Data Analysis and Visualization. Theory, Algorithms, and Applications (TopoInVis'09) (2011)
24. Zomorodian, A., Carlsson, G.: Computing persistent homology. Discrete Comput. Geom. **33**(2), 249–274 (2005)

Using Contour Trees in the Analysis and Visualization of Radio Astronomy Data Cubes

Paul Rosen, Anil Seth, Elisabeth Mills, Adam Ginsburg, Julia Kamenetzky, Jeff Kern, Chris R. Johnson, and Bei Wang

Abstract The current generation of radio and millimeter telescopes, particularly the Atacama Large Millimeter Array (ALMA), offers enormous advances in observing capabilities. While these advances represent an unprecedented opportunity to facilitate scientific understanding, the increased complexity in the spatial and spectral structure of these ALMA data cubes lead to challenges in their interpretation. In this paper, we perform a feasibility study for applying topological data analysis and visualization techniques never before tested by the ALMA community. Using techniques based on contour trees, we seek to improve upon existing analysis and visualization workflows of ALMA data cubes, in terms of accuracy and speed in feature extraction.

P. Rosen (✉)
University of South Florida, Tampa, USA
e-mail: prosen@usf.edu

A. Seth · C. R. Johnson · B. Wang
University of Utah, Salt Lake City, USA
e-mail: aseth@astro.utah.edu

C. R. Johnson
e-mail: crj@sci.utah.edu

B. Wang
e-mail: beiwang@sci.utah.edu

E. Mills
Brandeis University, Waltham, USA
e-mail: eacmills@brandeis.edu

A. Ginsburg · J. Kern
National Radio Astronomy Observatory, Charlottesville, USA
e-mail: aginsbur@aoc.nrao.edu

J. Kern
e-mail: jkern@nrao.edu

J. Kamenetzky
Westminster College, Salt Lake City, USA
e-mail: jkamenetzky@westminstercollege.edu

I. Hotz et al. (eds.), *Topological Methods in Data Analysis and Visualization VI*, Mathematics and Visualization, https://doi.org/10.1007/978-3-030-83500-2_6

We review our development process in building effective analysis and visualization capabilities for the astrophysicists. We also summarize effective design practices by identifying domain-specific needs of simplicity, integrability, and reproducibility, in order to best target and service the large astrophysics community.

1 Introduction

Radio astronomy is currently undergoing a revolution driven by new high spatial and spectral resolution observing capabilities. The current generation of radio and millimeter telescopes, particularly the Atacama Large Millimeter Array (ALMA), offers enormous advances in capabilities, including significantly increased sensitivity, resolution, and spectral bandwidth. While these advances represent an unprecedented opportunity to facilitate scientific understanding, they also pose a significant challenge. In some cases, the higher sensitivity and resolution they provide yield new detections of sources with well-ordered structure that is easy to interpret using current tools (e.g., [1]). However, these advances often lead to the detection of structure with increased spatial and spectral complexity, e.g., new molecules in the chemically-rich massive star-forming region Sgr B2, outflows in the nuclear region of the nearby galaxy NGC 253, and rich kinematic structure in the giant molecular cloud "The Brick" [6, 8, 49]. Visualization is a natural tool to study such data, which are typically modeled as 3D cubes, commonly referred to as *ALMA data cubes*, with two spatial dimensions and one spectral dimension (see Fig. 1). While visualizing volumes is not new to scientific visualization, ALMA data cubes present unique challenges. First of all, an ALMA data cube represents the complex interactions of radio signals produced by the bulk mixing and motion of various molecules deep in space. These data tend to have a high spectral resolution but low spatial resolution.

Yet these complex behaviors need precise examination. Second, the data have an extraordinarily low signal to noise ratio. This makes direct visualization impractical as the signal is difficult to extract. Third, the noise is spectrally varying and incoherent, therefore difficult to model and remove using conventional approaches. Due to these unique challenges, visualization alone is insufficient for analysis and exploration.

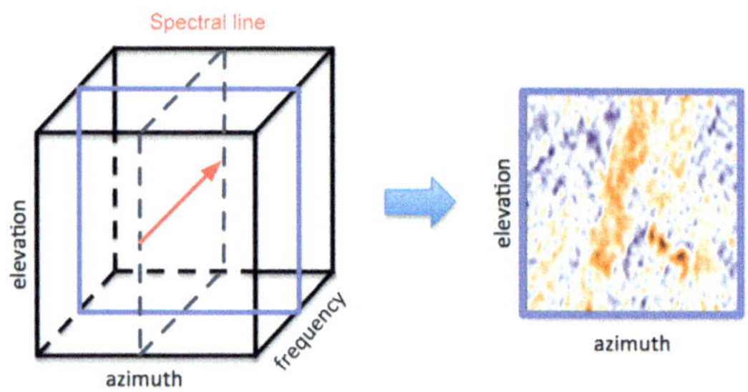

Fig. 1 An illustration of the ALMA data cube and a spectral line.

In this paper, we review our application development process in building effective analysis and visualization capabilities for ALMA data cubes. Our publicly available tool is called ALMA-TDA (https://github.com/SCIInstitute/ALMA-TDA).

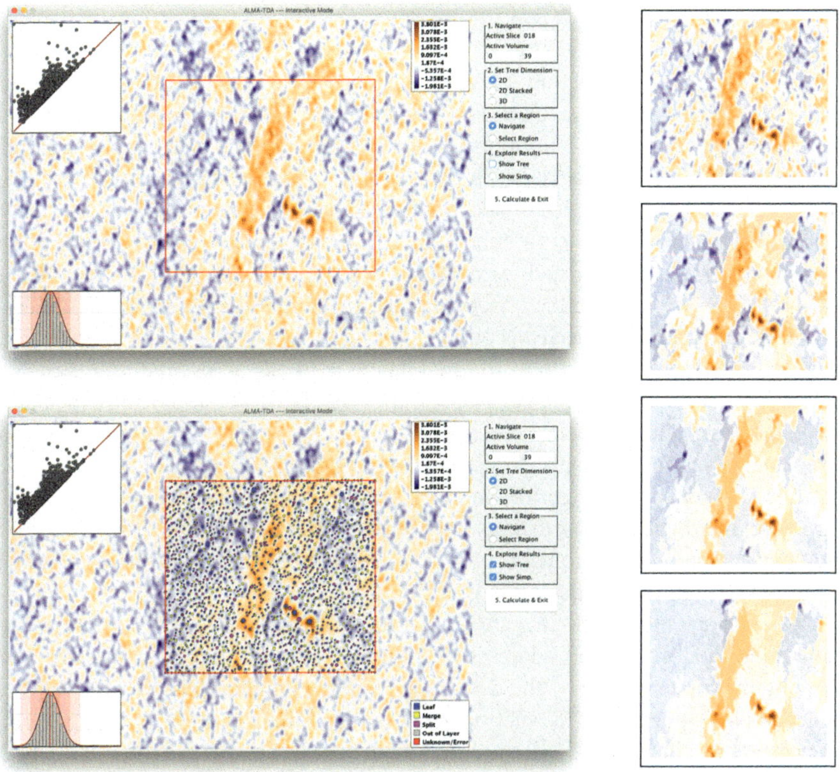

Fig. 2 An example of ALMA-TDA. The results of varying the simplification level on slice #18 of the Ghost of Mirach data set using a 2D contour tree. On the top left, a visualization of the original data is shown. On the bottom left, the contour tree computed on the region is shown with circles at critical point locations. Finally, on the right, the results of simplifying the data with simplification levels at 0.0005, 0.001, 0.0015, 0.002, from top to bottom, respectively

ALMA-TDA uses contour trees to extract and simplify the complex signals from noisy radio astronomy data. An example of our tool is shown in Fig. 2. We additionally summarize effective design practices targeting and servicing the large astrophysics community. In particular, we need to design tools with *simplicity* (i.e., light-weight), *integrability* (i.e., integrable within existing toolchains), and *reproducibility* (i.e., fully recorded analysis history via command-lines). We hope such learned design practices will provide guidelines toward the future development of tools and techniques that would benefit astrophysicists' scientific goals.

2 Science Case

The new complexity involving ALMA data cubes brought about by increased sensitivity, spatial and spectral resolution, and spectral bandwidth has become a significant bottleneck in science, as it not only challenges astrophysics analysis tools but also the users' ability to understand their data. For example, increased complexity in the spatial and velocity structure of spectral line emission makes even a single spectral line hard to interpret. When a cube contains the superposition of multiple spatial and kinematic structures, such as outflows and rotation and infall, each with their own relationship between the observed velocity and their actual position along the line of sight, traditional analysis and exploration tools do not perform well. Users (the astrophysicists) struggle to follow kinematic trends across multiple structures by examining movies or channel maps of the data. However, moment map analysis (e.g., integrated fluxes, mean velocities and mean line-widths), the most commonly used analysis tool for compressing this 3D information into a more easily parsed 2D form, no longer has a straightforward interpretation in the presence of such complex structure, in which mean velocities may be velocities at which no emission is actually present, and mean line widths may represent the distance between two velocity components, rather than the width of a single component.

Whether scientists can navigate and correctly interpret this new complexity will determine their success in addressing a number of important scientific questions. Among the topics driven by the detection of more complex structures are ISM turbulence [19, 48], the star formation process [32], filaments [46], molecular cloud structure and kinematics [49], and the kinematics of nearby galaxies [28, 31, 36] and high redshift galaxies [2, 58].

An even greater challenge arises from our ability to detect an increased number of spectral lines in more and more sources. There simply are no tools capable of simultaneously visualizing, comparing, and analyzing the dozens to hundreds of data cubes for all of the detected spectral lines in a given source. Standard methods that visualize the data as moment and channel maps, animate cubes as videos or 3D models, cannot scale up to the case involving large numbers of lines, even in non-complex, well-ordered cases, such as rotating disks or expanding stellar shells. Users become overwhelmed by, for example, comparing these typical diagnostics for two lines, side by side or one at a time. In the richest sources with thousands of lines, such comparisons will simply be impossible—it becomes necessary to resort to methods that entirely throw away either the spectral information of moment maps or the spatial information that requires model fitting of complex spectra (such as Principle Component Analysis). As a result, both exploration and analysis of the astronomy data becomes time consuming and potentially incomplete.

As we move into the future and the telescopes reach their full potential, complex spatial and velocity structures will no longer be a problem that typically occurs in a separate subset of sources than those exhibiting rich spectra behaviors—the two problems will coexist, compounding the highlighted issues. The visualization and analysis challenges currently facing radio astronomy will then only grow more pressing as the data volumes increase, and the instruments grow more sensitive.

Existing Tools. A critical aspect to the study of ALMA data cubes is the detection, extraction, and characterization of objects such as stars, galaxies, and blackholes. *Source finding* in radio astronomy is the process of detecting and characterizing objects in radio images (in the forms of data cubes), and returning a survey catalog of the extracted objects and their properties [26, 59]. A common practice is to use a computer program (i.e., a source finder) to search the data cubes, followed by manual inspection to confirm the sources of electromagnetic radiation [59]. An ideal source finder aims to determine the location and properties of these astronomical objects in a complete and reliable fashion [26], while manual inspection is often time-consuming and expensive.

Several existing tools have been used in the ALMA community in terms of source finding [27], including the popular ones such as clumpfind [60], dendrograms [54], cprops [53], and more recent ones such as FellWalker [7], SCIMES [18] and NeuroScope [37]. Clumpfind is designed for analyzing radio observations of molecular clouds obtained as 3D data cubes; it works by contouring the data, searching for local peaks of emission, and following them down to lower intensity levels [60]. The dendrograms of a data cube is an abstraction of the changing topology of the isosurfaces as a function of contour level, which captures the essential features of the hierarchical structure of the isosurfaces [54]. The FellWalker algorithm is a gradient-tracing watershed algorithm that segment images into regions surrounding local maxima [7]. FellWalker provides some ability to merge clumps, therefore simplify the underlying structures, and the merging criteria share some similarities with persistence-based simplification. However, these criteria are less mathematically rigorous compared to our approach. SCIMES (Spectral Clustering for Interstellar Molecular Emission Segmentation) considers the dendrogram of emission under graph theory and utilizes spectral clustering to find discrete regions with similar emission properties [18]. Finally, the most recent NeuroScope [37] (specifically targeted for ALMA data cubes) employs a set of neural machine learning tools for the identification and visualization of spatial regions with distinct patterns of motion.

However, the study of source finding for ALMA data cubes raises the following question: How can we help the astrophysicists to understand the *denoising* process? That is, how to best separate signals from noise, and to understand the effects of denoising on the original data? In other words, it is important for us to *quantify both signals and noise* as well as to *perform simplifications* of the underline data. This kind of study is current underdeveloped within the ALMA community.

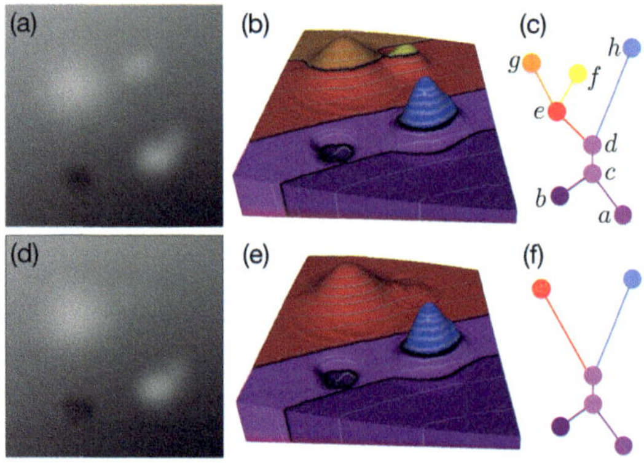

Fig. 3 **a** A grayscale image of a 2D scalar function before simplification. **b** height map of the contours corresponding to the scalar function shown in **a**. **c** The contour tree structures capture the evolution of terrain features (i.e., relations among local minima, local maxima, and saddles). **d**–**f**: The grayscale image, height map, and the contour tree after simplifying the features

3 Technical Background

From a technical perspective, we focus on performing data analysis and designing effective visualization of ALMA data cubes by employing the contour tree [11]. The contour tree is a mathematical object describing the evolution of the level sets of a scalar function defined on a simple, connected domain, such as the grayscale intensity defined on the 2D domain associated with a slice of a data cube (at a fixed frequency). There are two key properties associated with a contour tree, making it a feasible tool in the study of ALMA data cubes. First, a contour tree has a graph-based representation that captures the changes within the topology of a scalar function and provides a meaningful summarization of the associated data. Second, a contour tree can be easily simplified, in a quantifiable way, to remove noise while retaining important features in data.

Contour Trees. Scalar functions are ubiquitous in modeling scientific information. Topological structures, such as contour trees, are commonly utilized to provide compact and abstract representations for these functions. The contour tree of a scalar function $f : \mathbb{X} \rightarrow \mathbb{R}$ describes the connectivity of its *level sets* (isosurfaces) $f^{-1}(a)$ (for some $a \in \mathbb{R}$), whose connected components are referred to as *contours*. Given a scalar function defined within some Euclidean domain $\mathbb{X}$, the contour tree is constructed

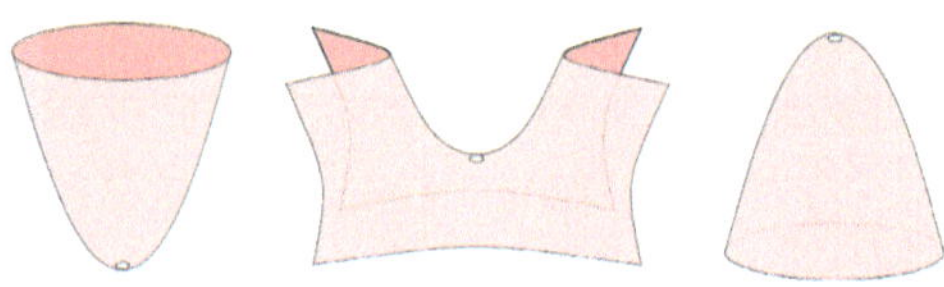

Fig. 4 Local structures of critical points. From left to right: a local minimum, a saddle point, and a local maximum

by collapsing the connected components of each level set to a point. The contour tree stores information regarding the number of components at any function value (isovalue) as well as how these components split and merge as the function value changes. Such an abstraction offers a global summary of the topology of the level sets and enables the development of compact and effective methods for modeling and visualizing scientific data. See Fig. 3(a)–(c) for an illustrative example. Vertices in the contour tree correspond to *critical points* of the 2D scalar function, namely, local minima, saddle points, and local maxima, whose local structures are illustrated in Fig. 4.

Persistence. To simplify a contour tree, we assign an importance measure to each edge of the tree and collapse (eliminate) edges of lower importance measures [10, 12]. Various geometric properties, such as persistence, volume, and surface area, can be used to compute the importance measure.

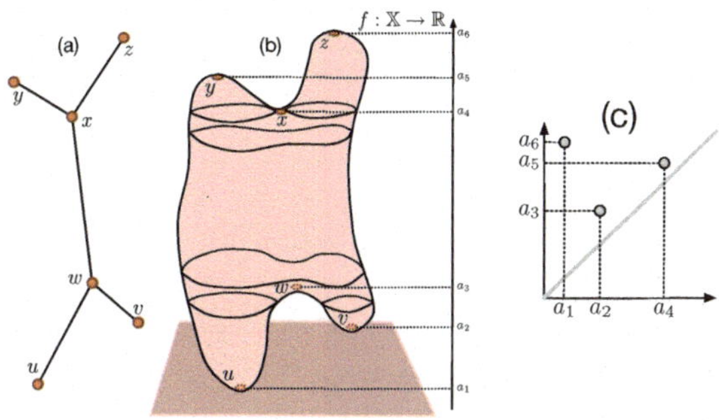

Fig. 5 Example of persistence pairing of critical points for **b** a 2D height function. The persistence pairing of **a** critical points from the contour tree gives rise to **c** a (scaled) persistence diagram

We apply ideas from topological persistence [21] in our feasibility study. We describe the idea of persistence using Fig. 5 as an illustrative example [20, Page 163]. Given a height function $f : \mathbb{X} \to \mathbb{R}$ defined on a 2D domain, let $\mathbb{X}_a$ denote the sublevel set of f, that is, $\mathbb{X}_a = f^{-1}(-\infty, a]$. Suppose we sweep a horizontal plane in the direction of increasing height values and keep track of the (connected) components of $\mathbb{X}_a$ while increasing a. A component of $\mathbb{X}_a$ *starts* at a local minimum and *ends* at a (negative) saddle point when it merges with an older component (i.e., a component that starts earlier). This defines a minimum-saddle persistence pair between critical vertices, and the *persistence* of such a pair is the height difference between them. Similarly, a hole/tunnel of $\mathbb{X}_a$ starts at a (positive) saddle point and ends at a local maximum (where it is capped off). This defines a saddle-maximum pair, with its persistence being the height difference between its vertices. In a nutshell, minima stars components, saddles merge components or create tunnels (complete loops), and maxima fill holes [20, Page 162].

Referring to Fig. 5: points u and v are local minima; y and z are local maxima; w is a negative saddle point; and x is a positive saddle point. Their corresponding height values are sorted as $a_1 < a_2 < \cdots < a_6$. We sweep a horizontal plane in the direction of increasing height value and keep track of the components in the sublevel set. The pair (v, w) forms a minimum-saddle persistence pair, as a component in the sublevel set starts at point v, and it merges with an older component that starts at point u. The pair (x, y) form a saddle-maximum pair. The pair (v, w) has a persistence of

$|a_2 - a_3|$, while the pair (x, y) has a persistence $|a_5 - a_4|$. The contour tree is shown in Fig. 5(a).

Persistence Diagram. The pairing of critical points also gives rise to a *persistence diagram* [16] that summarizes and visualizes topological features of a given function. A persistence diagram contains a multi-set of points in the plane; its x- and y-coordinates captures the start (*birth*) time and the end (*death*) time of a particular topological feature. The distance of the point to the diagonal captures the persistence of that feature. Points away from the diagonal have high persistence and correspond to signals of the data, while points that are close to the diagonal have low persistence, which is typically treated as noise.[1]

In the example of Fig. 5, the critical point pairs (x, y) and (v, w) give rise to points (a_4, a_5) and (a_2, a_3) in the persistence diagram, respectively. This persistence diagram also contains an additional off-diagonal point (a_1, a_6), which corresponds to the pairing of global minimum u with the global maximum z that captures the entire shape of data. This is a global feature that can not be simplified (see [17] for technical details).

In our context, we only care about minimum-saddle and saddle-maximum pairs.

Contour Tree Simplification. In the contour tree example of Fig. 3(c), the pair (b, c) is a minimum-saddle pair while the pairs (e, f) and (d, g) are saddle-maximum pairs. During a hierarchical simplification, the pair (e, f) has the smallest persistence. Therefore the edge connecting them is collapsed (simplified), as shown in Fig. 3(f); this can be achieved by a smooth deformation of the surface in Fig. 3(d). In this paper, we focus on the persistence-based simplification—other simplification schemes may be employed based on local geometric measures for individual contours [12], for instance, surface area and contained volume; we intend to use these geometric measures in the future to perform contour tree simplification that suppressing minor topological features of the astronomy data.

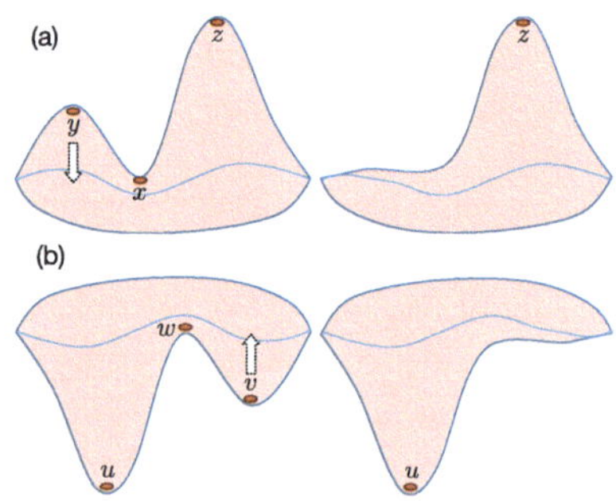

Fig. 6 Simplifying a saddle-maximum pair (x, y) in **a** and minimum-saddle pair (v, w) in **b** for a 2D scalar field. **a**: We reduce the height of the local maximum y to the level of saddle x, effectively flattening the region surrounding y. **b**: We increase the height of local minimum v to the level of the saddle w, again flattening the region surrounding v

Scalar Field Simplification. Given a contour tree simplification, we would like to compute its corresponding scalar field simplification. Simplifying a scalar function directly in a way that removes topological noise as determined by its persistence

[1] Based upon sublevel set filtration, topological features typically appear as points in the upper left corner of the persistence diagram; points in the lower right corner correspond to features in superlevel set filtration and/or extended persistence [17], which are not the focus of this paper.

diagram has been investigated extensively (e.g. [22]). As pointed out by Carr et al. [13], contour tree simplification has well-defined effects on the underlying scalar field: collapsing a leaf corresponds to leveling off (or flattening) regions surrounding a maximum or a minimum. This is a desirable simplification for the domain scientists, as they are interested in reducing noise to zero flux during the denoising process. Figure 3(b) and (c) demonstrate the result of edge collapsing: collapsing the edge (e, f) from the tree results in flattening the yellow region surrounding the local maximum f; this is equivalent to introducing a small perturbation to the neighborhoods of saddle-maximum pair (e, f) so that both critical points e and f are removed. Such a flattening process is further highlighted in Fig. 6.

Implementation. Both contour tree calculation and scalar field simplification use our own implementations of the associated algorithms.

Related Work. The contour tree was first introduced by van Kreveld et al. [30] to study contours on topographic terrain maps (i.e., curves containing sampled points with the same elevation values). It has since been widely used for both scientific and medical visualizations [3, 44, 55, 56]. Efficient algorithms for computing the contour tree [11, 15, 47] (and its variants, merge tree [41], and Reeb graph [45]) in arbitrary dimensions have been developed. The calculation of contour trees is theoretically $\mathscr{O}(n \log n)$. However, the actual running time is approximately $\mathscr{O}(n)$. The latest state-of-the-art regarding contour trees have been parallel or distributed implementations [9, 14, 25, 33, 38, 39]. We use an approach described in [52], which is implemented under the piecewise linear setting.

A related concept called dendrograms has been used in astronomical applications to segment data and to quantify hierarchical structure such as the amount of emission or turbulence at a given size scale, for example, to study the role of self-gravity in star formation [24]. A dendrogram is a tree-diagram typically used to characterize the arrangement of clusters produced by hierarchical clustering. It tracks how components (clusters) of the level sets merge as the function value changes, while a contour tree captures more complete topological changes (i.e., merge and split) of the level sets. The state-of-the-art Astronomical Dendrogram method [50] has limited capabilities in automatic data denoising, feature extraction, and interactive visualization.

4 Application Development Process

We revisit our application development process in building effective analysis and visualization capabilities of ALMA data cubes by reviewing the timeline of our project. We reflect on key activities with the goal of learning from experience and summarizing effective design targeting and serving the astrophysics community, and how different members of our team interact with one another, including computer scientists (both visualization and TDA experts) and radio astronomers. To give an overview of our design process, we describe the critical activities as identified in [34]: understand, ideate, make, and deploy.

The discussion of the project started in November 2014, when National Radio Astronomy Observatory (NRAO) scientist Dr. Jeff Kern, a coauthor of this paper, visited the Scientific Computing and Imaging (SCI) Institute and saw a talk on the topic of topological data analysis. Over the following months, follow-up conversations generated some initial excitement regarding the potentially applying topological techniques in understanding ALMA data cubes, which has never been done before.

To *understand* the problem domain and target users, we identified key opportunities, that is, applying emerging techniques from topological data analysis to the study of ALMA data cubes. The main motivation stemmed from Jeff's comments that "there simply are no tools capable of simultaneously visualizing, comparing, and analyzing the dozens to hundreds of data cubes for all of the detected spectral lines in a given source." We believed that introducing topological data analysis techniques to the ALMA community would potentially offer new insights regarding feature detection, as well as improve their workflow efficiency.

The *ideate* activity of the project started in May 2015, as the domain problems became better characterized and possible solutions via contour tree-based approaches appeared to have the greatest potential among the solution space. We externalized our ideas and expected technical challenges, while at the same time, formulating a potential analysis pipeline, visual encodings, and selecting interactive capabilities within a proposed system for ALMA data cubes.

By January 2016, we have already met with astrophysicists at the NRAO facility to learn their needs and conducted an on-campus interview with astrophysicist, Dr. Anil Seth, another collaborator of this project, who works with ALMA data cubes. We learned the typical pipeline in the analysis and visualization of ALMA data cubes, specifically, in Anil's case, via image editing tools or file viewers such as QFitsView [43] and SAOImage DS9 [29]. We also gave short tutorials regarding our proposed techniques to obtain comments and feedback from all our interactions.

We started our *make* activity by constructing a tangible prototype, specifically encompassing visualization decisions and interaction techniques. The process coupled the ideate and make activities in the design and refinement of our system. We identified that *quantification* (of signals and noise) and *simplification* are two of the most important aspects for our proposed framework. We went through multiple rounds of interface mock-ups and functionality discussions. We showcased our first prototype between June and August 2016, including one-on-one discussions with Anil and Dr. Julia Kamenetzky on our team, and through a number of talks given to the astrophysics community, with generally positive feedback.

Over the course of the next half a year, we rolled out multiple phases of *deploy* activities in order to put the prototype in a real-world setting to understand how to improve its effectiveness and performance. Our goal was to have a usable system that helps with the users' data-specific tasks. In January 2017, we organized a one-day workshop where we engaged in panel discussions on the current version of the prototype, gathered comments and suggestions, and discussed potential research and developmental directions moving forward. This workshop, in particular, helped to cement the lessons learned.

4.1 Designing to Serve the ALMA Community

Throughout the development process, we learned a few best practices for serving the ALMA community: simplicity, integrability, and reproducibility.

In terms of *simplicity*, the tool should have a sufficient but not overwhelming amount of visualization; and minimize GUI interactions. This philosophy is in sharp contrast with some of the common practices of many visualization tools, where we aim to create novel, exciting, and sometimes flashy visualizations. Our initial prototypes were full of many unnecessary functionalities and complex GUIs. We learned via feedback on user practices that a complex interface will distract or confuse the users to the point that they would not even try using the software. The tool should also be light-weight. That is, it should be easily installed on a desktop computer and not require extensive external dependencies or packages to be installed. For this reason, we chose Java as a platform from the beginning. Though not the most efficient, Java software is highly portable. This is well-aligned with properties of commonly used processing tools in the ALMA community.

In terms of *integrability*, the tool should be integrable with existing workflows and toolchains. This means that the core functionality of the software needs to be automatable. In addition to providing a GUI, we also provide a command-line interface for generating results, such that it can one day be integrated with other tools such as CASA (Common Astronomy Software Applications) [35], astropy [51], or SAMP (Simple Application Messaging Protocol) [57].

In terms of *reproducibility*, the analysis history using our tool should be recorded so that the results can be reproduced. This is supported in two ways. First, by enabling processing via the command-line, we can save parameters and automatically rerun the results later. Second, we minimize the amount of GUI interactions, as most of such interactions are exploratory and do not necessarily contribute to the final analysis. When the user is satisfied with their results using our visualization, the exact command required to reproduce the visualization results is output to the command-line for future reference.

5 Software Design

Our software is a visualization tool with both command-line only and interactive visualization operating modes.

The command-line mode provides a small set of options for complete reproducibility of any computation. Those options are:

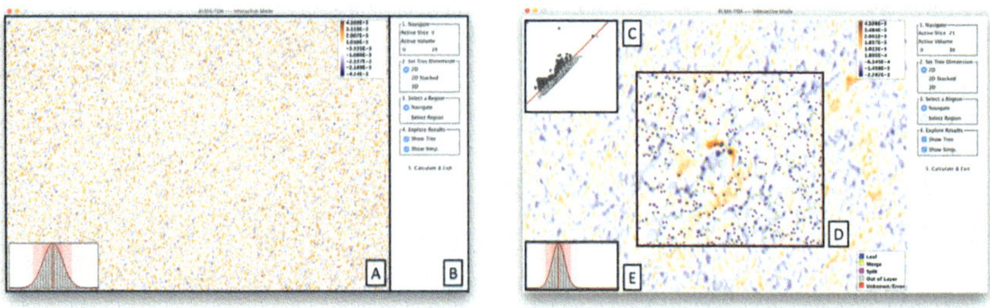

Fig. 7 Upon loading the software in the interactive mode, the user is presented with a view of the data. Left: Initial view of the software. Right: Visualizations are shown as the user selects a region of interest, and the contour tree is calculated (on the back end)

- **Input file** – Path to the file for processing.
- **X, Y, & Z range** – The dimensions of the region to be processed.
- **Simplification type** – Either 2D, for a single slice; 2D Stack, for a series of 2D slices; or 3D, for volume processing.
- **Simplification level** – Persistence level for feature simplification.
- **Output file** – Path to save results.

We also provide an interactive visualization mode to explore the capabilities of our approach and select these parameters. When starting the software, users need only add the "interactive" tag to the command-line, and the visualization launches.

The visualization initially opens to the interface seen in Fig. 7 (left). The interface is designed to include only minimal required capabilities. The main window, **A**, shows visualizations related to the loaded data cube. The GUI component, **B**, provides controls to set options for processing the data. The controls are placed in groups, numbered for steps 1–5. The GUI component is designed with both simplicity and functionality in mind to offer the users the most intuitive and yet fully-functional analysis capabilities.

5.1 Visual Elements

The visualization is composed of five main visual elements.

Scalar Field View (Fig. 7A). Being a sampling of radio waves, the 2D scalar field (a slice of ALMA data cube along the frequency axis) has both positive and negative amplitudes. It is therefore displayed using a divergent orange/purple colormap. By default, the first slice is selected and viewed by centering on the middle of the domain. The user can translate and zoom with the mouse. Different slices can be selected by changing the values in the controls in Fig. 7B.

Persistence Diagram (Fig. 7C). Once the contour tree is calculated, the data are displayed using a persistence diagram. Being that the distance from the diagonal is an analog to persistence, we use this visualization for interactively selecting the level of simplification by dragging the red simplification bar. Features below the bar are grayed out, indicating that those features will be simplified. Once released, a simplified contour tree (on the back-end) and a simplified scalar field (for the front-end) are calculated.

Contour Tree (Fig. 7D). Displaying the tree structure of the contour tree is not particularly meaningful, as it is both large and an abstract view of the data. However, seeing critical points and their persistence in the context of the data is valuable. The critical points are placed over the scalar field view at their respective spatial location. Their size is set based upon their persistence (higher persistence, larger point). Finally, their color is set by their type: local extrema (leaf of the tree) – blue, negative saddle points (merge) – yellow, and positive saddle points (split) – magenta. For 3D analysis, contour tree nodes off of the layer are colored gray. This view of the contour tree can be enabled or disabled on demand using the controls in Fig. 7B.

Simplified Scalar Field (Fig. 7D). Since users are, in large part, interested in the feature extraction power of this approach, we show the result of scalar field simplification in context. As the user adjusts the level of persistent simplification, the scalar field is simplified and overlaid with the original visualization. This view can be enabled or disabled on demand using the controls in Fig. 7B.

Histogram (Fig. 7E). A histogram is produced, indicating the distribution of (intensity) values of data cubes within the current view. In addition to showing histogram bins, this view shows the mean as a solid red line and ± 3 standard deviations as consecutively lighter red bars. This histogram is adapted as the user navigates their view or when the simplification level of the scalar field is adjusted. This view is important, as domain experts are interested in quantifying the total flux gained or lost during simplification. This is most observable by shifts in the mean.

5.2 *Interaction Process*

Though the use of our software requires some explanation, we strive to make it as simple to use as possible. Part of this effort is providing a simple and intuitive five-step approach to the users.

Step 1: Navigation. The users are first asked to navigate the view to the general region of interest. This includes translation and zooming, but it also includes selecting the slice or volume of interest.

Step 2: Tree Dimension. The dimension of the contour tree calculation must be selected next. The options include 2D, for a single slice; 2D stack, for 2D computation on a series of slices; and 3D for computation on a volume. These options will be discussed further in the case studies.

Step 3: Region Selection. Next, the specific region of interest is selected with the mouse. As soon as the mouse is released, computation begins. If the region is large, the

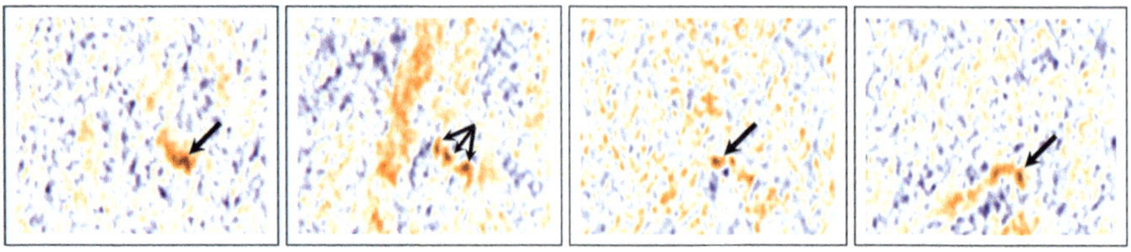

Fig. 8 Four slices, #16, 18, 20, & 22 from the Ghost of Mirach data set. The bright red spots (indicated by the arrows) in these images are the signal of interest

user is prompted with the option to cancel, due to computation time. We are actively investigating scalable contour tree computations to support larger data cubes with on-the-fly visualization.

Step 4: Exploration. Once the computation is completed, the user is invited to explore the domain. This includes navigation (translation, zooming, and changing slices) and adjustment of the simplification level. As simplification levels are adjusted, the user can observe changes in the scalar field, compare those changes to the original field, and look for changes in flux in the histogram.

Step 5: Compute and Exit. Steps 1–4 may be repeated as many times as necessary until the user is satisfied. Once done, the user clicks "Compute and Exit". This will trigger the processing of the data cube and the saving of the output. Finally, the precise command required to reproduce the results will be printed on the command-line.

6 Case Studies

We show the capabilities of our prototype with two case studies involving specific ALMA data cubes used by coauthors.

6.1 Ghost of Mirach Galaxy Data Set

NGC 404 (also known as Mirach's Ghost) is data of a molecular gas emission at the center of the nearby, low mass galaxy. The data was taken using ALMA on Oct. 31, 2015. A data cube is created using the default ALMA pipeline and involves a Fourier transformation of the interferometric data at each frequency. The data cube is approximate 4.5 GB with a resolution of 5400×5400 in the spatial domain and 30 in the spectral domain (i.e., 30 slices). However, the feature of interest is around 200×200 in size and covers around 10 slices. Scientists often sample cubes much larger than their feature of interest to reduce some structured errors, vignetting, for example.

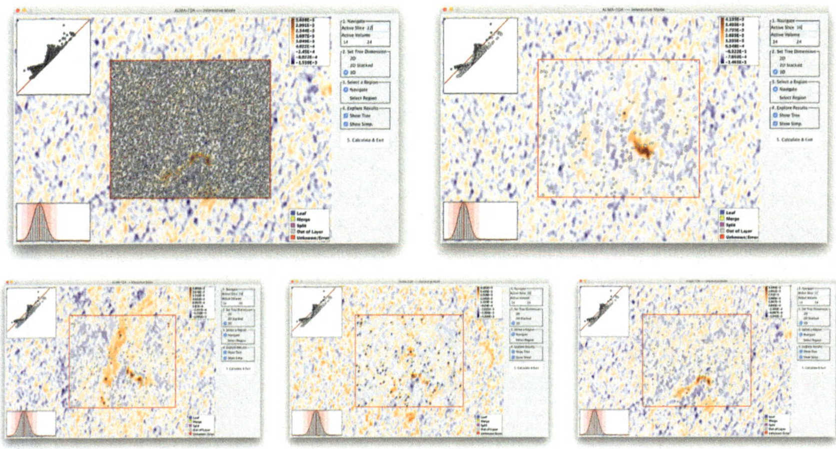

Fig. 9 The result of simplifying using the 3D contour tree on the Ghost of Mirach data set is worse than expected due to topological pants (tubes connecting through slices). Top left: Visualization of the 3D contour tree on slice 22. Top right: Simplification of slice 16. Bottom: Simplification of slices 18, 20, & 22, respectively. The persistent simplification level is 0.00128. At this level of simplification, many features (we visually consider) as noise remain and, in fact, will not be removed by multiple levels of simplification

Science Description. Excited molecular carbon monoxide gas emits light at 230 GHz. The Doppler shifts of this line emission can provide information on the motion of molecular gas in the galaxy. Visualization of the data of NGC404 shows a clear rotating disk located within the central 20 light-years of the galaxy (see Fig. 10). Similar rotating molecular gas disks have been used to measure the masses of supermassive black holes at the centers of galaxies (e.g. [4, 42]). However, the data is noisy, so coherent gas structures are hard to pick out. NGC 404 presents a special challenge due to the low mass of its black hole [40]. Fortunately, the high angular resolution of ALMA provides the highest sensitivity for measuring the black hole mass.

We can see an example of 4 spectral slices of the data set in Fig. 8. In these 4 slices, the bright red spots represent the signal, while most of the remaining patterns represent noise.

Varying Simplification Levels. Figure 2 shows an example of performing simplification on a single 2D spectra (i.e., a single slice along the frequency axis). The noisy structure is captured by the 2D contour tree as many low persistence features (bottom left). Increasing the level of simplification removes much of this noise (right). However, selecting a simplification level that is too aggressive may result in loss of signal (bottom right).

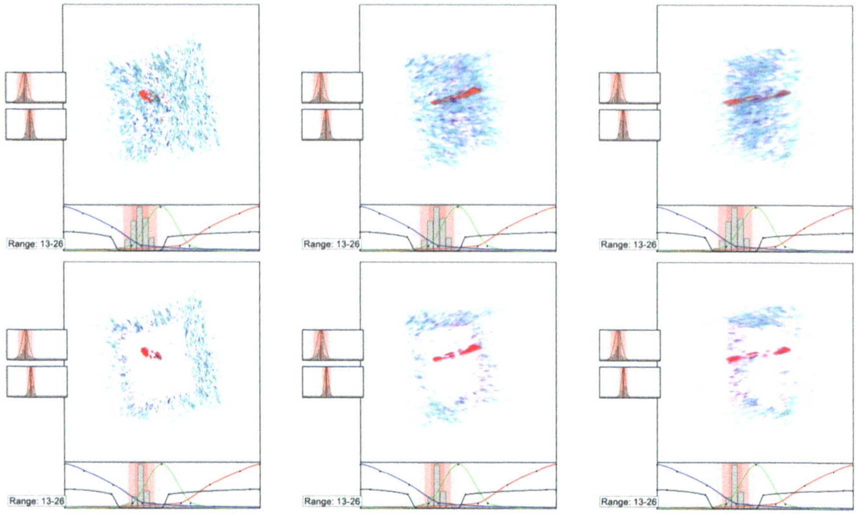

Fig. 10 Result of volume rendering the Ghost of Mirach data set before (top) and after (bottom) using a stack of images with 2D contour trees. The columns show 3 different viewing angles of slices 13–26. The persistent simplification level is 0.00138. Side views (middle and right) of the rendered volume are blurry due to lower resolution along the spectral dimension compared with the two spatial dimensions

3D Contour Trees. Since the spectral data are treated as cubes, our collaborators are interested in the structures that would be found using 3D contour trees. The result of capturing the 3D contour tree, shown in Fig. 9, is both a surprise and a disappointment. Although many critical points are found, the data suffer from *topological pants*—a sphere with three disjoint closed discs removed [5]. Essentially, the 3D contours of noisy features form a complex interconnect tubes through the volume that are not physically meaningful (see Fig. 11). This interferes with the kind of features that a contour tree can identify. The root cause of this is that each of the spectra is processed independently, and thus, there is no correlation between noise patterns across consecutive slices. Simplifying these temporal noise patterns as a whole is not physically meaningful, and they interfere with true features in the data. In addition, while (formally) wavelength (as the 3rd dimension) is continuous, phenomenon such as Rayleigh diffraction means that nearby spectra can differ radically; such a phenomenon, together with (essentially) independent noise on individual spectral channels, makes the assumption of a continuous 3rd dimension rather weak. The noise characteristic, therefore, drives our choice of 2D contour tree simplification instead of the 3D version.

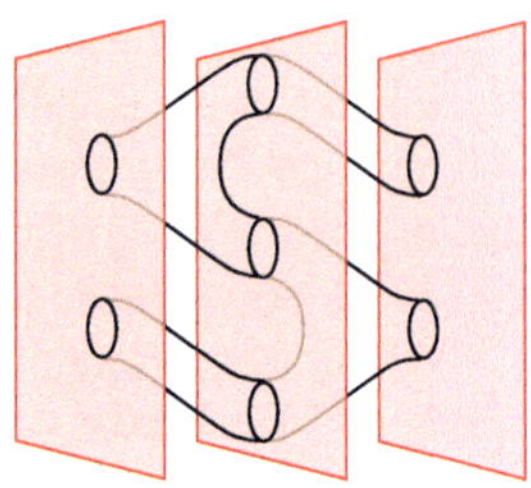

Fig. 11 Illustration of topological pants, a series of interconnected tubes, on 3 layers of a volume.

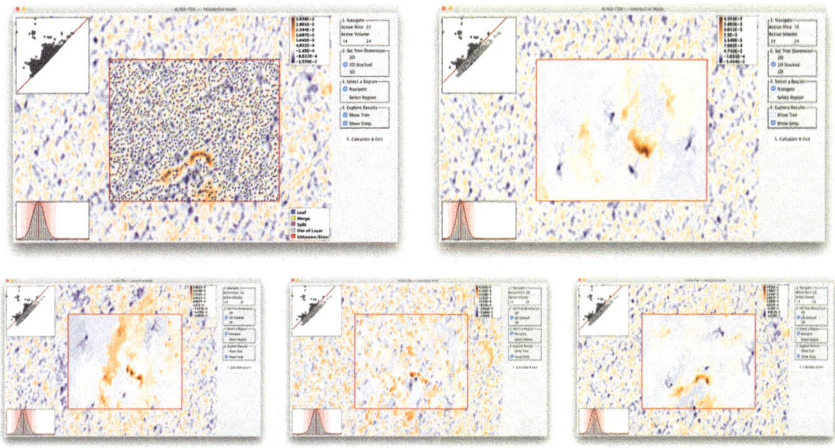

Fig. 12 Result of simplifying the Ghost of Mirach data set using a stack of images with 2D contour trees. Top left: Visualization of the 2D contour tree on slice 22. Top right: Simplification of slice 16. Bottom: Simplification of slices 18, 20, & 22, respectively. The persistent simplification level is 0.00138. The simplification level is good for all except slice 20, where a more aggressive level of simplification is called for

2D Contour Tree Stacks. On the other hand, the processing of 2D contour trees is highly successful. However, domain scientists still need the ability to process 3D cubes. The obvious solution is to use a series of 2D contour trees to control the simplification. Figure 12 shows the result of simplifying a stack of spectra (slices). This example uses a similar level of simplification to the 3D contour tree example in Fig. 9. In our implementation, the level of simplification is shared between all slices. This works well for slices 16, 18, and 22 (top right, bottom left, and bottom right, respectively). However, the level of simplification is not aggressive enough for slice 20 (bottom middle). At this point, the user could either select a more aggressive simplification or could choose to simplify slice 20 separately from the others. Figure 10 shows the stack of slices 13–26 drawn using a custom-built conventional volume renderer. Despite the natural denoising properties of volume rendering, the results without persistent simplification (Fig. 10 top) are difficult to interpret when compared to those with contour tree stack simplification (Fig. 10 bottom).

Moment 0 Analysis. Astrophysicists often use what is known as *moment analysis* to reduce the 3D spectrum to 2D images. Moment 0, 1, and 2 measure the mass of gas, the direction of gas movement, and the temperature of the gas, respectively. They are all integrals across the spectra. To demonstrate the noise reducing power of our approach, we show the result of the moment 0 analysis in Fig. 13 on the 2D stack simplification from Fig. 12. Moment 0 is calculated as $m_0 = \int I_v$, where I is the intensity for a given spectrum, v. By removing the noise from each of the layers, the resulting moment map is significantly less noisy, making the signal itself very apparent. Our collaborator also finds the dim feature pointed to by the arrow

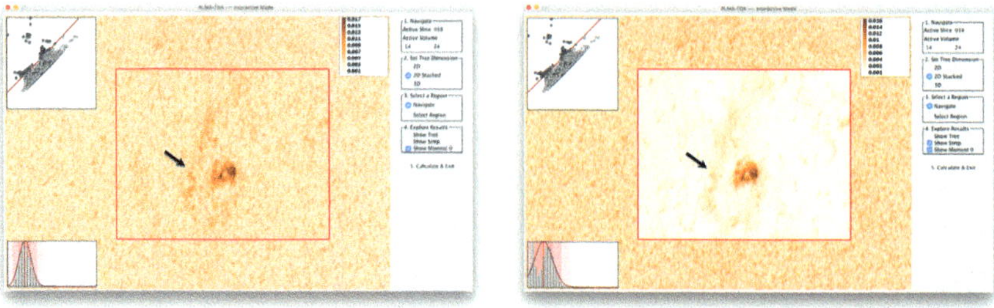

Fig. 13 Moment 0 analysis of Ghost of Mirach data set between slices 14 and 24 (the range of the signal) using a stack of 2D contour trees. Left: Visualization of moment 0 for the original data. Right: Moment 0 results using data with simplification level of 0.0020

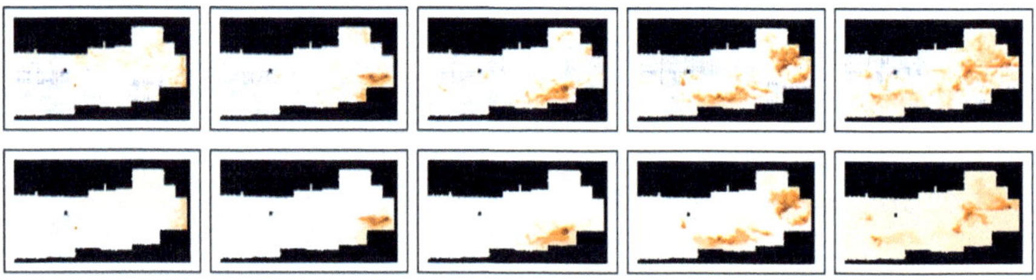

Fig. 14 Visualizations of selected slices from the range 100 to 200 of the CMZ data. Top: Slices 100, 120, 140, 160, and 180 before simplification, respectively. Bottom: Slices 100, 120, 140, 160, and 180 after simplification, respectively. The simplification level used is 3.45

very interesting. He and his collaborators have been actively debating whether this structure is a signal or a data processing artifact. Nevertheless, our approach retains it as a signal, and we are excited to see how our results generate further conversations regarding the data.

6.2 CMZ Data Set

The CMZ data are a ^{13}CO 2-1 image of the Central Molecular Zone (CMZ) of the galaxy (data are published in [23]). The data cube is approximately 500MB with a resolution of 1150 $\times$ 200 in the spatial domain and 500 in the spectral domain (i.e., 500 slices). We look at 100 slices of a region with a resolution of about 300 $\times$ 200.

Science Description. The cube shows the low-density molecular gas in the Galaxy's center, with higher intensities generally indicating that there is more gas moving at a particular velocity along each line of sight. It contains highly turbulent gas with properties that are very different than the rest of the Galaxy. Domain scientists use these data to measure the structure of the interstellar medium, which is important for determining how stars are formed and how galaxies evolve. Because the gas they are

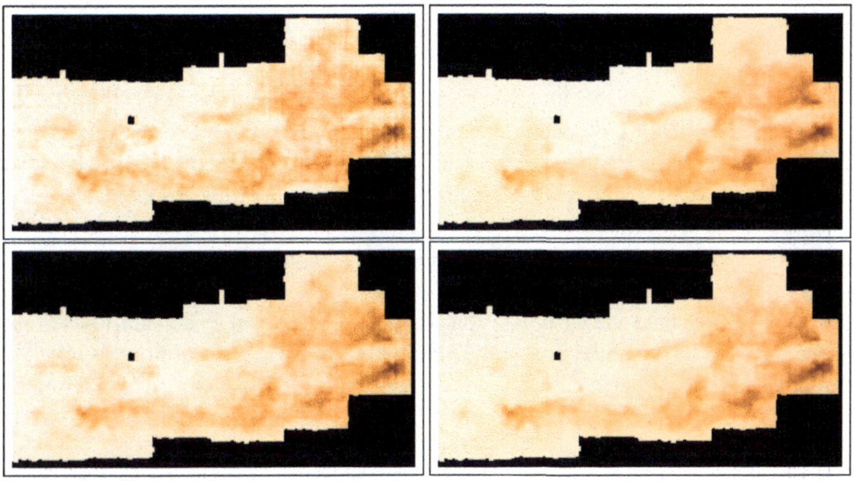

Fig. 15 Visualizations of moment 0 for slices 100 to 200 of the CMZ data set computed using 2D contour trees. Top left: Moment 0 on the original data. Top right: Moment 0 on all 100 simplified slices. Bottom left: Moment 0 only using every 4th slice. Bottom right: Moment 0 only using every 8th slice. The simplification level used is 3.45

seeing is in diffuse clouds that do not have well-defined edges, signal identification is a critical component in improving their understanding of how the gas changes state. Identifying structures in the gas is useful for determining how turbulent it is on different scales, which plays a key role in may star formation theories.

Denoising Slices. Figure 14 shows a number of slices denoised. The signal to noise ratio in this data set is much better than the previous ones. Nevertheless, many low persistent features have been removed using our approach.

Denoising for Moment Analysis. Deep cubes (those with many slices) such as this one are often created in order to mitigate the impact of noise during moment analysis—by more densely sampling the frequency domain, noise from any single slice has a smaller impact on the output. However, creating deep cubes such as this is computationally and manpower expensive. NRAO has significant human and computational infrastructure dedicated to generating data cubes from the raw data captured by radio telescopes. By providing strong denoising capabilities, data cubes can be sampled at lower spectral frequencies and still produce similar moment maps. See Fig. 15 for an example. Here, the top shows the moment map on the original data. Then, moment maps are shown that are calculated using every slice (100 total), every 4th slices (25 total), and every 8th slice (12 total). The results using fewer slices are virtually indistinguishable from the version using all 100 slices.

7 Discussion

In this feasibility study, we focus on persistence-based simplification of ALMA data cubes. Our application development process focuses on the usability objectives of collaborators, *simplicity*, *integrability*, and *reproducibility*, and we recommend these design objectives for anyone else wishing to collaborate with astrophysicists.

Despite our initial inclination to build a large scale visualization system, we find that this is unnecessary given the existing array of visualization options. Instead, what is needed is a simple and compact tool to understand the impact of parameter selection on the data via visualization. Parameter selection is not intuitive to new users. Without the visualization of the parameter selection, that intuition is relatively difficult to build. Nevertheless, once the selection is complete, the visualization and data processing can be easily reproduced using the information retained via the command-line interface.

Thus far, the reception of our approach has been good. Virtually everyone who has seen the results are impressed, for some, almost to the point of skepticism. Public outreach with such a new tool using unfamiliar techniques remains challenging. Among astrophysicists, there is a desire to understand both the tool and the underlying technique. Given the complexities of topological data analysis, this can be a challenging, but potentially transformative undertaking.

Acknowledgements This work was funded in part by a NRAO-NSF ALMA Development Grant titled *Feature Extraction & Visualization of ALMA Data Cubes through Topological Data Analysis*.

References

1. Partnership, A.L.M.A., Brogan, C.L., et al.: First results from high angular resolution alma observations toward the HL Tau region. ArXiv:1503.02649 (2015)
2. Partnership, A.L.M.A., Vlahakis, C., et al.: ALMA long baseline observations of the strongly lensed submillimeter galaxy HATLAS $J090311.6 + 003906$ at $z = 3.042$. ArXiv:1503.02652 (2015)
3. Bajaj, C.L., Pascucci, V., Schikore, D.R.: The contour spectrum. In: Proceedings of the 8th Conference on Visualization, pp. 167–ff (1997)
4. Barth, A.J., et al.: Toward precision black hole masses with ALMA: NGC 1332 as a case study in molecular disk dynamics. Astrophys. J. **823**(1), 51 (2016)
5. Basmajian, A., Saric, D.: Geodesically complete hyperbolic structures. arxiv.org/abs/1508.02280 (2016)
6. Belloche, A., Garrod, R.T., Müller, H.S.P., Menten, K.M.: Detection of a branched alkyl molecule in the interstellar medium: iso-propyl cyanide. Science **345**, 1584–1587 (2014)
7. Berry, D.: Fellwalker - a clump identification algorithm. Astron. Comput. **10**, 22–31 (2015)
8. Bolatto, A.D., et al.: Suppression of star formation in the galaxy NGC 253 by a starburst-driven molecular wind. Nature **499**, 450–453 (2013)
9. Carr, H., Sewell, C., Lo, L.T., Ahrens, J.: Hybrid data-parallel contour tree computation. In: Turkay, C., Wan, T.R. (eds.) Computer Graphics and Visual Computing. The Eurographics Association (2016)
10. Carr, H., Snoeyink, J.: Path seeds and flexible isosurfaces using topology for exploratory visualization. In: Proceedings of the Symposium on Data Visualisation (2003)

11. Carr, H., Snoeyink, J., Axen, U.: Computing contour trees in all dimensions. Comput. Geom. Theory Appl. **24**(3), 75–94 (2003)
12. Carr, H., Snoeyink, J., van de Panne, M.: Simplifying flexible isosurfaces using local geometric measures. In: Proceedings of the 15th IEEE Visualization, pp. 497–504 (2004)
13. Carr, H., Snoeyink, J., van de Panne, M.: Flexible isosurfaces: simplifying and displaying scalar topology using the contour tree. Comput. Geom. Theory Appl. **43**(1), 42–58 (2010)
14. Carr, H., Weber, G., Sewell, C., Ahrens, J.: Parallel peak pruning for scalable SMP contour tree computation. In: IEEE Symposium on Large Data Analysis and Visualization (2016)
15. Chiang, Y.J., Lenz, T., Lu, X., Rote, G.: Simple and optimal output-sensitive construction of contour trees using monotone paths. Comput. Geom. **30**(2), 165–195 (2005)
16. Cohen-Steiner, D., Edelsbrunner, H., Harer, J.: Stability of persistence diagrams. Discrete Comput. Geom. **37**(1), 103–120 (2007)
17. Cohen-Steiner, D., Edelsbrunner, H., Harer, J.: Extending persistence using Poincaré and Lefschetz duality. Found. Comput. Math. **9**(1), 79–103 (2009)
18. Colombo, D., Rosolowsky, E., Ginsburg, A., Duarte-Cabral, A., Hughes, A.: Graph-based interpretation of the molecular interstellar medium segmentation. Mon. Not. R. Astron. Soc. **454**(2), 2067–2091 (2015)
19. De Breuck, C., et al.: ALMA resolves turbulent, rotating [CII] emission in a young starburst galaxy at $z = 4.8$. Astron. Astrophys. **565**, A59 (2014)
20. Edelsbrunner, H., Harer, J.: Computational Topology: An Introduction. American Mathematical Society, Providence (2010)
21. Edelsbrunner, H., Letscher, D., Zomorodian, A.J.: Topological persistence and simplification. Discrete Comput. Geom. **28**, 511–533 (2002)
22. Edelsbrunner, H., Morozov, D., Pascucci, V.: Persistence-sensitive simplification of functions on 2-manifolds. In: Proceedings of the 22nd Annual ACM Symposium on Computational Geometry, pp. 127–134 (2006)
23. Ginsburg, A., et al.: Dense gas in the galactic central molecular zone is warm and heated by turbulence. Astron. Astrophys. **586**, A50 (2016)
24. Goodman, A.A., et al.: A role for self-gravity at multiple length scales in the process of star formation. Nature **457**(7225), 63–66 (2009)
25. Gueunet, C., Fortin, P., Jomier, J., Tierny, J.: Contour forests: fast multi-threaded augmented contour trees. In: IEEE Symposium on Large Data Analysis and Visualization (2016)
26. Hancock, P.J., Murphy, T., Gaensler, B.M., Hopkins, A., Curran, J.R.: Compact continuum source-finding for next generation radio surveys. Mon. Not. R. Astron. Soc. **422**(2), 1812–1824 (2012)
27. Hopkins, A.M., et al.: The ASKAP/EMU source finding data challenge. Publ. Astron. Soc. Aust. **32**, e037 (2015)
28. Johnson, K.E., et al.: The physical conditions in a pre-super star cluster molecular cloud in the antennae galaxies. Astrophys. J. **806**, 35 (2015)
29. Joye, W., Mandel, E.: New features of SAOImage DS9. In: Astronomical Data Analysis Software and Systems XII, vol. 295, p. 489 (2003)
30. van Kreveld, M., van Oostrum, R., Bajaj, C.L., Pascucci, V., Schikore, D.: Contour trees and small seed sets for isosurface traversal. In: Proceedings of the 19th Annual Symposium on Computational Geometry, pp. 212–220 (1997)
31. Leroy, A.K., et al.: ALMA reveals the molecular medium fueling the nearest nuclear starburst. Astrophys. J. **801**, 25 (2015)
32. Liu, H.B., et al.: ALMA resolves the spiraling accretion flow in the luminous *ob* cluster-forming region *g*33.92+0.11. Astrophys. J. **804**, 37 (2015)
33. Maadasamy, S., Doraiswamy, H., Natarajan, V.: A hybrid parallel algorithm for computing and tracking level set topology. In: Proceedings of the 19th International Conference on High Performance Computing, pp. 1–10 (2012)
34. McKenna, S., Mazur, D., Agutter, J., Meyer, M.: Design activity framework for visualization design. IEEE Trans. Vis. Comput. Graph. **20**, 2191–2200 (2014)

35. McMullin, J., Waters, B., Schiebel, D., Young, W., Golap, K.: Casa architecture and applications. In: Astronomical Data Analysis Software and Systems XVI, vol. 376, p. 127 (2007)
36. Meier, D.S., et al.: ALMA multi-line imaging of the nearby starburst NGC 253. Astrophys. J. **801**, 63 (2015)
37. Merényi, E., Taylor, J., Isella, A.: Mining complex hyperspectral alma cubes for structure with neural machine learning. In: IEEE Symposium Series on Computational Intelligence (2016)
38. Morozov, D., Weber, G.: Distributed merge trees. In: Proceedings of the 18th ACM SIGPLAN Symposium on Principles and Practice of Parallel Programming, pp. 93–102 (2013)
39. Morozov, D., Weber, G.H.: Distributed contour trees. In: Topological Methods in Data Analysis and Visualization III: Theory. Algorithms, and Applications, pp. 89–102. Springer, Cham (2014)
40. Nguyen, D.D., et al.: Improved dynamical constraints on the mass of the central black hole in NGC 404. Astrophys. J. **836**(2), 237 (2017)
41. Oesterling, P., Heine, C., Weber, G., Morozov, D., Scheuermann, G.: Computing and visualizing time-varying merge trees for high-dimensional data. In: Carr, H., Garth, C., Weinkauf, T. (eds.) Topological Methods in Data Analysis and Visualization IV. Springer, Cham (2015)
42. Onishi, K., et al.: WISDOM project - I: black hole mass measurement using molecular gas kinematics in NGC 3665. Mon. Not. R. Astron. Soc. **468**(4), 4663–4674 (2017)
43. Ott, T.: QFitsView. http://www.mpe.mpg.de/~ott/QFitsView/
44. Pascucci, V., Cole-McLaughlin, K., Scorzelli, G.: Multi-resolution computation and presentation of contour trees. Technical report, UCRL-PROC-208680, Lawrence Livermore National Laboratory (2005)
45. Pascucci, V., Scorzelli, G., Bremer, P.T., Mascarenhas, A.: Robust on-line computation of Reeb graphs: simplicity and speed. ACM Trans. Graph. **26**(3), 58.1–58.9 (2007)
46. Peretto, N., et al.: Global collapse of molecular clouds as a formation mechanism for the most massive stars. Astron. Astrophys. **555**, A112 (2013)
47. Raichel, B., Seshadhri, C.: A mountaintop view requires minimal sorting: a faster contour tree algorithm. ArXiv:1411.2689 (2014)
48. Rathborne, J.M., et al.: Turbulence sets the initial conditions for star formation in high-pressure environments. Astrophys. J. Lett. **795**, L25 (2014)
49. Rathborne, J.M., et al.: A cluster in the making: ALMA reveals the initial conditions for high-mass cluster formation. Astrophys. J. **802**, 125 (2015)
50. Robitaille, T., Beaumont, C., McDonald, B., Rosolowsky, E.: Astrodendro, a Python package to compute dendrograms of astronomical data. http://www.dendrograms.org/
51. Robitaille, T.P., et al.: Astropy: a community python package for astronomy. Astron. Astrophys. **558**, A33 (2013)
52. Rosen, P., Tu, J., Piegl, L.: A hybrid solution to calculating augmented join trees of 2D scalar fields in parallel. In: CAD Conference and Exhibition (2017)
53. Rosolowsky, E., Leroy, A.: Bias-free measurement of giant molecular cloud properties. Publ. Astron. Soc. Pac. **118**(842), 590–610 (2006)
54. Rosolowsky, E.W., Pineda, J.E., Kauffmann, J., Goodman, A.A.: Structural analysis of molecular clouds: dendrograms. Astrophys. J. **679**(2), 1338–1351 (2008)
55. Schneider, D., Wiebel, A., Carr, H., Hlawitschka, M., Scheuermann, G.: Interactive comparison of scalar fields based on largest contours with applications to flow visualization. IEEE Trans. Vis. Comput. Graph. **14**(6), 1475–1482 (2008)
56. Sohn, B.S., Bajaj, C.: Time-varying contour topology. IEEE Trans. Vis. Comput. Graph. **12**(1), 14–25 (2006)
57. Taylor, M., Boch, T., Taylor, J.: SAMP, the simple application messaging protocol: letting applications talk to each other. Astron. Comput. **11**, 81–90 (2015)
58. Wang, R., et al.: Star formation and gas kinematics of quasar host galaxies at $z \sim 6$: new insights from ALMA. Astrophys. J. **773**, 44 (2013)
59. Westerlund, S., Harris, C., Westmeier, T.: Assessing the accuracy of radio astronomy source finding algorithms. Publ. Astron. Soc. Aust. **29**, 301–308 (2012)
60. Williams, J.P., de Geus, E.J., Blitz, L.: Determining structure in molecular clouds. Astrophys. J. **428**(2), 693–712 (1994)

Part II: Topological Methods in Complex Fields – Flow Fields, Tensor Fields, and Multi-fields

Objective Finite-Time Flow Topology from Flowmap Expansion and Contraction

Roxana Bujack, Soumya Dutta, Duan Zhang, and Tobias Günther

Abstract We extend the definition of the classic instantaneous vector field saddles, sinks, and sources to the finite-time setting by categorizing the domain based on the behavior of the flow map w.r.t. contraction or expansion. Since the intuitive Lagrangian approach turns out to be unusable in practice because it requires advection in unstable regions, we provide an alternative, sufficient criterion that can be computed in a robust way. We show that both definitions are objective, relate them to existing approaches, and show how the generalized critical points and their separatrices can be visualized.

1 Introduction

The topological analysis of time-dependent vector fields remains to this day a very active research area in flow visualization. Similar to the classic steady case, we expect that particle motion is guided by a number of topological elements that have mainly

---

Electronic supplementary material The online version of this chapter (https://doi.org/10.1007/978-3-030-83500-2_7) contains supplementary material, which is available to authorized users.

R. Bujack (✉) · S. Dutta · D. Zhang
Los Alamos National Laboratory, Los Alamos, USA
e-mail: bujack@lanl.gov

S. Dutta
e-mail: sdutta@lanl.gov

D. Zhang
e-mail: dzhang@lanl.gov

T. Günther
Department of Computer Science, ETH Zurich, Zürich, Switzerland
e-mail: tobias.guenther@inf.ethz.ch

I. Hotz et al. (eds.), *Topological Methods in Data Analysis and Visualization VI*, Mathematics and Visualization, https://doi.org/10.1007/978-3-030-83500-2_7

been investigated individually, such as vortices [16, 20, 57], separating structures [22, 38, 55] and attractors [60]. In this paper, we introduce a finite-time generalization of the classic 2D vector field topology that maintains physical meaning in time-varying flows. In particular, we request the following properties for the topological structures to be meaningful over finite-time windows:

- In steady flows, the method is consistent with classic vector field topology.
- The definition of topological elements is objective, i.e. invariant w.r.t. Galilean transformations of the frame of reference.
- The feature definition is pathline-oriented and therefore in accordance with particle movement.

In a nutshell, the contributions of this work are as follows

- A coherent theoretical framework of an objective Lagrangian finite-time flow topology that ties together approaches from the literature.
- A non-Lagrangian sufficient definition that exceeds its Lagrangian counterpart in robustness.
- A simple algorithm for the extraction based on first-order approximation.
- Efficient visualizations of the finite-time topology.

Reviewing related work (Sect. 2), suggests a Lagrangian definition of finite-time topology as a logical consequence, because it bridges the gap between several approaches. Unfortunately, we will see quickly that it is practically useless because of its lack of robustness (Sect. 3). Therefore, we will dedicate most of this paper to the theoretical analysis of a non-Lagrangian alternative, which forms a sufficient criterion for the intuitive Lagrangian definition (Sect. 4). Finally, we will showcase results and suggest visualizations.

2 Related Work

The recent survey [6] on time dependent flow topology provides an overview on the goals, challenges, and state of the art.

2.1 *Classic Steady Vector Field Topology*

Classic steady vector field topology provides us with a compact description of the asymptotic motion of particles [26, 42]. Governing the asymptotic motion are a number of topological elements, which were described by Helman and Hesselink [27], including critical points (sinks, sources, centers, saddles), boundary elements (attachment and detachment points), the manifolds that separate flow regions of homogeneous asymptotic behavior (separatrices), and periodic orbits [2]. The extension to the 3D case [28] gave rise to a broader variety of elements, such as bifurcation lines [40] (lines to which nearby streamlines are asymptotically drawn to or repelled away from at an exponential rate) or saddle connectors [53] (individual streamlines that connect saddles). Aside from characterizations as extremal lines [33] of vortex-related scalar fields [47, 48], vortex corelines have also been expressed as lines along which the velocity vector aligns with the single real-valued eigenvector of the Jacobian matrix [52]. The parallel vectors operator [39] became a very powerful descriptor for such line features. In fact, both vortex corelines and bifurcation lines can be expressed in this way, with the only difference being that vortex corelines require swirling motion [39] (complex eigenvalues in the Jacobian) and bifurcation lines require attracting and repelling behavior [38, 44] (negative determinant in the plane orthogonal to the flow). Extensions include the characterization of higher-order critical points into sectors of elliptic, parabolic, or hyperbolic behavior [8, 49, 59] and higher-order bent vortex corelines [45].

2.2 *Streamlines vs. Pathlines*

More recent research concentrated on the definition and extraction of topological structures in time-dependent flows [41], in which we face two major challenges. First, aside from periodic flows, the temporal domain is usually bounded, which does not permit the observation of asymptotic motion. Second, the topology of streamlines (i.e. the observation of individual time slices) is irrelevant for pathlines, which was for instance demonstrated for vortex corelines [57]. The difference between streamline-oriented and pathline-oriented topology was discussed by Theisel et al. [54] in detail. Wiebel et al. [60] demonstrated in a simple 2D rotating petri-dish example that most existing techniques fail to detect the attracting vortex center that moves on a circular path. In the literature, this flow is sometimes also referred to as the Beads problem [58]. Integration-based methods can find the coreline, including the particle density estimate to extract the preferential particle settling [60] and the vortex coreline in the vector field in which streaklines are tangent curves [58]. Local methods failed due to lack of rotation invariance in the feature definitions, which can be obtained by a deformation from Cartesian to polar coordinates [18].

2.3 *Reference Frames*

A number of methods suggest to reduce the time-dependent topology back to the steady case by a suitable choice of the reference frame. Wiebel et al. [61] and Bhatia et al. [3] used flow decompositions to subtract a flow component that is irrotational and incompressible, i.e. harmonic. Fuchs et al. [14] selected a reference frame in which the velocity vanishes at locations at which the acceleration is zero. Bujack et al. [5] selected extrema in the determinant of the Jacobian to determine the reference frame. To determine a reference frame in which the vector field becomes steady [37,

43], reference frames have been calculated by local [16, 19] and global [20] linear optimizations, as well as by deep learning [32]. Alternatively, several local feature definitions possess a certain reference frame invariance. However, most of them, are only invariant to equal-speed translations, e.g., vorticity, λ_2 [30], and the Q-criterion [29]. Objectivity is achieved only by a few, such as by the instantaneous vorticity deviation [24].

2.4 Lagrangian Coherent Structures

In contrast to the local approaches, a large body of research searched for structures that behave coherently over a finite-time window. This research includes region-based vortex methods [24], coherent sets [13], and coherent line and surface structures, typically called Lagrangian coherent structures (LCS) [23]. The latter results in material lines that order the flow, including jet cores (parabolic LCS), vortex boundaries (elliptic LCS) and separating structures (hyperbolic LCS). As approximation to hyperbolic LCS, Haller [25] suggested to use the finite-time Lyapunov exponent (FTLE) [50], which measures the separation of nearby-released particles over a finite-time window. A number of approaches to compute FTLE exist, including a discretization of the flow map [25], localized FTLE [31], timeline cell tracking [34], a direct sampling of an advected circle [56] and Monte Carlo rendering [17]. Later, Haller [22] suggested to extract hyperbolic LCS by looking for the biggest separation orthogonal to a material surface. Similarly, Friederici et al. [11, 12] analyzed the finite-time separation orthogonal to a separatrix in steady flows.

2.5 Time-Dependent Saddles

Theisel et al. [54] categorized pathlines into attractors, repellors, and saddle-like trajectories based on whether their surrounding pathlines converge toward it in forward integration, in backward integration, or neither. In the fluid dynamics community, Haller [21] defined uniformly hyperbolic trajectories as pathlines with the property that half of their neighboring pathlines converge toward them in forward direction and the other in backward direction. Further, he introduced the concept of hyperbolicity time as the maximal amount of time a pathline spends in a region in which the Jacobian determinant is strictly negative and shows that the local maxima of hyperbolicity time are a first approximation to the uniformly hyperbolic trajectories. Inspired by Haller's hyperbolic trajectories [21], Sadlo and Weiskopf [46] generalized the concept of saddle-type critical points to time-dependent vector fields using the intersections of forward and backward FTLE ridges. The motivation behind this choice is that just like saddles, these areas show divergent behavior in forward as well as backward direction in time. As introduced by Wiebel et al. [62], they used these points as seeds for generalized streaklines, which form a generalization of sep-

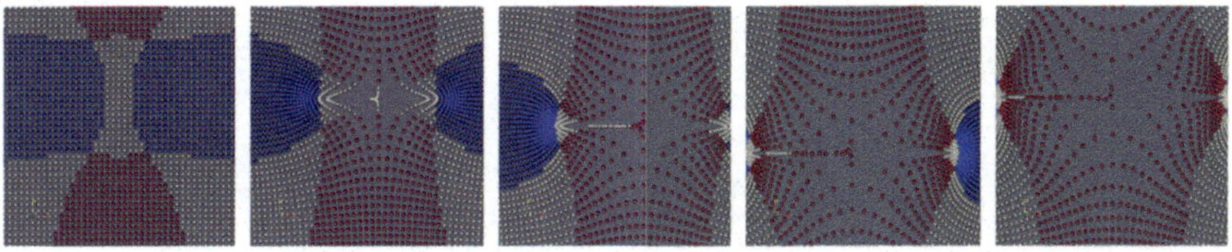

(a) Timestep 0. (b) Timestep 5. (c) Timestep 10. (d) Timestep 15. (e) Timestep 20.

Fig. 1 The Lagrangian definition is not robust. Categorization of the pathlines of the accelerated translation of a steady saddle is visualized through color coding: red–source; blue–sink; white–saddle. Theoretically, the saddle in the center of timestep 0 should move once around the origin of coordinates on a circle. Instead it is fully driven away by the expanding regions

aratrices to time-dependent flows. Later, Üffinger et al. [55] extended the concept to 3D. To approximate the path of a saddle, i.e. a bifurcation line in 2D space-time, Machado et al. [38] applied the reduced velocity criterion [39, 52] and iteratively aligned the extracted bifurcation line with the flow to obtain a pathline. In his recent survey on LCS, Haller [23] formulated four desirable properties: objectivity, finite-time nature, Lagrangian invariance, and spatial convergence. He points out that most classic definitions of material stability look strictly in forward direction to assess repelling behavior and strictly in backward direction to assess repelling behavior. Instead, repelling and attracting behavior should be assessed over the full time window, i.e. both forward and backward from the current point in time. He rejecteds Shadden's definition of LCS as second derivative ridges [50] and suggests shrink lines and stretch lines as LCSs [9, 10].

3 Intuitive Approach

Many attempts to generalize classic vector field topology to a time-dependent setting are based on translating the convergence and divergence properties of the classic critical points to pathlines. Most approaches deal with saddles [4, 21, 22, 46]. A few take into account sources or sinks, too [54, 60]. In this paper, we also define our categorization by pulling together existing work into one coherent framework. Intuitively speaking, we consider a pathline a **Lagrangian finite-time saddle** if part of its neighborhood has attracting behavior and part of its neighborhood has repelling behavior. We consider it a **Lagrangian finite-time sink** if all of its neighborhood has attracting behavior, and a **Lagrangian finite-time source** if all of its neighborhood has repelling behavior. The term Lagrangian or Lagrangian invariant refers to the ability of a structure to move with the flow, i.e. to be invariant w.r.t. advection [23]. We translate this into a concise mathematical definition.

Definition 1 (Lagrangian Finite-time Topological Categories). We consider a point and time $(x_0, t) \in \mathbb{R}^d \times \mathbb{R}$ a Lagrangian finite-time saddle for a given time interval $t \in [t_0, t_1] \subset \mathbb{R}$ if for any $\epsilon > 0$, we can find a plane containing 4 points

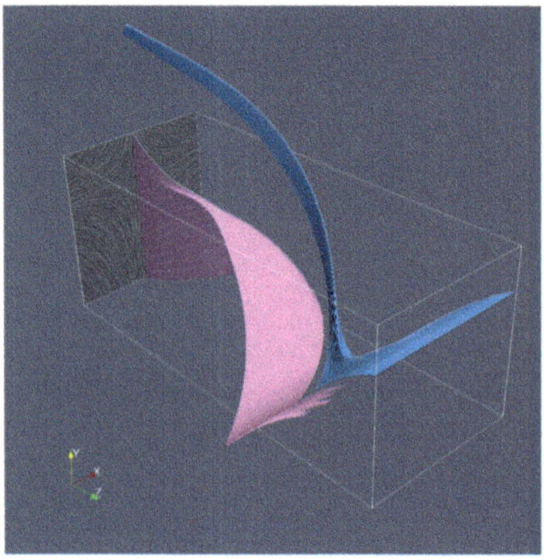

(a) FTLE ridges for the saddle under accelerated translation do not intersect.

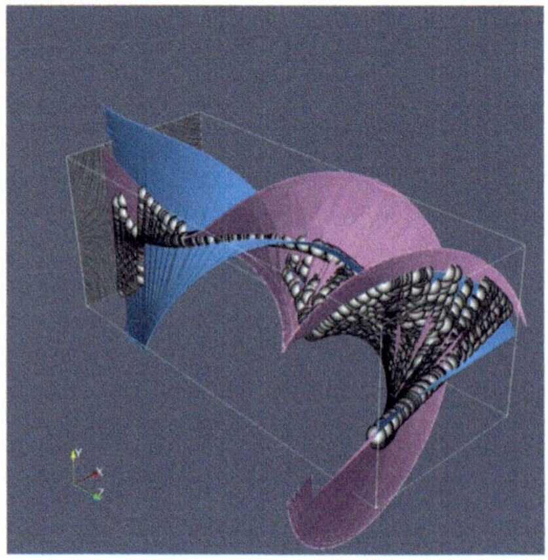

(b) FTLE ridges intersect almost everywhere for the saddle under accelerated translation.

Fig. 2 Lagrangian intersection of forward and backward FTLE is not robust

$x_1, ..., x_4 \in B_\epsilon(x_0)$ in its ϵ-neighborhood (numbered in positive orientation around x_0) so that the pathlines starting at (x_1, t) and (x_3, t) will expand from x_0 forward in time until t_1 while (x_2, t) and (x_4, t) contract. We consider it a Lagrangian finite-time sink if there is an $\epsilon_0 > 0$ such that for all $\epsilon : \epsilon_0 > \epsilon > 0$, a pathline starting at any point $x \in B_\epsilon(x_0)$ in its ϵ-neighborhood will contract to x_0 and a finite-time source if it expands. □

To categorize the steady flow behavior in finite-time, we define contraction and expansion as follows:

Definition 2 (Finite-time Contraction and Expansion). We consider two trajectories $x_0(t), x_i(t) : \mathbb{R} \to \mathbb{R}^d$ expanding in forward time for a given finite-time interval $t \in [t_0, t_1]$ if $\|x_0(t_0) - x_i(t_0)\| < \|x_0(t_1) - x_i(t_1)\|$ and contracting if $\|x_0(t_0) - x_i(t_0)\| > \|x_0(t_1) - x_i(t_1)\|$. Expansion in forward time is equivalent to contraction in backward time and vice versa. □

Definition 1 is objective [51] and Lagrangian invariant [23], i.e. it is advected by the flow. It is not able to classify centers and it does not always coincide with the steady topology, for example for linear fields. It is straightforward and very intuitive and nicely ties together different related work, but it suffers from a significant drawback. In practice, it is pretty much unusable because it is not robust. The categorization of the different pathlines at time t_0 works just fine, but to determine where these areas of a category go, we have to integrate along unstable manifolds that strongly deflect the pathlines, Fig. 1.

Impossibility of integration purely along stable manifolds. To advect in a robust way, an idea would be to make use of the backward integration [21, 22, 46]. But advecting the forward time attracting regions (sinks and part of the saddles) in the forward time direction and advecting the backward time attracting regions (sources and the other part of the saddles) in the backward time direction does not work either, because the saddle lies on a repelling manifold for both directions. This part will be deflected no matter from where we integrate. Figure 2 illustrates the problem in space-time. Theoretically, the saddle lies on the intersection line of the attracting manifolds in forward and backward directions. But due to the strong deflection, the surfaces may not intersect at all or become aligned.

4 Theory

In this section, we will provide a definition of a non-Lagrangian finite-time topology, which is a necessary condition for the intuitive Definition 1, but allows for a robust extraction. We will study its properties and derive an algorithm for its efficient computation based on its first-order approximation.

4.1 *Mathematical Definition*

Analogously to the Lagrangian Definition 1, we state a concise mathematical definition that concisely describes the intuitive physical categorization of the domain into contracting and expanding regions. The first of the two main differences is that we no longer require these regions to be Lagrangian, which means that instead of categorizing pathlines, we categorize points in space and time. Second, we explicitly consider these point's contracting and expanding behavior (Definition 2) in forward and also in backward time.

Definition 3 (Finite-time Topological Categories). We consider a point in space and time $(x_0, t) \in \mathbb{R}^d \times \mathbb{R}$ a **finite-time saddle** for a given time interval $t \in [t_0, t_1] \subset \mathbb{R}$ if for any $\epsilon > 0$, we can find 4 points $x_1, ..., x_4 \in B_\epsilon(x_0)$ in its ϵ-neighborhood (numbered in positive orientation around x_0) so that the pathlines starting at (x_1, t) and (x_3, t) will expand from x_0 forward in time until t_1 and contract backward until t_0 while (x_2, t) and (x_4, t) do the opposite. We consider it a **finite-time sink** if there is an $\epsilon_0 > 0$ such that all $\epsilon : \epsilon_0 > \epsilon > 0$, so that a pathline starting at any point $x \in B_\epsilon(x_0)$ in its ϵ-neighborhood will contract to x_0 forward in time until t_1 and expands backward until t_0 and a **finite-time source** for the opposite. □

4.2 *Relation to the Lagrangian Definition*

Definition 3 is sufficient for Definition 1, which means that every point in space and time that is classified as a finite-time saddle/source/sink lies on a pathline that is classified as a Lagrangian finite-time saddle/source/sink.

To see that, let

$$F_{t_0}^{t_1} : \mathbb{R} \times \mathbb{R} \times \mathbb{R}^d \to \mathbb{R}^d, \quad t \times t_0 \times x_0 \mapsto F_{t_0}^{t_1}(x_0), \tag{1}$$

with

$$\begin{aligned} F_{t_0}^{t_0}(x_0) &= x_0, \\ F_{t_1}^{t_2}(F_{t_0}^{t_1}(x_0)) &= F_{t_0}^{t_2}(x_0), \end{aligned} \tag{2}$$

denote the flow map describing how a flow parcel at (x_0, t_0) moves to $F_{t_0}^{t_1}(x_0)$ in the time interval $t_1 - t_0$. Then, we can compactly write the conditions in Definition 3. For a saddle, there exist x_{odd}, x_{even} such that:

$$\begin{aligned} \| F_t^{t_1}(x_{odd}) - F_t^{t_1}(x_0) \| &> \|x_{odd} - x_0\|, \\ \| F_t^{t_1}(x_{even}) - F_t^{t_1}(x_0) \| &< \|x_{even} - x_0\|, \\ \| F_t^{t_0}(x_{odd}) - F_t^{t_0}(x_0) \| &< \|x_{odd} - x_0\|, \\ \| F_t^{t_0}(x_{even}) - F_t^{t_0}(x_0) \| &> \|x_{even} - x_0\|, \end{aligned} \tag{3}$$

for a sink for all x_i holds:

$$\begin{aligned} \| F_t^{t_1}(x_i) - F_t^{t_1}(x_0) \| &< \|x_i - x_0\|, \\ \| F_t^{t_0}(x_i) - F_t^{t_0}(x_0) \| &> \|x_i - x_0\|, \end{aligned} \tag{4}$$

and for a source for all x_i holds:

$$\begin{aligned} \| F_t^{t_1}(x_i) - F_t^{t_1}(x_0) \| &> \|x_i - x_0\|, \\ \| F_t^{t_0}(x_i) - F_t^{t_0}(x_0) \| &< \|x_i - x_0\|. \end{aligned} \tag{5}$$

From this, we can directly derive the properties of the pathline through (x_0, t). We will show this for the case of a source. Assume (5) holds then at time t_0, all points $(F_t^{t_0}(x_i), t_0)$ in the neighborhood of the starting location of this pathline $(F_t^{t_0}(x_0), t_0)$ satisfy

$$\| F_t^{t_0}(x_i) - F_t^{t_0}(x_0) \| \overset{(5)}{<} \|x_i - x_0\| \overset{(5)}{<} \| F_t^{t_1}(x_i) - F_t^{t_1}(x_0) \|, \tag{6}$$

which is the condition for the pathline to be a Lagrangian finite-time source. The size of ϵ_0 in Definition 1 depends on the respective flow field, but its existence is guaranteed if it is continuous, because the flowmap is as many times differentiable as the vector field [1]. The other cases work analogously.

4.3 Objectivity

We consider objectivity [51] important because this property ensures that two observers do not get different answers from looking at the same physical phenomenon. Within the flow, smaller features get advected by larger ones, which results as a mixture of different 'best' reference frames to look at the flow. Definition 3 is objective, i.e. invariant w.r.t a Euclidean transformation of the reference frame $x' = Q(t)x + c(t)$ with a time-dependent orthogonal matrix $Q : \mathbb{R} \rightarrow \in SO(d)$ and a translation $c : \mathbb{R} \rightarrow \mathbb{R}^d$. This follows from the transformation properties of the flowmap under Euclidean transformations $F'^{t_1}_{t_0}(x'_0) = Q(t_1)F^{t_1}_{t_0}(x_0) - c(t_1)$ [36], because of which the difference suffices

$$\begin{aligned} F'^{t_1}_{t_0}(x'_0) - F'^{t_1}_{t}(x'_i) &= Q(t_1)F^{t_1}_{t_0}(x_0) - c(t_1) - Q(t_1)F^{t_1}_{t}(x_i) + c(t_1) \\ &= Q(t_1)(F^{t_1}_{t_0}(x_0) - F^{t_1}_{t}(x_i)) \end{aligned} \tag{7}$$

and the distance

$$\begin{aligned} \|F'^{t_1}_{t_0}(x'_0) - F'^{t_1}_{t}(x'_i)\|^2 &= (F'^{t_1}_{t_0}(x'_0) - F'^{t_1}_{t}(x'_i))^T(F'^{t_1}_{t_0}(x'_0) - F'^{t_1}_{t}(x'_i)) \\ &\overset{(7)}{=} (Q(t_1)(F^{t_1}_{t_0}(x_0) - F^{t_1}_{t}(x_i)))^T(Q(t_1)(F^{t_1}_{t_0}(x_0) - F^{t_1}_{t}(x_i))) \\ &= (F^{t_1}_{t_0}(x_0) - F^{t_1}_{t}(x_i))^T Q(t_1)^T Q(t_1)(F^{t_1}_{t_0}(x_0) - F^{t_1}_{t}(x_i)) \\ &= (F^{t_1}_{t_0}(x_0) - F^{t_1}_{t}(x_i))^T(F^{t_1}_{t_0}(x_0) - F^{t_1}_{t}(x_i)) \\ &= \|F^{t_1}_{t_0}(x_0) - F^{t_1}_{t}(x_i)\|. \end{aligned} \tag{8}$$

4.4 Linear Approximation

The difference between two close points can be approximated using Taylor's theorem. In our case, the conditions (3) to (5) can be expressed using the deformation gradient $\nabla F : \mathbb{R}^{d\times d}$

$$F^{t_1}_{t}(x_0) - F^{t_1}_{t}(x_i) = \nabla F^{t_1}_{t}(x_0)(x_i - x_0) + O(\|x_i - x_0\|^2). \tag{9}$$

For the limit $\epsilon \rightarrow 0$, we can write its magnitude as

$$\begin{aligned} \|F^{t_1}_{t}(x_0) - F^{t_1}_{t}(x_i)\|^2 &= (F^{t_1}_{t}(x_0) - F^{t_1}_{t}(x_i))^T(F^{t_1}_{t}(x_0) - F^{t_1}_{t}(x_i)) \\ &= (x_0 - x_i)^T(\nabla F^{t_1}_{t}(x_0))^T \nabla F^{t_1}_{t}(x_0)(x_0 - x_i). \end{aligned} \tag{10}$$

The first part of condition (5) can be rewritten as the ratio

$$\frac{\| F_t^{t_1}(x_0) - F_t^{t_1}(x_i) \|}{\| x_0 - x_i \|} > 1. \tag{11}$$

With the unit vector

$$n = \frac{x_0 - x_i}{\| x_0 - x_i \|} \tag{12}$$

and the Cauchy-Green strain tensor $C_{t_0}^{t_1}(x_0) = (\nabla F_{t_0}^{t_1}(x_0))^T \nabla F_{t_0}^{t_1}(x_0)$ from continuum mechanics, the ratio (11) can be estimated through

$$\frac{\| F_t^{t_1}(x_0) - F_t^{t_1}(x_i) \|^2}{\| x_0 - x_i \|^2} = n^T (\nabla F_t^{t_1}(x_0))^T \nabla F_t^{t_1}(x_0) n = n^T C_t^{t_1}(x_0) n. \tag{13}$$

Because of

$$\frac{\| F_t^{t_1}(x_1) - F_t^{t_1}(x_3) \|}{\| x_1 - x_3 \|} > 1 \Leftrightarrow \frac{\| F_t^{t_1}(x_1) - F_t^{t_1}(x_3) \|^2}{\| x_1 - x_3 \|^2} > 1 \quad \Leftrightarrow n^T C_t^{t_1}(x_0) n > 1, \tag{14}$$

the conditions in (3) to (5) can be expressed through the eigenvalues of C. Since the eigenvectors maximize $\max_{\|n\|=1} |n^T C n|$, the conditions are transferred to the eigenvalues. In particular, for a point (x_0, t) to be a first-order approximation to the finite-time saddle in the interval $[t_0, t_1]$, the tensors $C_t^{t_1}(x_0)$ and $C_t^{t_0}(x_0)$ must each have eigenvalues greater as well as smaller than 1. The eigenvalues need to be both smaller than 1 for $C_t^{t_1}(x_0)$ and both greater than 1 for $C_t^{t_0}(x_0)$ for a point to be a first-order approximation of a finite-time sink and the opposite for a finite-time source.

The linear approximation is also objective. Because of $\nabla_{x'} x = \frac{\mathrm{d}x}{\mathrm{d}x'} = Q^T$ and the chain rule [36], the deformation gradient suffices:

$$\nabla F_{t_0}^{t_1\prime}(x') = Q(t_1) \nabla F_{t_0}^{t_1}(x) Q(t_0)^T. \tag{15}$$

and the Cauchy-Green strain tensor

$$\begin{aligned} C_{t_0}^{t_1\prime}(x') &= (\nabla F_{t_0}^{t_1\prime}(x'))^T \nabla F_{t_0}^{t_1\prime}(x') \\ &= (Q(t_1) \nabla F_{t_0}^{t_1}(x) Q(t_0)^T)^T Q(t_1) \nabla F_{t_0}^{t_1}(x) Q(t_0)^T \\ &= Q(t_0)((\nabla F_{t_0}^{t_1}(x))^T \nabla F_{t_0}^{t_1}(x)) Q(t_0)^T \\ &= Q(t_0) C_{t_0}^{t_1}(x) Q(t_0)^T. \end{aligned} \tag{16}$$

This approximation is not necessarily objective, because it has two time dependencies that the definition of objectivity does not encompass, but its eigenvalues are objective. Let v be an eigenvector of C' with eigenvalue λ, i.e. $Cv = \lambda v$, then $\tilde{v} = Q(t_0) v$ is an eigenvector of C with the same eigenvalue

$$C' \tilde{v} \overset{(16)}{=} Q(t_0) C_{t_0}^{t_1}(x) Q(t_0)^T \tilde{v} = Q(t_0) C_{t_0}^{t_1}(x) v = Q(t_0) \lambda v = \lambda Q(t_0) v = \lambda \tilde{v}. \tag{17}$$

4.5 Strength

As can be seen in Fig. 1, Definitions 1 and 3 usually do not produce isolated points but areas of coherent classification. For each connected component of one category, we can choose a point as a representative through demanding that it shows the corresponding contracting or expanding behavior in locally the strongest way, for example for the saddle through maximizing

$$M_{t_0}^{t_1}(x_0,t) := \max_{x_{odd},x_{even}\in B_\epsilon(x_0)} \min(\frac{\|F_t^{t_1}(x_{odd}) - F_t^{t_1}(x_0)\|}{\|x_{odd} - x_0\|}, \frac{\|x_{even} - x_0\|}{\|F_t^{t_1}(x_{even}) - F_t^{t_1}(x_0)\|}, \frac{\|x_{odd} - x_0\|}{\|F_t^{t_0}(x_{odd}) - F_t^{t_0}(x_0)\|}, \frac{\|F_t^{t_0}(x_{even}) - F_t^{t_0}(x_0)\|}{\|x_{even} - x_0\|}), \tag{18}$$

for the sink through maximizing

$$M_{t_0}^{t_1}(x_0,t) := \min_{x_i\in B_\epsilon(x_0)} \min(\frac{\|F_t^{t_1}(x_i) - F_t^{t_1}(x_0)\|}{\|x_i - x_0\|}, \frac{\|x_i - x_0\|}{\|F_t^{t_0}(x_i) - F_t^{t_0}(x_0)\|}), \tag{19}$$

and for the source through maximizing

$$M_{t_0}^{t_1}(x_0,t) := \min_{x_i\in B_\epsilon(x_0)} \min(\frac{\|x_i - x_0\|}{\|F_t^{t_1}(x_i) - F_t^{t_1}(x_0)\|}, \frac{\|F_t^{t_0}(x_i) - F_t^{t_0}(x_0)\|}{\|x_i - x_0\|}). \tag{20}$$

The inner most min refers to the minimum of the forward and backward terms to avoid the detection of examples that only exhibit the behavior in one direction. The second min avoids line sinks and sources, which do not have expanding or contracting behavior in one direction. The outer most maximization refers to the candidate points x_0 that exhibit the respective behavior in locally the strongest way.

4.6 Weighting Related to FTLE

The first-order approximation shows that our measures of strength are related to FTLE, where the largest eigenvalue λ_{max} of the Cauchy-Green strain tensor C is evaluated. To consider the dependence on the size of the time interval and the potentially rapid growth of the expansion, λ_{max} is weighted via

$$\tilde{\lambda}_{max}(C_{t_0}^{t_1}(x_0)) := FTLE_{t_0}^{t_1}(x_0) = \frac{\log\sqrt{\lambda_{max}(C_{t_0}^{t_1}(x_0))}}{t_1 - t_0}, \tag{21}$$

Analogously, we can weight the largest and smallest eigenvalue $\lambda_{max}, \lambda_{min}$ of the Cauchy-Green strain tensor for weighted first-order approximations of our measures of strength. The logarithm changes the limit where changes between the categories happen to 0, which leads to the following cases

$$(x_0, t) \text{ is a } \begin{cases} \text{saddle} & \text{if } \tilde{\lambda}_{max}(C_t^{t_1}) > 0 \wedge \tilde{\lambda}_{min}(C_t^{t_1}) < 0 \wedge \tilde{\lambda}_{max}(C_t^{t_0}) > 0 \\ & \wedge \tilde{\lambda}_{min}(C_t^{t_0}) < 0 \wedge v_{max}(C_t^{t_1}) \nparallel v_{min}(C_t^{t_0}), \\ \text{source} & \text{if } \wedge \tilde{\lambda}_{min}(C_t^{t_1}) > 0 \wedge \tilde{\lambda}_{max}(C_t^{t_0}) < 0, \\ \text{sink} & \text{if } \wedge \tilde{\lambda}_{max}(C_t^{t_1}) < 0 \wedge \tilde{\lambda}_{min}(C_t^{t_0}) > 0 \\ \text{neither} & \text{else.} \end{cases} \tag{22}$$

Please note that for a first-order saddle, we additionally have to make sure that the directions of the strongest expansion in forward and backward time do not coincide to guarantee that there are really 4 separate points x_i as demanded in Definition 3 instead of x_{odd} as suggested in forward time to coincide with x_{even} as suggested by backward time, which would occur for example for a blue sky bifurcation, i.e. a flow that is first a sink and then turns into a source or vice versa.

That also means that we can directly use the absolute value of the weighted eigenvalues to determine how strong the contracting or expanding properties of each point are. In particular, we use the minimum over all four $\tilde{\lambda}$

$$\tilde{M}_{t_0}^{t_1}(x_0, t) := \min_{i \in \{0,1\}} \min_{j \in \{min, max\}} \frac{|\tilde{\lambda}_j(C_t^{t_i}(x_0))|}{|t_i - t|}. \tag{23}$$

If a point does not fall into a category (for example, it is a source in forward time and saddle in backward time), we set the strength to zero. If a point is a saddle, we additionally weight it by the scalar product across the eigenvectors to exclude areas where they coincide in forward and backward direction. All in all, we get the measure of strength

$$\mathcal{M}_{t_0}^{t_1}(x_0, t) = \begin{cases} \tilde{M}_{t_0}^{t_1}(x_0, t) & \text{if source or sink,} \\ |v_{max}(C_t^{t_1})^T v_{min}(C_t^{t_0})| \tilde{M}_{t_0}^{t_1}(x_0, t) & \text{if saddle,} \\ 0 & \text{else.} \end{cases} \tag{24}$$

We compute this scalar measure of strength for the whole domain, which will allow us to determine strong representatives for coherent regions of the same behavior and to remove weak occurrences for reducing clutter in the visualizations. Since, the eigenvectors are orthogonal, we do not need to consider the other pair.

The measure of strength is also objective. We already know that the eigenvalues are objective from (17) and we can see that the product of the eigenvectors $v_1(C_t^{t_1}(x_0))^T v_2(C_t^{t_0}(x_0))$ is objective, too, because their transformed equivalents suffice

$$\begin{aligned} v_1'(C_t^{t_1 \prime}(x_0'))^T v_2'(C_t^{t_0 \prime}(x_0')) &\overset{(17)}{=} (Q(t) v_1(C_t^{t_1}(x_0)))^T Q(t) v_2(C_t^{t_0}(x_0)) \\ &= v_1(C_t^{t_1}(x_0))^T v_2(C_t^{t_0}(x_0)). \end{aligned} \tag{25}$$

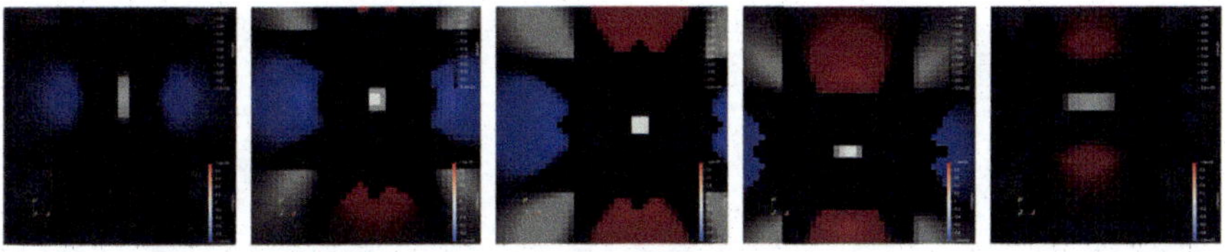

(a) Timestep 1. (b) Timestep 5. (c) Timestep 10. (d) Timestep 15. (e) Timestep 19.

Fig. 3 The non-Lagrangian definition correctly categorizes the pathlines of the accelerated translation of a steady saddle. Red: source, blue: sink, white: saddle, black: neither. We show the strength of each region by overlaying (24) ranging from black for low strength to transparent for high strength

4.7 Separatrices

It is common practice to use generalized streaklines [62] seeded around the locally strongest saddles, sometimes also called bifurcation lines in space-time, [38, 46, 55]. In particular, we seed pathlines with a small offset in both directions of the eigenvector $v_{max}(C_t^{t_1}(x_0))$ corresponding to the bigger eigenvalue for the forward separatrix advection and analogously with a small offset in both directions of the eigenvector $v_{max}(C_t^{t_0}(x_0))$ for the backward separatrix. Then, we generate surfaces from them in space-time. Figure 5 shows a visualization. The temporally local separatrices can be produced from slicing the volume at one timestep.

5 Experiments

In this section, we concentrate mainly on experiments for which we actually know the ground truth to demonstrate the correctness of the proposed method. For this purpose, we use two analytic data sets. The first one is a steady saddle

$$v(x) = v(x, y) = 2\begin{pmatrix} x + 0.5 \\ -y \end{pmatrix} e^{-2\sqrt{(x+0.5)^2+y^2}} \tag{26}$$

that is moved through an accelerated translation. A Euclidean transformation

$$x' = Q(t)x + c(t) \tag{27}$$

changes a velocity field via

$$v'(x, t) = Q(t)v(Q^T(t)(x - c(t))) + \dot{Q}(t)(x - c(t)) + \dot{c}(t). \tag{28}$$

We use $c(t) = \frac{1}{2}(\sin(\theta) + 1, \cos(\theta))^T$ with $\theta = 2\pi t^2/|T|^2$ and $|T|$ denoting the number of time steps, which moves the saddle clockwise on the circle with radius

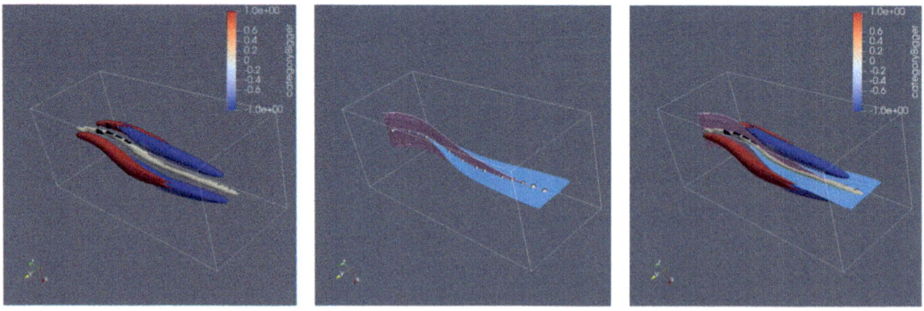

(a) Categories of saddle under accelerated translation. (b) Separatrices of saddle under acc. translation. (c) Categories and separatrices of translation.

(d) Categories of saddle under accelerated rotation. (e) Separatrices of saddle under acc. rotation. (f) Categories and separatrices of rotation.

Fig. 4 Visualizations of the finite-time topology in space-time for two examples of Euclidean transformations. Left: isosurfaces of the strength colored by category. Red– source; blue– sink; white– saddle. Middle: separatrices, i.e. streak surfaces forward (red) and backward (blue) in time seeded at the strongest saddle offset in the direction of the eigenvectors of the Cauchy Green strain tensor. Right: both. The path of the saddle is one full circle in both cases. The shapes of the sinks, sources, and separatrices reveal that the top movement is a pure translation, whereas the bottom is a rotation

0.5 around $(0, 0)^T$ starting at $(0, 0.5)^T$. The motivation of using accelerated moving reference frames is that this is the most complicated case. If a method detects this one correctly, it will also work for constant movements. We have already seen the results of the Lagrangian categorization for this dataset in Fig. 1. The results of the robust categorization using the suggested sufficient first-order approximation suggested in this work can be found in Fig. 3. This figure shows the expected behavior with the accelerated movement around the origin. On top of the category, we also encode the strength of the occurrence fading out weak areas into black. This approach is consistent with color-coding black areas that belong in no category. Please note that the two sources (red) and two sinks (blue) around the saddle (white) are a result of the Gaussian weighting in combination with the saddle. The actual expanding and contracting character of these regions can be well perceived in the particle view Fig. 1 offers.

Fig. 5 Quad gyre: separatrices (pink: fw, blue: bw) of strongest saddle (white) in spacetime

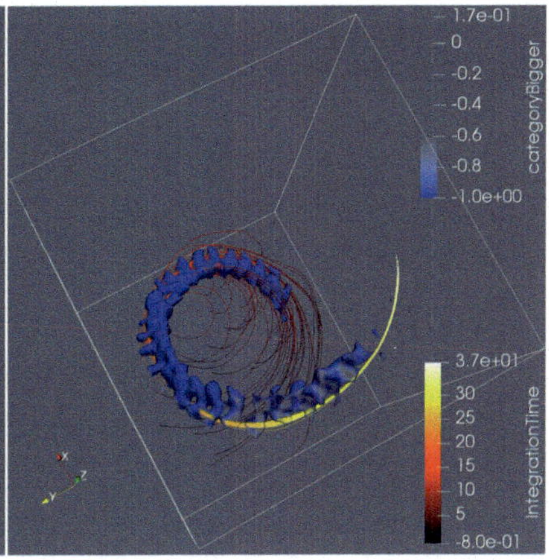

Fig. 6 Petri-dish: the rotating sink (blue) and pathlines for comparison in spacetime

The second analytic dataset is the same saddle (26) performing an accelerated rotation with $Q(t) \in SO_2$ being the rotation matrix by $\theta = 2\pi t^2/|T|^2$. For both flows, we use the spatial domain $[-2, 2]^2$ with resolution 81^2 and the full 21 time steps $[t_1, t_1] = [0, 20]$. To avoid boundary artifacts, we computed the flowmap on a bigger domain. Both transformations are purely Euclidean. They can be interpreted as a change of the reference frame of the observer and an objective method should be able to detect the saddle on the circle.

Figure 4 shows the results of the classification and the separatrices for both transformations in space-time. We visualize the different categories using the same color coding. Saddles are white, sources are red, sinks are blue, and points that fall in no category are black. For the reduction of weak occurrences to gain a less cluttered, more expressive visualization, we applied isocontours on the scalar strength field (24) and colored the result using the scalar field of the categories (22). Storing the two fields makes the visualization of the method easy in any common visualization tool. To get the separatrices for each time slice, we first chose representatives for the saddle-type regions by selecting the locations with the global maximum of the strength. Then, we seeded pathlines as described in Sect. 4.7. Our method detects the true locations of the saddle up to the accuracy of one cell. It cannot find the exact location within a cell, because the maximum always lies on a gridpoint in a piecewise linear field.

For these datasets, the intersection of forward and backward FTLE as suggested by Sadlo [46] produces the same results. The Lagrangian forward and backward FTLE produces no result for the detection of the saddles. Even though, the ridges are detected correctly at the first and last time step, the surfaces are deflected so strongly that they do not intersect at all for the translation and almost everywhere for the rotation, Fig. 1.

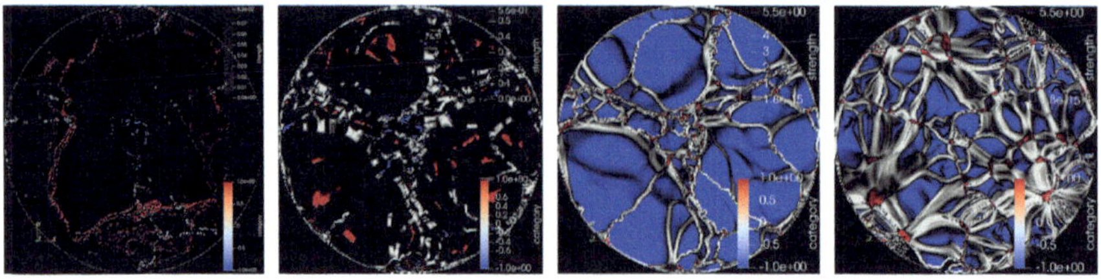

(a) Full time [0, 120]. (b) POI [65, 75]. (c) Bw POI [65, 75]. (d) Fw POI [65, 75].

Fig. 7 Categorization in viscous fingers dataset for timestep $t = 70$

Figure 5 shows the extraction of the strongest saddle and the separatrices of the quad gyre, which extends the double gyre [50] to the domain $[0, 1]^2$. We used the resolution 201^2 and one full period in time. Here, the global maximum of our measure (24) coincides with the intersection of forward and backward FTLE of the adjacent intervals [46]. The double gyre is an incompressible flow, which is why it does not have sources and sinks.

Figures 6 and 7 show results of the categorization for two flow simulations using red for sources, blue for sinks, white for saddles, and black for neither. The rotating sink in the last 40 timesteps of the petri-dish dataset [5, 7, 60] is nicely extracted but the complicated topology in the viscous fingers dataset [35] from the SciVis contest http://sciviscontest.ieeevis.org is harder to interpret. A limit of our method is reached if the data shows strong contraction and spans a long period of time. Once all particles in the flowmap are accumulated in one point, nothing is left for it to capture in coming time steps, which leads to detail getting lost and most points not belonging in either category 7a. We show the strength of each region by overlaying (24), ranging from black for low strength to transparent for high strength. This issue is a known problem of Lagrangian methods and can be overcome by guaranteeing Lagrangian invariance for a time period of interest (POI) only [15]. Figures 7a and 7b show the difference of the global $[t_0, t_1] = [0, 120]$ and the POI 70 approach for time step 70. Figures 7c and 7d show the partial POI categories considering only backward and only forward information, which together form Fig. 7b. The comparison shows that the chosen time interval influences how a point is categorized.

6 Discussion

The extension of vector field topology to time-dependent flows has been extensively studied not only in the scientific visualization community. Our results are based on many approaches that have been published previously.

The closest related work w.r.t. saddle is [4]. The definition of finite-time saddle is identical to ours, but sources and sinks were not treated. There, the connection of the saddle part to Lagrangian coherent structures based on FTLE is treated. Sadlo and

Weiskopf [46] suggested to intersect forward and backward FTLE ridges, which corresponds to half of the constraints in Definition 3. Approaches of this kind, where different time intervals are combined, were criticized by Haller [23], because they are not Lagrangian w.r.t. the total time interval. But as we have seen, the Lagrangian equivalent in Definition 1 is infeasible in practice. Definition 3 bridges the gap between the Lagrangian approach and the FTLE intersection of adjacent intervals [46] providing a categorization that is both Lagrangian and robust.

The closest work w.r.t. sinks is probably by Wiebel et al. [60]. They used the density maximum of particles that were seeded equidistantly in space and repeatedly over time. Up to the exact evaluation of the density maxima, our sink definition is in accordance with theirs because dense particle positions correspond to contracting flowmap behavior. The main difference is that we consider a concrete finite-time interval $[t_0, t_1]$, while they seed repeatedly in time. Their method can be interpreted as averaging the results of ours over the intervals $[t_i, t_1]$ with , $i \in [t_0, t_1]$. They do not consider saddles in their work.

Probably the closest related work overall is by Theisel et al. [54]. In their pathline-based approach, they also categorize pathlines into attractors, repellors, and saddle-like trajectories based on whether their surrounding pathlines converge toward it in forward integration, in backward integration, or both. There are three main differences to our work. First, their approach is local in time. They use the instantaneous orientation of the pathlines in spacetime, which means it cannot encompass the finite-term behavior of the flow. Second for the actual computation, they categorize a point using the Jacobian of the vector field that results from projection of these directions on the plane through spacetime that is orthogonal to the pathline through it. This approach is not objective. Finally, there is no notion of strength or the extraction of representatives, or separatrices.

Our notion of separatrices uses generalized streaklines [62] seeded on the saddles, which is identical to related work on hyperbolic trajectories, saddle core lines, and bifurcation lines [38, 46, 55].

7 Conclusion

We have presented an intuitive Lagrangian extension of the classic 2D vector field critical points saddle, source, and sink to finite-time in Definition 1. It is objective and reflects particle movement in a physically meaningful way. Since it is not robust in practice, we also provide a sufficient criterion in Definition 3 and a first-order approximation for the computation of the category and the strength. We show its independence on changes of the reference frame and point out its relations to existing approaches in the literature.

Looking at the discussion, we do not necessarily consider Definitions 1 and 3 a huge leap over existing methods. We consider the main contribution of this paper to be how this theoretical framework encompasses saddles, sources, sinks, and separatrices and therefore ties together multiple valuable approaches from the literature.

Limitations are that our method is not able to detect all classic critical points, e.g., in linear steady fields because there, the Cauchy Green strain tensor is constant. In addition, just like FTLE, it may detect shear as saddles and may require a high resolution and long computation times for the generation of the flowmap. Furthermore, it loses its ability to capture details in long simulation runs with strong contraction when all particles gather in one point. The categorization is always tied to a given time interval. The same point in space and time could be classified differently for different intervals. In the future, we will analyze strategies to choose meaningful time intervals. Finally, the categorization is undefined at the boundary where particles leave the domain and at the boundary times t_0 and t_1. Analysis of how the method extends to 3D flow will be future work.

Acknowledgements We gratefully acknowledge the support of the U.S. Department of Energy through the LANL Laboratory Directed Research Development Program under project number 20190143ER for this work published under LA-UR-19-23386. Further, we thank Dr. Reymond Chan and Professor Angela Stevens for providing the petri-dish dataset.

References

1. Arnold, V.I.: Geometrical Methods in the Theory of Ordinary Differential Equations, vol. 250. Springer, Heidelberg (2012)
2. Asimov, D.: Notes on the topology of vector fields and flows. Technical report RNR-93-003, NASA Ames Research Center (1993)
3. Bhatia, H., Pascucci, V., Kirby, R.M., Bremer, P.-T.: Extracting features from time-dependent vector fields using internal reference frames. In: Computer Graphics Forum, volume 33, pp. 21–30. Wiley Online Library (2014)
4. Bujack, R., Dutta, S., Baeza Rojo, I., Zhang, D., Günther, T.: Objective finite-time saddles and their connection to FTLE. In: Johansson, J., Sadlo, F., Marai, G.E. (eds.) EuroVis 2019 - Short Papers, pp. 49–53. The Eurographics Association (2019)
5. Bujack, R., Hlawitschka, M., Joy, K.I.: Topology-inspired Galilean invariant vector field analysis. In: Proceedings of the IEEE Pacific Visualization Symposium, PacificVis 2016, Taipei, Taiwan, pp. 72–79 (2016)
6. Bujack, R., Yan, L., Hotz, I., Garth, C., Wang, B.: State of the art in time-dependent flow topology: interpreting physical meaningfulness through mathematical properties. In: Computer Graphics Forum (2020)
7. Chan, R.: A biofluid dynamic model for centrifugal accelerated cell culture systems. Ph.D. dissertation, Leipzig University, Germany (2008)
8. Effenberger, F., Weiskopf, D.: Finding and classifying critical points of 2D vector fields: a cell-oriented approach using group theory. Comput. Vis. Sci. **13**(8), 377–396 (2010)
9. Farazmand, M., Haller, G.: Computing Lagrangian coherent structures from their variational theory. Chaos Interdisc. J. Nonlinear Sci. **22**(1), 013128 (2012)
10. Farazmand, M., Haller, G.: Attracting and repelling Lagrangian coherent structures from a single computation. Chaos Interdisc. J. Nonlinear Sci. **23**(2), 023101 (2013)
11. Friederici, A., Günther, T., Rössl, C., Theisel, H.: Finite time steady vector field topology - theoretical foundation and 3D case. In: Vision, Modeling and Visualization, pp. 95–102 (2017)
12. Friederici, A., Rössl, C., Theisel, H.: Finite time steady 2D vector field topology. In: Topological Methods in Data Analysis and Visualization, pp. 253–266. Springer (2015)
13. Froyland, G.: An analytic framework for identifying finite-time coherent sets in time-dependent dynamical systems. Physica D **250**, 1–19 (2013)

14. Fuchs, R., Kemmler, J., Schindler, B., Waser, J., Sadlo, F., Hauser, H., Peikert, R.: Toward a Lagrangian vector field topology. In: Computer Graphics Forum, volume 29, pp. 1163–1172. Wiley Online Library (2010)
15. Germer, T., Otto, M., Peikert, R., Theisel, H.: Lagrangian coherent structures with guaranteed material separation. In: Computer Graphics Forum, volume 30, pp. 761–770. Wiley Online Library (2011)
16. Günther, T., Gross, M., Theisel, H.: Generic objective vortices for flow visualization. ACM Trans. Graph. (TOG) **36**(4), 141 (2017)
17. Günther, T., Kuhn, A., Theisel, H.: MCFTLE: Monte Carlo rendering of finite-time Lyapunov exponent fields. Comput. Graph. Forum **35**(3), 381–390 (2016). (Proceedings of EuroVis)
18. Günther, T., Schulze, M., Theisel, H.: Rotation invariant vortices for flow visualization. IEEE Trans. Vis. Comput. Graph. **22**(1), 817–826 (2016). (Proceedings of the IEEE Scientific Visualization)
19. Günther, T., Theisel, H.: Hyper-objective vortices. IEEE Trans. Vis. Comput. Graph. **26**(3), 1532–1547 (2018)
20. Hadwiger, M., Mlejnek, M., Theußl, T., Rautek, P.: Time-dependent flow seen through approximate observer killing fields. IEEE Trans. Vis. Comput. Graph. **25**(1), 1257–1266 (2019). (Proceedings of the IEEE Scientific Visualization)
21. Haller, G.: Finding finite-time invariant manifolds in two-dimensional velocity fields. Chaos Interdisc. J. Nonlinear Sci. **10**(1), 99–108 (2000)
22. Haller, G.: A variational theory of hyperbolic Lagrangian coherent structures. Physica D **240**(7), 574–598 (2011)
23. Haller, G.: Lagrangian coherent structures. Ann. Rev. Fluid Mech. **47**, 137–162 (2015)
24. Haller, G., Hadjighasem, A., Farazmand, M., Huhn, F.: Defining coherent vortices objectively from the vorticity. J. Fluid Mech. **795**, 136–173 (2016)
25. Haller, G., Yuan, G.: Lagrangian coherent structures and mixing in two-dimensional turbulence. Physica D **147**(3–4), 352–370 (2000)
26. Heine, C., Leitte, H., Hlawitschka, M., Iuricich, F., De Floriani, L., Scheuermann, G., Hagen, H., Garth, C.: A survey of topology-based methods in visualization. Comput. Graph. Forum **35**(3), 643–667 (2016)
27. Helman, J.L., Hesselink, L.: Representation and display of vector field topology in fluid flow data sets. Computer **22**(8), 27–36 (1989)
28. Helman, J.L., Hesselink, L.: Visualizing vector field topology in fluid flows. IEEE Comput. Graph. Appl. **11**, 36–46 (1991)
29. Hunt, J.C.R.: Vorticity and vortex dynamics in complex turbulent flows. Trans. Can. Soc. Mech. Eng. **11**(1), 21–35 (1987). (Proceedings of CANCAM)
30. Jeong, J., Hussain, F.: On the identification of a vortex. J. Fluid Mech. **285**, 69–94 (1995)
31. Kasten, J., Petz, C., Hotz, I., Noack, B.R., Hege, H.-C.: Localized finite-time Lyapunov exponent for unsteady flow analysis. In: Vision, Modeling and Visualization, pp. 265–276 (2009)
32. Kim, B., Günther, T.: Robust reference frame extraction from unsteady 2D vector fields with convolutional neural networks. Comput. Graph. Forum (2019). (Proceedings of EuroVis)
33. Kindlmann, G., Chiw, C., Huynh, T., Gyulassy, A., Reppy, J., Bremer, P.-T.: Rendering and extracting extremal features in 3D fields. Comput. Graph. Forum **37**(3), 525–536 (2018)
34. Kuhn, A., Engelke, W., Rössl, C., Hadwiger, M., Theisel, H.: Time line cell tracking for the approximation of Lagrangian coherent structures with subgrid accuracy. Comput. Graph. Forum **33**(1), 222–234 (2014)
35. Kuhnert, J., Sudarshan, S.: Meshfree numerical schemes for time dependent problems in fluid and continuum mechanics. In: Advances in PDE Modeling and Computation, pp. 119–136 (2014)
36. Liu, I.-S.: On the transformation property of the deformation gradient under a change of frame. J. Elast. **71**(1–3), 73–80 (2003)
37. Lugt, H.J.: The dilemma of defining a vortex. In: Recent Developments in Theoretical and Experimental Fluid Mechanics, pp. 309–321. Springer (1979)

38. Machado, G., Boblest, S., Ertl, T., Sadlo, F.: Space-time bifurcation lines for extraction of 2D Lagrangian coherent structures. Comput. Graph. Forum **35**(3), 91–100 (2016)
39. Peikert, R., Roth, M.: The parallel vectors operator: a vector field visualization primitive. In: Proceedings of the Conference on Visualization 1999: Celebrating Ten Years, VIS 1999, Los Alamitos, CA, USA, pp. 263–270. IEEE Computer Society Press (1999)
40. Perry, A.E., Chong, M.S.: A description of eddying motions and flow patterns using critical-point concepts. Ann. Rev. Fluid Mech. **19**(1), 125–155 (1987)
41. Pobitzer, A., Peikert, R., Fuchs, R., Schindler, B., Kuhn, A., Theisel, H., Matkovic, K., Hauser, H.: The state of the art in topology-based visualization of unsteady flow. Comput. Graph. Forum **30**(6), 1789–1811 (2011)
42. Post, F.H., Vrolijk, B., Hauser, H., Laramee, R.S., Doleisch, H.: The state of the art in flow visualisation: feature extraction and tracking. Comput. Graph. Forum **22**(4), 775–792 (2003)
43. Robinson, S.K.: Coherent motions in the turbulent boundary layer. Ann. Rev. Fluid Mech. **23**(1), 601–639 (1991)
44. Roth, M.: Automatic Extraction of Vortex Core Lines and Other Line Type Features for Scientific Visualization, vol. 2. Hartung-Gorre (2000)
45. Roth, M., Peikert, R.: A higher-order method for finding vortex core lines. In: Proceedings of the Conference on Visualization 1998, pp. 143–150. IEEE Computer Society Press (1998)
46. Sadlo, F., Weiskopf, D.: Time-dependent 2-D vector field topology: an approach inspired by Lagrangian coherent structures. In: Computer Graphics Forum, volume 29, pp. 88–100. Wiley Online Library (2010)
47. Sahner, J., Weinkauf, T., Hege, H.-C.: Galilean invariant extraction and iconic representation of vortex core lines. In: Proceedings of the Eurographics/IEEE VGTC Symposium on Visualization (EuroVis), pp. 151–160 (2005)
48. Sahner, J., Weinkauf, T., Teuber, N., Hege, H.-C.: Vortex and strain skeletons in Eulerian and Lagrangian frames. IEEE Trans. Vis. Comput. Graph. **13**(5), 980–990 (2007)
49. Scheuermann, G., Hagen, H., Krüger, H., Menzel, M., Rockwood, A.: Visualization of higher order singularities in vector fields. In: Proceedings of the 8th Conference on Visualization 1997, pp. 67–74. IEEE Computer Society Press (1997)
50. Shadden, S.C., Lekien, F., Marsden, J.E.: Definition and properties of Lagrangian coherent structures from finite-time Lyapunov exponents in two-dimensional aperiodic flows. Physica D **212**(3), 271–304 (2005)
51. Song, Y.: A note on Galilean invariants in semi-relativistic electromagnetism. arXiv preprint arXiv:1304.6804 (2013)
52. Sujudi, D., Haimes, R.: Identification of swirling flow in 3-D vector fields. In: 12th Computational Fluid Dynamics Conference, p. 1715 (1995)
53. Theisel, H., Weinkauf, T., Hege, H.-C., Seidel, H.-P.: Saddle connectors - an approach to visualizing the topological skeleton of complex 3D vector fields. In: Proceedings of the IEEE Visualization, pp. 225–232 (2003)
54. Theisel, H., Weinkauf, T., Hege, H.-C., Seidel, H.-P.: Topological methods for 2D time-dependent vector fields based on stream lines and path lines. IEEE Trans. Vis. Comput. Graph. **11**(4), 383–394 (2005)
55. Üffinger, M., Sadlo, F., Ertl, T.: A time-dependent vector field topology based on streak surfaces. IEEE Trans. Vis. Comput. Graph. **19**(3), 379–392 (2013)
56. Üffinger, M., Sadlo, F., Kirby, M., Hansen, C.D., Ertl, T.: FTLE computation beyond first-order approximation. In: Eurographics (Short Papers), pp. 61–64 (2012)
57. Weinkauf, T., Sahner, J., Theisel, H., Hege, H.-C.: Cores of swirling particle motion in unsteady flows. IEEE Trans. Vis. Comput. Graph. **13**(6), 1759–1766 (2007)
58. Weinkauf, T., Theisel, H.: Streak lines as tangent curves of a derived vector field. IEEE Trans. Vis. Comput. Graph. **16**(6), 1225–1234 (2010)
59. Weinkauf, T., Theisel, H., Shi, K., Hege, H.-C., Seidel, H.-P.: Extracting higher order critical points and topological simplification of 3D vector fields. In: VIS 2005: IEEE Visualization 2005, pp. 559–566. IEEE (2005)

60. Wiebel, A., Chan, R., Wolf, C., Robitzki, A., Stevens, A., Scheuermann, G.: Topological flow structures in a mathematical model for rotation-mediated cell aggregation. In: Topological Methods in Data Analysis and Visualization, pp. 193–204. Springer (2011)
61. Wiebel, A., Garth, C., Scheuermann, G.: Computation of localized flow for steady and unsteady vector fields and its applications. IEEE Trans. Vis. Comput. Graph. **13**(4), 641–651 (2002)
62. Wiebel, A., Tricoche, X., Schneider, D., Jaenicke, H., Scheuermann, G.: Generalized streak lines: analysis and visualization of boundary induced vortices. IEEE Trans. Vis. Comput. Graph. **13**(6), 1735–1742 (2007)

Coreline Criteria for Inertial Particle Motion

Irene Baeza Rojo and Tobias Günther

Abstract Dynamical systems, such as the second-order ODEs that govern the motion of finite-sized objects in fluids, describe the evolution of a state by a trajectory living in a high-dimensional phase space. The high dimensionality leads to visualization challenges and, for the case of inertial particles, multiple models exist that pose different assumptions. In this paper, we thoroughly address the extraction of a specific feature, namely the vortex corelines of inertial particles. Based on a general template model that comprises two of the most commonly used inertial particle ODEs, we first transform their high-dimensional tangent vector field into a Galilean reference frame in which the observed inertial particle flow becomes as steady as possible. In the optimal frame, we derive first-order and second-order vortex coreline criteria, allowing us to extract straight and bent inertial vortex corelines using 3D and 6D parallel vectors operators, respectively. With this, we generalize existing work in multiple ways: not only do we handle two inertial particle models at once, we extend the concept of second-order vortex corelines to the inertial case and make them Galilean-invariant by deriving the criteria from a steady reference frame, rather than from a geometric characterization.

1 Introduction

In this paper, we study the vortical motion of finite-sized objects in time-dependent fluids, such as sand particles in air or bubbles in water. Such vortex dynamics occur for instance when helicopters approach the ground [1–3], in the detection of marine debris [4], and in the formation of rain [5]. Previous methods [6, 7] used an aerosol par-

Present Address:
I. B. Rojo · T. Günther (✉)
Department of Computer Science, ETH Zürich, Zürich, Switzerland
e-mail: tobias.guenther@inf.ethz.ch

I. B. Rojo
e-mail: irene.baeza@inf.ethz.ch

T. Günther
FAU Erlangen-Nürnberg, Erlangen, Germany

I. Hotz et al. (eds.), *Topological Methods in Data Analysis and Visualization VI*, Mathematics and Visualization, https://doi.org/10.1007/978-3-030-83500-2_8

ticle model that had only one degree of freedom. We extend the vortex coreline extraction in unsteady flows to a more general inertial particle model that captures a wider spectrum of density ratios between the particle and the surrounding medium. Using a generalized description that contains both models as special cases, we develop a first-order and a second-order feature extractor based on Sujudi-Haimes [8] and Roth and Peikert [9], respectively. While the first-order extraction can be carried out with standard 3D parallel vectors extractors [10], the second-order criterion requires the search for 6D parallel vectors in a high-dimensional 6D space, for which we use a Bézier-based subdivision with subsequent Newton iterations. Instead of deriving the vortex criteria only from a geometric point of view in space-time [6], we extract the vortex corelines in local Galilean reference frames in which the flow field becomes as steady as possible [11, 12]. When it comes to vortex coreline extraction, there are two orthogonal concepts: the shape of the coreline, and the motion of the coreline. While previous work [7] concentrated on the motion, our focus is on the shape of the coreline, as we search for straight and curved corelines. Assuming that vortices perform Galilean transformations, we transform our flow into a reference frame in which the time partial derivative vanishes by optimizing for a Galilean transformation. In the optimal frame, the resulting vortex extraction methods therefore become Galilean invariant. In summary, we contribute:

- a generalization of the inertial vortex coreline criterion of Günther and Theisel [6] to a more general particle model,
- and a second-order criterion that extracts bent inertial vortex corelines, for which we generalize Roth and Peikert [9].

Notation. In the following, we denote scalars s in italic letters, vectors $\mathbf{v}$ are bold and matrices $\mathbf{J}$ are bold upper-case letters. Throughout the chapter, $\mathbf{I}$ denotes the identity matrix. Quantities in the space-velocity domain are denoted with a tilde $\tilde{\mathbf{v}}$.

2 Related Work

2.1 Galilean Invariance

Galilean invariance is a desirable formal property that the measure of a feature might have. If a measure (such as a vortex measure) is Galilean invariant, then it does not change under Galilean transformations of the reference frame [23]. Formally, a Galilean transformation maps a point $(\mathbf{x}, t)$ to the point $(\mathbf{x}^*, t^*)$:

$$\mathbf{x}^* = \mathbf{x} + \mathbf{c} + t\,\mathbf{d}, \qquad t^* = t + a, \tag{1}$$

where $\mathbf{c}$ and $\mathbf{d}$ are constant vectors and a is a constant scalar. Accordingly, a vector field $\mathbf{u}(\mathbf{x}, t)$ is transformed to:

$$\mathbf{u}^*(\mathbf{x}^*, t^*) = \mathbf{u}(\mathbf{x}, t) + \mathbf{d} \tag{2}$$
$$= \mathbf{u}(\mathbf{x}^* - \mathbf{c} - t\,\mathbf{d}, t^* - a) + \mathbf{d}, \tag{3}$$

which follows from differentiation of Eq. (1) to consider how the tangent of a pathline $\frac{d\mathbf{x}^*(t^*)}{dt} = \mathbf{u}^*(\mathbf{x}^*, t^*)$ is transformed. A measure $\mathcal{M}$ is Galilean invariant if it gives the same result at both locations in the accordingly transformed vector fields, i.e., $\mathcal{M}(\mathbf{x}, \mathbf{u}, t) = \mathcal{M}(\mathbf{x}^*, \mathbf{u}^*, t^*)$. For example, applying the ∇ operator to both sides of Eq. (2) shows that the Jacobian is Galilean invariant, since $\nabla\mathbf{u}^*(\mathbf{x}^*, t^*) = \nabla\mathbf{u}(\mathbf{x}, t)$, given that $\nabla\mathbf{d} = \mathbf{0}$ because $\mathbf{d}$ is constant.

Since the motion of the reference frame and the motion of the feature are relative to each other, Galilean invariance not only guarantees that measures do not change under motions of the observer; they also do not change when the feature itself is moving. In other words, a Galilean invariant vortex measure will give the same result if vortices move with constant speed in a constant direction, which ultimately allows us to extract moving vortices.

2.2 *Inertial Particle Motion*

The motion of finite-sized particles in fluids can be described by a second-order ODE, which can be rephrased into a coupled first-order ODE [6]. Depending on the possible simplifying assumptions, different equations of motion are possible, which resulted in a number of different particle models [13–16]. Throughout this work, we use two inertial particle models. The underlying fluid flow is described by the $n-$d unsteady vector field $\mathbf{u}(\mathbf{x}, t) : \mathbb{R}^n \times \mathbb{R} \rightarrow \mathbb{R}^n$.

Model 1. The first model was described by Crowe et al. [13] and considers small particles in air. With $\mathbf{g}$ being the gravity vector, it reads:

$$\tilde{\mathbf{v}}(\mathbf{x}, \mathbf{v}, t) = \begin{pmatrix} \mathbf{v} \\ \frac{\mathbf{u}(\mathbf{x},t)-\mathbf{v}}{r} + \mathbf{g} \end{pmatrix}. \tag{4}$$

The particle response time $r = \frac{d_p^2 \rho_p}{18\mu}$ determines how quickly an inertial particle aligns its own velocity $\mathbf{v}$ with the underlying air flow $\mathbf{u}(\mathbf{x}, t)$, where d_p is thc particle size, ρ_p is the particle density and μ is the viscosity of the underlying air flow. The smaller r is, the lighter the particle, with tracer particles arising in the limit for $r \rightarrow 0$. This particle model has frequently been used to model the motion of sand particles in air [1, 3, 17–19], and assumes that particles are spherical, very small, and have a much higher density than the surrounding air.

Model 2. The second model was used by Haller [14] and distinguishes between aerosols and bubbles, based on the density ratio $R = 2\rho_f/(\rho_f + 2\rho_p)$, where ρ_f is the fluid density and ρ_p is the particle density. For $R < 2/3$ we have aerosols ($\rho_p > \rho_f$, e.g., sand particles in air), for $R = 2/3$ we obtain neutrally buoyant particles

($\rho_p = \rho_f$) and for $R > 2/3$ the motion of bubbles ($\rho_p < \rho_f$) is modeled:

$$\tilde{\mathbf{v}}(\mathbf{x}, \mathbf{v}, t) = \begin{pmatrix} \mathbf{v} \\ \frac{R}{St}(\mathbf{u}(\mathbf{x}, t) - \mathbf{v}) + \frac{3R}{2}\frac{D\mathbf{u}(\mathbf{x},t)}{Dt} + \left(1 - \frac{3R}{2}\right)\mathbf{g} \end{pmatrix}, \tag{5}$$

where $\frac{D\mathbf{u}(\mathbf{x},t)}{Dt} = \mathbf{J}\mathbf{u} + \mathbf{u}_t$ is the material derivative of the flow, i.e., the acceleration, and $\mathbf{u}_t = \frac{\partial \mathbf{u}(\mathbf{x},t)}{\partial t}$ is the time partial derivative of the flow. The Stokes number St determines the amount of inertia, with $St \to 0$ approaching the behavior of tracer particles. This model and its variations, such as gravity-free environments or neutrally buoyant particles, have been used extensively in the fluid dynamics literature [14, 15, 20–22].

2.3 Vortex Corelines of Massless Flows

Vortex coreline definitions have evolved over the last two decades, adding more and more generality. We mainly focus on the methods that led up to this paper. For a more comprehensive overview, we refer to the recent survey of Günther and Theisel [23].

Corelines in Steady Flows. The earliest influential algorithms were built for steady flows. Globus et al. [24] traced streamlines from attracting and repelling foci and Sujudi and Haimes [8] introduced the reduced velocity criterion. In the presence of complex eigenvalues in the Jacobian $\mathbf{J} = \nabla\mathbf{u}$, let $\mathbf{e}$ be the single eigenvector with a real eigenvalue. Sujudi and Haimes requested that the projection of the velocity along vector $\mathbf{e}$ is zero: $\mathbf{u} - (\mathbf{u}^T\mathbf{e})\,\mathbf{e} = \mathbf{0}$. In other words, the flow exactly on the vortex coreline is not in the swirling plane, but only moves forward. Peikert and Roth [10] introduced the parallel vectors (PV) operator, which returns for two vector fields the set of locations at which the two vector fields are parallel. With $\mathbf{u} \parallel \mathbf{e}$, the method of Sujudi-Haimes [8] can be rephrased in PV notation, which is often computed as $\mathbf{u} \parallel \mathbf{J}\,\mathbf{u}$. Other vortex coreline conditions, such as helicity extrema, vorticity extrema and λ_2 extrema, can likewise be expressed by the PV operator [10, 25].

Corelines in Unsteady Flows. Fuchs et al. [26] extended this reduced velocity criterion to unsteady flows by extracting $\mathbf{u} \parallel \mathbf{J}\,\mathbf{u} + \mathbf{u}_t$. Another approach was taken by Weinkauf et al. [27], who applied the reduced velocity criterion in space-time. Since there are, in space-time, two eigenvectors with real eigenvalue (one of them is zero), the flow vector $\mathbf{u}$ must be in the plane that is spanned by these two eigenvectors to not take part in the swirling plane. This co-planar vectors condition simplifies to the PV condition $\mathbf{u} - \mathbf{f} \parallel \mathbf{J}(\mathbf{u} - \mathbf{f})$, with $\mathbf{f}$ being the feature flow field [28]. Günther et al. [29] introduced a rotation invariant coreline condition. By linear optimization, rotating and translating reference frames have recently been extracted, in which the flow becomes steady [30, 31]. The idea is that, in the optimal frame, vortex features are no longer obscured by ambient motion. The solution must vary spatially, since no global observer exists for the entire domain [11, 12, 32]. While Günther and

Theisel optimized the reference frame locally at each point in space-time, Hadwiger et al. [31] formulated the search for optimal reference frames as a global optimization problem, in which the motion is described by an approximate Killing field. Note that a global optimization is infeasible for inertial particle motion, since this would require to discretize the entire seven-dimensional phase space. More recently, reference frame optimization has been extended to spatially-varying transformations, which opened a path to topology-based methods beyond vortices [33]. Günther and Theisel [34] have shown that the method of Weinkauf et al. [27] is optimal for vortices performing Galilean transformations. Other reference frame adjustments used flow decompositions [35–37], adjusted to Galilean-invariant extremal features [38] or used machine learning [39].

Second-Order Corelines. The local coreline conditions above can all be classified as first-order methods, since they only involve first-order derivatives. For steady flows $\mathbf{u}(x, y, z)$, the method of Roth and Peikert [9] computes bent vortex corelines with

$$\mathbf{u} \parallel \mathbf{b} \quad \text{with} \quad \mathbf{b} = \frac{D}{Dt}(\mathbf{J}\mathbf{u}) = (\nabla\mathbf{J}\mathbf{u} + \mathbf{J}\mathbf{J})\,\mathbf{u}, \tag{6}$$

where $\nabla\mathbf{J}\mathbf{u} = \mathbf{J}_x u + \mathbf{J}_y v + \mathbf{J}_z w$. Note that this method is not Galilean invariant, even when calculating $\mathbf{b} = \frac{D}{Dt}(\mathbf{J}\mathbf{u} + \mathbf{u}_t)$, since the material derivative entails a multiplication with $\mathbf{u}$, which is not invariant. In this paper, we extend this approach to the high-dimensional vector field of inertial particles. When considered in a reference frame in which the flow is steady, the approach becomes reference frame invariant.

2.4 *Vortex Corelines of Inertial Particles*

For inertial particles, Günther and Theisel [6] proposed an inertial first-order vortex coreline criterion that is Galilean invariant. However, their condition is tailored to Model 1 and can only guarantee to correctly extract straight vortex corelines. To arrive at this condition, they followed a geometric construction [27].

The method was extended by Günther and Theisel [7] to handle more kinds of vortex motions, namely all smooth rotations and translations. For this, a linear optimization is needed that finds a reference frame in which the high-dimensional flow becomes as steady as possible. Then, the above method [6] is applied to find straight inertial vortex corelines. As before, the extractor was tailored to Model 1.

In this paper, we further extend the method of Günther and Theisel [6] in three ways. First, we derive vortex criteria not only for Model 1, but for a more general template that includes Model 1 and Model 2 as special cases. Second, we extend the method of Roth and Peikert [9] to derive a second-order vortex coreline criterion that extracts bent inertial vortex corelines. Third, instead of deriving the vortex criteria geometrically, we follow a reference frame centered approach, which is for both the first-order and second-order case Galilean invariant, and arrives for the special case of Model 1 at the same condition as [6]. This gives new insights on the mathematical properties of their solution.

3 Vortex Coreline Criteria for Inertial Particles

Our inertial vortex coreline criteria are fundamentally based on the observation of the time-dependent high-dimensional vector field in a reference frame in which the flow becomes steady. Since we concentrate on the shape of the corelines instead of the motion, we assume a Galilean transformation. Therefore, we first introduce Galilean reference frame transformations of inertial particles and derive the analytic solution to the optimal reference frame for both inertial particle models. Afterwards, we derive the inertial vortex coreline conditions that are applied in the optimal frame. We begin with first-order criteria in 2D, in 2D space-time, and in 3D. Afterwards, we introduce the second-order vortex coreline criterion that allows us to find bent corelines.

3.1 *Generalized Inertial Particle Motion*

Similar to Günther and Gross [40], we utilize a generalized template-based description of the particle motion, which allows us to express local properties of inertial particles for multiple models at once. Their template-based description introduces abstract variables that are assigned dependent on the particle model. While Günther and Gross generalized the high-dimensional Jacobian matrix $\tilde{\mathbf{J}}$, we generalize the underlying high-dimensional vector field $\tilde{\mathbf{v}}$ instead. Let κ be a constant scalar and $\mathbf{k}(\mathbf{x}, t)$ be an n-d vector field. The change in particle position $\mathbf{x}$ and particle velocity $\mathbf{v}$ of an inertial particle are described by the high-dimensional vector field $\tilde{\mathbf{v}}(\mathbf{x}, \mathbf{v}, t)$:

$$\tilde{\mathbf{v}}(\mathbf{x}, \mathbf{v}, t) = \frac{\mathrm{d}}{\mathrm{d}t}\begin{pmatrix}\mathbf{x}\\ \mathbf{v}\end{pmatrix} = \begin{pmatrix}\mathbf{v}\\ \mathbf{k}(\mathbf{x}, t) - \frac{\mathbf{v}}{\kappa}\end{pmatrix}. \tag{7}$$

Intuitively speaking, position $\mathbf{x}$ and velocity $\mathbf{v}$ are the two properties that are stored per particle. Then, the high-dimensional vector field $\tilde{\mathbf{v}}$ is the vector field in which inertial particle trajectories are traced as tangent curves. The high-dimensional Jacobian matrix $\tilde{\mathbf{J}} = \nabla\tilde{\mathbf{v}}$ contains the $\mathbf{x}$ and $\mathbf{v}$ partials as column vectors and becomes:

$$\tilde{\mathbf{J}}(\mathbf{x}, \mathbf{v}, t) = \left(\frac{\partial\tilde{\mathbf{v}}(\mathbf{x},\mathbf{v},t)}{\partial\mathbf{x}}, \frac{\partial\tilde{\mathbf{v}}(\mathbf{x},\mathbf{v},t)}{\partial\mathbf{v}}\right) = \begin{pmatrix}\mathbf{0} & \mathbf{I}\\ \nabla\mathbf{k}(\mathbf{x}, t) & -\frac{1}{\kappa}\mathbf{I}\end{pmatrix}, \tag{8}$$

which can be used to locally analyze the behavior of inertial particles. By a suitable choice of κ and $\mathbf{k}(\mathbf{x}, t)$, Eqs. (7) and (8) can describe inertial particle motion:

$$\text{Model 1}: \quad \kappa = r, \quad \mathbf{k}(\mathbf{x},t) = \frac{\mathbf{u}(\mathbf{x},t)}{r} + \mathbf{g}, \tag{9}$$

$$\text{Model 2}: \quad \kappa = \frac{St}{R}, \quad \mathbf{k}(\mathbf{x},t) = \frac{R}{St}\mathbf{u}(\mathbf{x},t) + \frac{3R}{2}\frac{D\mathbf{u}(\mathbf{x},t)}{Dt} + \left(1 - \frac{3R}{2}\right)\mathbf{g}. \tag{10}$$

Galilean Transformation. Since inertial particle motion is described by a higher-dimensional vector field, we now consider how an inertial particle in the high-dimensional space changes under Galilean transformations. From Eqs. (1) and (2), we see how position $\mathbf{x}(t)$ and velocity $\mathbf{v}(t)$ of an inertial particle $\tilde{\mathbf{p}}(t) = \begin{pmatrix}\mathbf{x}(t)\\ \mathbf{v}(t)\end{pmatrix}$ are transformed to $\tilde{\mathbf{p}}^*(t)$:

$$\tilde{\mathbf{p}}^*(t) = \begin{pmatrix}\mathbf{x}^*(t)\\ \mathbf{v}^*(t)\end{pmatrix} = \begin{pmatrix}\mathbf{x}(t) + \mathbf{c} + t\,\mathbf{d}\\ \mathbf{v}(t) + \mathbf{d}\end{pmatrix}. \tag{11}$$

The above particle $\tilde{\mathbf{p}}^*$ moves tangentially in the high-dimensional vector field $\tilde{\mathbf{v}}^*$ that governs its motion. By differentiating Eq. (11), we consider the tangent of our particle at $(\mathbf{x}^*, \mathbf{v}^*)$ to see how the governing high-dimensional vector field is transformed

$$\tilde{\mathbf{v}}^* = \frac{\mathrm{d}\tilde{\mathbf{p}}^*}{\mathrm{d}t}_{|\mathbf{x}^*,\mathbf{v}^*} = \begin{pmatrix}\frac{\mathrm{d}\mathbf{x}(t)}{\mathrm{d}t} + \mathbf{d}\\ \frac{\mathrm{d}\mathbf{v}(t)}{\mathrm{d}t}\end{pmatrix} = \frac{\mathrm{d}\tilde{\mathbf{p}}}{\mathrm{d}t}_{|\mathbf{x},\mathbf{v}} + \begin{pmatrix}\mathbf{d}\\ \mathbf{0}\end{pmatrix} = \tilde{\mathbf{v}} + \begin{pmatrix}\mathbf{d}\\ \mathbf{0}\end{pmatrix}. \tag{12}$$

By inserting our generalized particle model from Eq. (7) into the Galilean transformed high-dimensional flow in Eq. (12), we obtain the transformed high-dimensional vector field that governs the inertial particle motion for both our considered models:

$$\tilde{\mathbf{v}}^* = \begin{pmatrix}\mathbf{v} + \mathbf{d}\\ \mathbf{k}(\mathbf{x},t) - \frac{\mathbf{v}}{\kappa}\end{pmatrix}, \quad \tilde{\mathbf{J}}^* = \tilde{\mathbf{J}}, \quad \tilde{\mathbf{v}}_t^* = \begin{pmatrix}\mathbf{0}\\ \mathbf{k}_t - \nabla\mathbf{k}\cdot\mathbf{d}\end{pmatrix}. \tag{13}$$

The Jacobian $\tilde{\mathbf{J}}^* = \nabla\tilde{\mathbf{v}}^*$ and the time partial derivative of the high-dimensional flow $\tilde{\mathbf{v}}_t^* = \frac{\partial\tilde{\mathbf{v}}^*}{\partial t}$ follow directly by differentiation. Note that the calculation of the derivatives requires application of the chain rule, since with Eq. (1) position $\mathbf{x}$ is $\mathbf{x} = \mathbf{x}^* - \mathbf{c} - t\,\mathbf{d}$.

3.2 *Inertial Motion in Steady Frame*

To find the most-steady reference frame for inertial particle motion, Günther and Theisel [7] considered all smooth rotations and translations of the reference frame. The final vortex extraction eventually resulted in a parallel vectors operation in 6D. Since we concentrate in this paper on coreline shapes, we assume a Galilean transformation for simplicity. For the first-order criteria, this will result in a 3D parallel vectors problem that can be solved efficiently.

To find a steady reference frame, we rearrange the velocity subspace of Eq. (13) to select the translation parameter $\mathbf{d}$ for which the time partial of the high-dimensional flow $\tilde{\mathbf{v}}^*$ vanishes, i.e., $\tilde{\mathbf{v}}_t^* = \mathbf{0}$:

$$\boxed{\mathbf{d} = (\nabla \mathbf{k})^{-1} \mathbf{k}_t}. \tag{14}$$

When inserting Eq. (14) in Eq. (13), the flow becomes steady.

For the specific models from Eqs. (9) and (10) we get:

$$\text{Model 1}: \quad \mathbf{d} = \mathbf{J}^{-1}\,\mathbf{u}_t. \tag{15}$$

$$\text{Model 2}: \quad \mathbf{d} = \left(\mathbf{J} + \frac{3St}{2}\frac{\nabla D\mathbf{u}}{Dt}\right)^{-1} \left(\mathbf{u}_t + \frac{3St}{2}\frac{D^2\mathbf{u}}{Dt^2}\right). \tag{16}$$

In case of Model 1, we have $\mathbf{d} = -\mathbf{f}$, where $\mathbf{f} = -\mathbf{J}^{-1}\mathbf{u}_t$ is the feature flow field [28] of massless particles, cf. Eq. (26) in [23]. Next, we extract inertial vortex corelines in the Galilean reference frame that is as steady as possible.

3.3 *First-Order Corelines*

3.3.1 2D Conditions

In the 2D steady vector fields of tracer particles, vortex centers are critical points with complex eigenvalues in the Jacobian [41]. Thus, for inertial flows we set $\tilde{\mathbf{v}}^* = \mathbf{0}$ in the high-dimensional inertial flow in Eq. (7). The position subspace gives a condition for $\mathbf{v}$, namely $\mathbf{v} = -\mathbf{d}$, which we insert in the velocity subspace, resulting in:

$$\mathbf{k} + \frac{\mathbf{d}}{\kappa} = \mathbf{0} \quad \Leftrightarrow \quad \boxed{\kappa\mathbf{k} + \mathbf{d} = \mathbf{0}} \quad \Leftrightarrow \quad \kappa(\nabla\mathbf{k})\mathbf{k} + \mathbf{k}_t = \mathbf{0}. \tag{17}$$

Inserting Model 1 from Eq. (9) or Model 2 from Eq. (10) into the boxed condition in Eq. (17) gives 2D conditions for locating the vortex center.

$$\text{Model 1}: \quad \mathbf{u} + r\,\mathbf{g} + \mathbf{d} = \mathbf{0}, \tag{18}$$

$$\text{Model 2}: \quad \mathbf{u} + \frac{3St}{2}\frac{D\mathbf{u}}{Dt} + \left(\frac{St}{R} - \frac{3St}{2}\right)\mathbf{g} + \mathbf{d} = \mathbf{0}. \tag{19}$$

To illustrate the above conditions, Fig. 1 shows space-time visualizations of the analytic MOVING CENTER flow (cf. Eq. (44) in Sect. 5) for Model 1. In this flow, the vortex moves with constant speed over time, as illustrated by inertial pathlines (left). The motion results in a feature flow that is also constant and moves in the same direction as the coreline (right). Using line integral convolution (LIC), we visualize

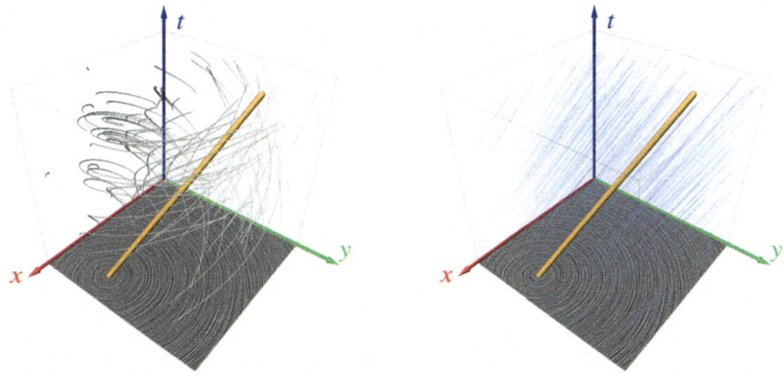

Fig. 1 Inertial vortex coreline in the Moving Center flow for Model 1 using $\mathbf{g} = (-2, 1)^{\mathrm{T}}$ and $r = 0.1$. The flow $\kappa\mathbf{k} + \mathbf{d}$ is shown with a LIC slice at the bottom, and the coreline (•) shows the center of the vortex over time. Left, inertial pathlines (•) depict the behavior of inertial particles in the flow. Right, the ambient flow field $\mathbf{d}$ (•) is constant and evolves parallel to the vortex coreline

a slice of the vector field $\kappa\mathbf{k} + \mathbf{d}$, where 2D vortex centers can be found to obtain a seed point for the space-time tracking.

Space-Time Extraction. The above condition in Eq. (17) determines vortex centers for a given moment in time. By lifting the condition into space-time, the paths of vortex centers can be extracted at once using a parallel vectors operator [10]. In space-time, the following PV condition arises, where the last component is time:

$$\begin{pmatrix} \kappa\mathbf{k} \\ 1 \end{pmatrix} \parallel \begin{pmatrix} -\mathbf{d} \\ 1 \end{pmatrix}. \tag{20}$$

The space-time condition has interesting properties and can be reformulated as shown next. First, let $\bar{\mathbf{k}}$ be the vector field on the left side and let $\bar{\mathbf{K}} = \nabla\bar{\mathbf{k}}$ be its space-time Jacobian:

$$\bar{\mathbf{k}} = \begin{pmatrix} \kappa\mathbf{k} \\ 1 \end{pmatrix}, \qquad \bar{\mathbf{K}} = \left(\frac{\partial\bar{\mathbf{k}}}{\partial\mathbf{x}}, \frac{\partial\bar{\mathbf{k}}}{\partial t}\right) = \kappa \begin{pmatrix} \nabla\mathbf{k} & \mathbf{k}_t \\ \mathbf{0} & 0 \end{pmatrix}. \tag{21}$$

Since $\bar{\mathbf{K}} \begin{pmatrix} -\mathbf{d} \\ 1 \end{pmatrix} = \mathbf{0}$, we can see that $\begin{pmatrix} -\mathbf{d} \\ 1 \end{pmatrix}$ is an eigenvector of $\bar{\mathbf{K}}$ with the corresponding eigenvalue 0. With Eq. (20) it follows that on a vortex coreline, $\bar{\mathbf{k}}$ is also an eigenvector of $\bar{\mathbf{K}}$ and thus the reduced velocity criterion of Sujudi-Haimes [8] applies. When using the parallel vectors version [10] of the criterion on $\bar{\mathbf{k}}$, we get:

$$\kappa \begin{pmatrix} \nabla\mathbf{k} & \mathbf{k}_t \\ \mathbf{0} & 0 \end{pmatrix} \begin{pmatrix} \kappa\mathbf{k} \\ 1 \end{pmatrix} \parallel \begin{pmatrix} \kappa\mathbf{k} \\ 1 \end{pmatrix} \Leftrightarrow \boxed{\begin{pmatrix} \kappa(\nabla\mathbf{k})\mathbf{k} + \mathbf{k}_t \\ 0 \end{pmatrix} \parallel \begin{pmatrix} \kappa\mathbf{k} \\ 1 \end{pmatrix}}. \tag{22}$$

Note that the condition on the right can be computed without having to linearly solve for $\mathbf{d}$ in Eq. (14). The criterion on the right of Eq. (22) can only be fulfilled if its left side becomes zero, due to the time components (only the zero vector can be parallel to $\bar{\mathbf{k}}$). Thus, the following conditions are equivalent:

$$\mathbf{k} + \frac{\mathbf{d}}{\kappa} = \mathbf{0} \quad \Leftrightarrow \quad \kappa(\nabla\mathbf{k})\mathbf{k} + \mathbf{k}_t = \mathbf{0}. \tag{23}$$

The latter asks for the acceleration of vector field $\mathbf{k}$ to be zero.

3.3.2 3D Conditions

To find 3D corelines, we apply the method of Sujudi and Haimes [8] to the high-dimensional flow in the optimal steady frame, i.e., to $\tilde{\mathbf{v}}^*$ in Eq. (13). We search for locations at which the high-dimensional velocity $\tilde{\mathbf{v}}^*$ aligns with its acceleration:

$$\tilde{\mathbf{v}}^* \parallel \tilde{\mathbf{J}}^*\tilde{\mathbf{v}}^* \quad \Rightarrow \quad \begin{pmatrix} \mathbf{v} + \mathbf{d} \\ \mathbf{k} - \frac{\mathbf{v}}{\kappa} \end{pmatrix} \parallel \begin{pmatrix} \mathbf{k} - \frac{\mathbf{v}}{\kappa} \\ \nabla\mathbf{k}(\mathbf{v} + \mathbf{d}) - \frac{\mathbf{k} - \frac{\mathbf{v}}{\kappa}}{\kappa} \end{pmatrix}. \tag{24}$$

The above parallel vectors condition in Eq. (24) is six-dimensional and requires the search for both position and velocity. By requiring parallelism in the space subspace and the velocity subspace individually, the condition is simplified, since the dependence on the velocity $\mathbf{v}$ disappears, which was demonstrated for Model 1 [6]. In the Appendix 1, we show this for the general case, arriving at:

$$\boxed{\kappa\mathbf{k} + \mathbf{d} \parallel \nabla\mathbf{k}\,(\kappa\mathbf{k} + \mathbf{d})}, \tag{25}$$

which is now independent of the particle velocity $\mathbf{v}$. Thus, the parallel vectors condition in Eq. (25) can be searched in the position subspace using standard extractors, as introduced by Peikert and Roth [10]. Inserting Model 1 from Eq. (9) or Model 2 from Eq. (10) into Eq. (25) gives:

$$\text{Model 1}: \quad \mathbf{u} + r\,\mathbf{g} + \mathbf{d} \quad \parallel \quad \mathbf{J}\,(\mathbf{u} + r\,\mathbf{g} + \mathbf{d})\,, \tag{26}$$

$$\text{Model 2}: \quad \mathbf{w} \parallel \left[\mathbf{J} + \frac{3St}{2}\nabla\left(\frac{D\mathbf{u}}{Dt}\right)\right]\mathbf{w}, \tag{27}$$

$$\text{with } \mathbf{w} = \mathbf{u} + \frac{3St}{2}\frac{D\mathbf{u}}{Dt} + \left(\frac{St}{R} - \frac{3St}{2}\right)\mathbf{g} + \mathbf{d}. \tag{28}$$

Note that Eq. (26) is the 3D condition that was derived by Günther and Theisel [6] (Eq. (32) in their paper for $\mathbf{d} = -\mathbf{f}$ as shown above), which appears here as special case.

Curvature of the Coreline. In the optimal steady reference frame, the condition $\tilde{\mathbf{v}}^* \parallel \tilde{\mathbf{J}}^*\tilde{\mathbf{v}}^*$ in Eq. (24) determines locations at which the high-dimensional flow is parallel to the acceleration. If both are parallel, the curvature of the resulting coreline vanishes. The curvature of a parametric curve $\mathbf{x}(t)$ is given in any dimension by: $\kappa = \frac{\sqrt{\|\dot{\mathbf{x}}\|^2\|\ddot{\mathbf{x}}\|^2-(\dot{\mathbf{x}}^{\mathrm{T}}\ddot{\mathbf{x}})^2}}{\|\dot{\mathbf{x}}\|^3}$. The numerator vanishes if the enclosed angle θ between $\dot{\mathbf{x}}$ and $\ddot{\mathbf{x}}$ is zero, due to the dot product $\dot{\mathbf{x}}^{\mathrm{T}}\ddot{\mathbf{x}} = \|\dot{\mathbf{x}}\|\,\|\ddot{\mathbf{x}}\|\cos(\theta)$. In Eq. (24), we have velocity $\dot{\mathbf{x}} = \tilde{\mathbf{v}}^*$ and steady acceleration $\ddot{\mathbf{x}} = \tilde{\mathbf{J}}^*\tilde{\mathbf{v}}^*$. Thus, the first-order vortex coreline criterion models vortices as straight line structures.

3.4 Second-Order Corelines

Roth and Peikert [9] introduced an extension of Sujudi and Haimes [8] that models bent vortex corelines. Applied to our high-dimensional vector field in the optimal steady reference frame, vortex corelines are identified as locations that fulfill:

$$\tilde{\mathbf{v}}^* \parallel \tilde{\mathbf{b}}^* \quad \text{with} \quad \tilde{\mathbf{b}}^* = \frac{D}{Dt}(\tilde{\mathbf{J}}^*\tilde{\mathbf{v}}^*). \tag{29}$$

The acceleration $\tilde{\mathbf{a}}^*$ in the transformed reference frame is

$$\tilde{\mathbf{a}}^* = \frac{D\tilde{\mathbf{v}}^*}{Dt} = \tilde{\mathbf{J}}^*\tilde{\mathbf{v}}^* = \begin{pmatrix} \mathbf{k} - \frac{\mathbf{v}}{\kappa} \\ \nabla\mathbf{k}(\mathbf{v}+\mathbf{d}) + \frac{\mathbf{v}}{\kappa^2} - \frac{\mathbf{k}}{\kappa} \end{pmatrix}, \tag{30}$$

which follows from insertion of Eqs. (13) into the left hand side of Eq. (30). Computing the material derivative of Eq. (30) as $\tilde{\mathbf{b}}^* = \frac{D\tilde{\mathbf{a}}^*}{Dt} = (\nabla\tilde{\mathbf{a}}^*)\tilde{\mathbf{v}}^*$ gives that the change in acceleration $\tilde{\mathbf{b}}^*$ is:

$$\tilde{\mathbf{b}}^* = \begin{pmatrix} \nabla\mathbf{k}(\mathbf{v}+\mathbf{d}) + \frac{\mathbf{v}}{\kappa^2} - \frac{\mathbf{k}}{\kappa} \\ \nabla(\nabla\mathbf{k})(\mathbf{v}+\mathbf{d})(\mathbf{v}+\mathbf{d}) - \frac{\nabla\mathbf{k}}{\kappa}(\mathbf{v}+\mathbf{d}) + \nabla\mathbf{k}(\mathbf{k}-\frac{\mathbf{v}}{\kappa}) - \frac{\mathbf{v}}{\kappa^3} + \frac{\mathbf{k}}{\kappa^2} \end{pmatrix}, \tag{31}$$

with $\nabla(\nabla\mathbf{k})(\mathbf{v}+\mathbf{d}) = \frac{\partial(\nabla\mathbf{k})}{\partial x}\cdot(u+d_1) + \frac{\partial(\nabla\mathbf{k})}{\partial y}\cdot(v+d_2) + \frac{\partial(\nabla\mathbf{k})}{\partial z}\cdot(w+d_3)$ for $\mathbf{d} = (d_1, d_2, d_3)^{\mathrm{T}}$. With Eq. (29), this leads to the following parallel vectors problem:

$$\boxed{\begin{pmatrix} \mathbf{v}+\mathbf{d} \\ \mathbf{k} - \frac{\mathbf{v}}{\kappa} \end{pmatrix} \parallel \begin{pmatrix} \nabla\mathbf{k}(\mathbf{v}+\mathbf{d}) + \frac{\mathbf{v}}{\kappa^2} - \frac{\mathbf{k}}{\kappa} \\ \nabla(\nabla\mathbf{k})(\mathbf{v}+\mathbf{d})(\mathbf{v}+\mathbf{d}) - \frac{\nabla\mathbf{k}}{\kappa}(\mathbf{v}+\mathbf{d}) + \nabla\mathbf{k}(\mathbf{k}-\frac{\mathbf{v}}{\kappa}) - \frac{\mathbf{v}}{\kappa^3} + \frac{\mathbf{k}}{\kappa^2} \end{pmatrix}} \tag{32}$$

This is a 6D parallel vectors problem, since both position $\mathbf{x}$ (for the evaluation of $\mathbf{k}$) and velocity $\mathbf{v}$ need to be searched. The extraction algorithm is described in Sect. 4.

4 Implementation

In the following, we explain the numerical extraction algorithms used to locate inertial critical points and inertial vortex corelines.

First-order Method. To find inertial critical points in 2D according to Eq. (17), we use the numerical subdivision approach of Globus et al. [24]. The line-based extraction of inertial critical paths in 2D space-time in Eq. (20) and the first-order extraction of inertial 3D corelines in Eq. (25) are formulated as 3D parallel vectors problems [10]. We used a Bézier-based subdivision to find the roots of the cross product [42, 43], which we refined using Newton iterations [44]. Several other existing algorithms would also be applicable [10, 45].

Second-order Method. The second-order method requires a 6D parallel vectors extraction in a 6D space. Recently, Hofmann and Sadlo [46] introduced the dependent vectors operator which extends the parallel vector operator to arbitrary dimensions, where our 6D PV problem is a special case. While their method provides a general framework for high dimensional features, their proposed algorithm does not scale well for our problem. The direct computation of the two 6D vector fields in Eq. (32) is too expensive, both in terms of computation time and memory consumption. To solve this problem, we follow the approach of Günther and Theisel [7], which we adapt to our vector configuration.

By writing the position subspace and the velocity subspace of Eq. (32) as functions in $\mathbf{v}$, we can simplify the search. By introducing the spatially-varying 3D vector fields $\underline{\mathbf{a}}$, $\underline{\mathbf{b}}$, $\underline{\mathbf{c}}$, $\underline{\mathbf{d}}$, matrix fields $\underline{\mathbf{B}}$, $\underline{\mathbf{C}}$, $\underline{\mathbf{D}}$, and tensor field $\underline{\underline{\mathbf{D}}}$, we can calculate the 6D vectors in Eq. (32) for a certain $\mathbf{v}$ on demand using only data stored in 3D:

$$\tilde{\mathbf{v}}^* \parallel \tilde{\mathbf{b}}^* \quad \Leftrightarrow \quad \begin{pmatrix} \mathbf{v} + \underline{\mathbf{a}} \\ \underline{\mathbf{B}}\,\mathbf{v} + \underline{\mathbf{b}} \end{pmatrix} \parallel \begin{pmatrix} \underline{\mathbf{C}}\,\mathbf{v} + \underline{\mathbf{c}} \\ \underline{\underline{\mathbf{D}}}\mathbf{v} \cdot (\mathbf{v} + \underline{\mathbf{a}}) + \underline{\mathbf{D}}\,\mathbf{v} + \underline{\mathbf{d}} \end{pmatrix}, \tag{33}$$

with the vector fields, matrix fields and tensor field:

$$\underline{\mathbf{a}} = \mathbf{d}, \qquad \underline{\mathbf{b}} = \mathbf{k}, \tag{34}$$

$$\underline{\mathbf{c}} = \nabla\mathbf{k} \cdot \mathbf{d} - \frac{\mathbf{k}}{\kappa}, \qquad \underline{\mathbf{C}} = \nabla\mathbf{k} + \frac{1}{\kappa^2}\mathbf{I}, \tag{35}$$

$$\underline{\mathbf{d}} = \nabla(\nabla\mathbf{k})\mathbf{d} \cdot \mathbf{d} - \frac{\nabla\mathbf{k}\mathbf{d}}{\kappa} + \nabla\mathbf{k}\mathbf{k} + \frac{\mathbf{k}}{\kappa^2}, \qquad \underline{\mathbf{B}} = -\frac{1}{\kappa}\mathbf{I}, \tag{36}$$

$$\underline{\mathbf{D}} = \nabla(\nabla\mathbf{k})\mathbf{d} - \frac{2\nabla\mathbf{k}}{\kappa} - \frac{1}{\kappa^3}\mathbf{I}, \qquad \underline{\underline{\mathbf{D}}} = \nabla(\nabla\mathbf{k}). \tag{37}$$

In practice, we discretize the above fields onto a piecewise linear tetrahedral 3D grid. At three spatial grid points $\mathbf{x}_i$ with $i \in \{1, 2, 3\}$ of a triangle, we therefore have the quantities $\underline{\mathbf{a}}_i$, $\underline{\mathbf{b}}_i$, $\underline{\mathbf{c}}_i$, $\underline{\mathbf{d}}_i$, $\underline{\mathbf{B}}_i$, $\underline{\mathbf{C}}_i$, $\underline{\mathbf{D}}_i$ and $\underline{\underline{\mathbf{D}}}_i$, which can be linearly interpolated with the barycentric weights a, b, c, subject to $a + b + c = 1$:

$$\underline{\mathbf{a}} = a\ \underline{\mathbf{a}}_1 + b\ \underline{\mathbf{b}}_2 + c\ \underline{\mathbf{c}}_3, \tag{38}$$

$$\underline{\mathbf{b}} = a\ \underline{\mathbf{b}}_1 + b\ \underline{\mathbf{b}}_2 + c\ \underline{\mathbf{b}}_3, \tag{39}$$

$$\vdots \tag{40}$$

$$\underline{\underline{\mathbf{D}}} = a\ \underline{\underline{\mathbf{D}}}_1 + b\ \underline{\underline{\mathbf{D}}}_2 + c\ \underline{\underline{\mathbf{D}}}_3. \tag{41}$$

At the same time, the velocity subspace is discretized onto a tetrahedral grid, for which we likewise assume barycentric interpolation with barycentric weights d, e, f, g inside the tetrahedra, subject to $d + e + f + g = 1$:

$$\mathbf{v} = d\ \mathbf{v}_1 + e\ \mathbf{v}_2 + f\ \mathbf{v}_3 + g\ \mathbf{v}_4. \tag{42}$$

Thus, the 6D PV condition $\tilde{\mathbf{v}}^* \parallel \tilde{\mathbf{b}}^*$ in Eq. (33) can now be expressed with Eqs. (36)–(42) in barycentric coordinates $(a, \ldots, g)$. To search the entire 6D space, each pair of position triangle and velocity tetrahedra must be tested. The PV solutions are points on the position triangles, which are connected to 3D lines in a post-process.

Two 6D vector fields $\tilde{\mathbf{v}}$ and $\tilde{\mathbf{w}}$ are parallel iff in generalization of the cross product the anti-symmetric matrix $\tilde{\mathbf{X}}$ is zero, with

$$\tilde{\mathbf{v}} \parallel \tilde{\mathbf{w}} \Leftrightarrow \tilde{\mathbf{X}} = \mathbf{0}_{6\times 6} \quad \text{with} \quad \tilde{\mathbf{X}}_{i,j} = \tilde{\mathbf{u}}_i\ \tilde{\mathbf{w}}_j - \tilde{\mathbf{w}}_i\ \tilde{\mathbf{u}}_j, \tag{43}$$

and the element indices $i, j \in \{1, \ldots, 6\}$. Thus, the search for parallel vectors becomes a root finding problem in all entries of matrix $\tilde{\mathbf{X}}$, which is quadratic in barycentric coordinates $\tilde{\mathbf{X}}(a, b, c; d, e, f, g)$.

After converting matrix $\tilde{\mathbf{X}}$ into Bernstein-Bezier form [42, 43], we use the convex hull property to quickly decide whether a root may exist inside a pair of position triangle and velocity tetrahedron. The conversion into Bernstein-Bezier form was described by Günther and Theisel [7], and we refer to their implementation section for the details. If a solution could exist in a tested pair, they performed a recursive subdivision, similar to Oster et al. [51], until the solution was found numerically. Their computation time is in the order of multiple hours. To speed up convergence, we instead use multi-variate Newton iterations [44] to locate the barycentric coordinate at which the matrix $\tilde{\mathbf{X}}$ vanishes. After uniformly scaling the linear 6D vector fields $\tilde{\mathbf{v}}$ and $\tilde{\mathbf{w}}$ such that the longest vector has unit length, we iteratively minimize the Frobenius norm of $\tilde{\mathbf{X}}$. Since we do not recursively subdivide, significantly less pairs of triangles and tetrahedra need to be tested. For example, the computation time in the VORTEX RING at an initial spatial resolution of 32^3 voxels reduces from 70 min to 1.5 min. In practice, we use higher grid resolutions, as shown later in Table 1, as our algorithm assumes linear interpolation on the tetrahedral elements covering the voxels of the grid. Note however that Newton iterations will only find one solution and can miss others if there a multiple solutions on the face of a cell. It should therefore only be applied when the bilinear face of a voxel is small enough, either because the initial grid resolution is fine enough, or after a certain number of recursive subdivisions have been performed. Our implementation of the multi-variate Newton

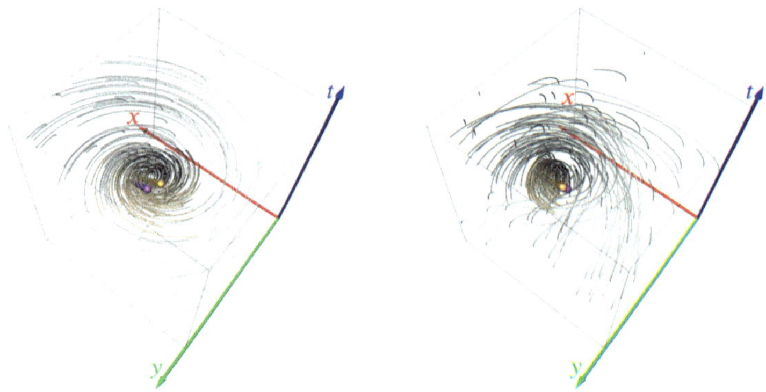

Fig. 2 Inertial (●) and massless (●) vortex coreline in the MOVING CENTER flow for Model 2 using $\mathbf{g} = (0, 2)^T$. On the left, $R = 0.5$ and $St = 0.1$, which models the behavior of aerosols such as sand particles in air. On the right, $R = 0.7$ and $St = 0.7$, which models the motion of bubbles. The corelines show the center of the vortex, while inertial pathlines (●) depict the overall movement of particles, showing that inertial particles close to the coreline revolve around the inertial vortex

iterations does not include further optimizations, such as line search or relaxation. Thus, there is room for further performance improvements.

5 Results

In the following, we test our extractors in several 2D and 3D flows, both analytic and numerical. For each experiment, we specify the inertial parameters as well as the applied extraction methods, that is, first-order or second-order coreline extraction.

5.1 Comparison of Inertial Particle Parameters

The first 2D flow contains a vortex center translating with constant speed along a straight line over time [27]:

$$\mathbf{u}(x, y, t) = \begin{pmatrix} -y + \frac{t}{2} - 1 \\ x + \frac{t}{2} - 1 \end{pmatrix}. \tag{44}$$

We consider the vector field in the space-time domain $D \times T = [-2, 2]^2 \times [0, 4]$. Figure 2 shows the effect of the parameters St and R on the vortex position for Model 2, showing the behavior of both aerosols and bubbles. Pathlines (●) close to the vortex center (●) revolve around it over time, which indicates that the coreline extraction is accurate. In contrast to massless flows, inertial pathlines seeded from the coreline

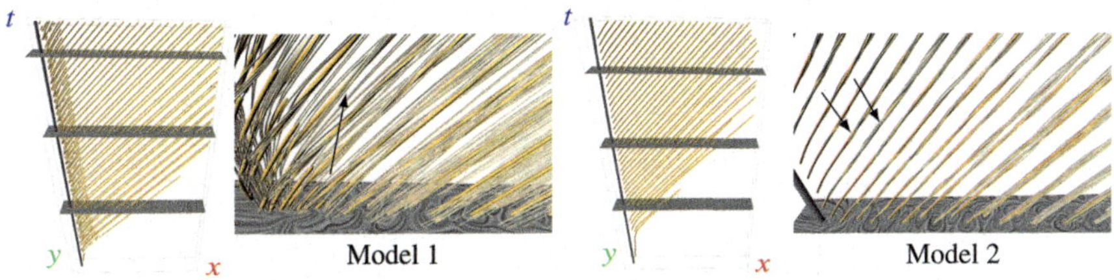

Fig. 3 Inertial vortex corelines in the CYLINDER flow for Model 1 (left) and 2 (right) using $\mathbf{g} = (0, 1)^T$, $r = 0.05$, $R = 2/3$ and $St = 0.01$. Inertial pathlines (●) show that particles rotate around inertial corelines (●), giving evidence of correctness

do not necessarily stay exactly on the coreline due to the inertia that carries them outward. Nevertheless, the general motion around the coreline can be observed.

5.2 Comparison of Inertial Particle Models

Cylinder Flow in 2D. The numerical 2D CYLINDER flow contains a von-Kármán vortex street in the wake of an obstacle. The viscous fluid was injected from the left into a channel bounded by solid walls with a slip boundary condition. Figure 3 shows two examples of Model 1 and 2 applied to this flow. In both cases, the symmetric flow creates translating vortices that remain coherent over time, which is visible in the space-time overview of the corelines (top). Depending on the particle model parameters, inertial particles can get trapped in the vortices while they are rotating around the corelines over time, which is visualized with inertial pathlines in the zoomed images. It becomes apparent that aerosol trajectories of Model 1 spiral away from the corelines, whereas the smaller neutrally buoyant particles of Model 2 remain close to their coreline. For this data set, the behavior of different inertial model parameters has previously been studied by Baeza Rojo et al. [47]. In our work, we automatically extract the vortex corelines, which compactly summarizes the loci of rotating motion.

Delta Wing in 3D. This numerical 3D flow was provided by Markus Rütten and contains a simulation of a triangular surface in upstream flow that generates two large wake vortices. We used a gravity-free environment, i.e., $\mathbf{g} = \mathbf{0}$, to extract a set of vortex corelines for both particle models. Figure 4 shows the difference between Model 1 and Model 2. Although the location of vortex corelines is similar here, Model 1 trajectories exhibit significantly more inertia when using a response time of $r = 0.01$, compared to trajectories in Model 2 when using $R = 0.1$ and $St = 0.001$ as parameters. The Model 2 particles are smaller and thus follow the flow more tangentially than the Model 1 particles. Similar to the massless case, inertial particles of Model 2 revolve very closely around the vortex corelines.

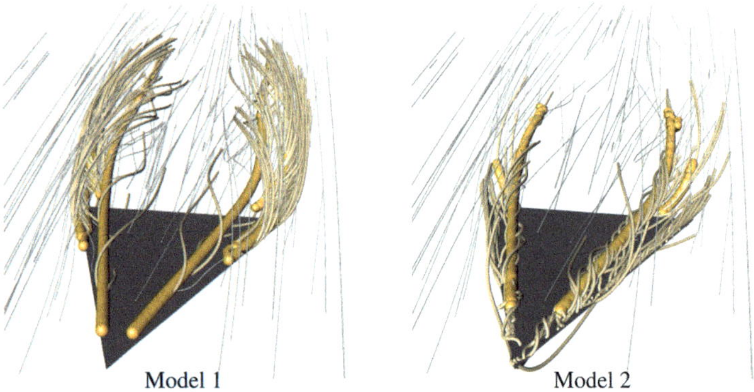

Fig. 4 Inertial vortex corelines for model 1 (left) and 2 (right) in the Delta Wing using $\mathbf{g} = (0, 0, 0)$, $r = 0.01$, $R = 0.1$ and $St = 0.001$

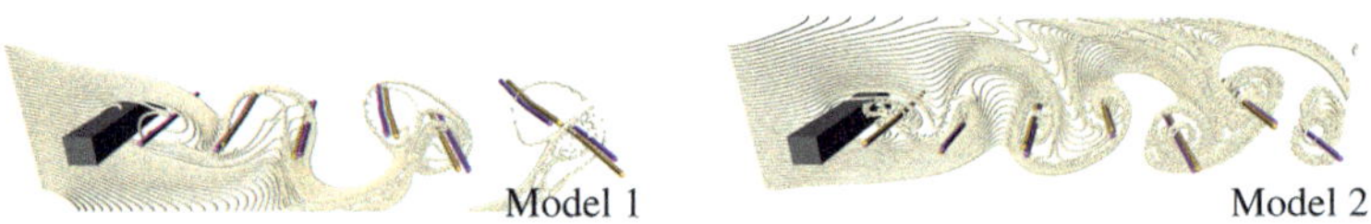

Fig. 5 Comparison of inertial (•) and massless (•) vortex corelines in the flow behind a Square Cylinder for Model 1 (top, $r = 0.25$) and Model 2 (bottom, $R = 0.8$ and $St = 0.4$). The heavy aerosol particles of Model 1 are dragged down by gravity $\mathbf{g} = (0, -1, 0)$ while the parameters of Model 2 represent bubbles, which rise upward

Square Cylinder in 3D. This 3D unsteady fluid flow sequence shows the development of a von-Kármán vortex street. The obstacle, which is a squared cylinder, is positioned between two parallel walls, producing a periodic shedding of vortices over time. Figure 5 shows selected frames of an accompanying animation, where we can see how inertial particles rotate around the extracted corelines. The heavy particles in Model 1 are affected by gravity, which drags particles down. The inertial vortex corelines (•) are shifted horizontally compared to the massless case (•) towards the updraft direction of the vortex, since this is where gravity cancels to zero. The shift therefore depends on the rotation direction of the vortex. Particles from Model 2, on the other hand, resemble bubbles (i.e., they have lower density than the liquid) and thus slowly rise up. It is also apparent that the behavior of particles in and around the vortices differs, since the inertia of particles of Model 1 carries particles further out. The low particle density inside vortices is characteristic for aerosol particles.

5.3 *Second-Order Corelines in 3D*

Vortex Ring. Our next vector field contains a translating Vortex Ring, which serves as synthetic test case for the second-order extractor. The velocity magnitude

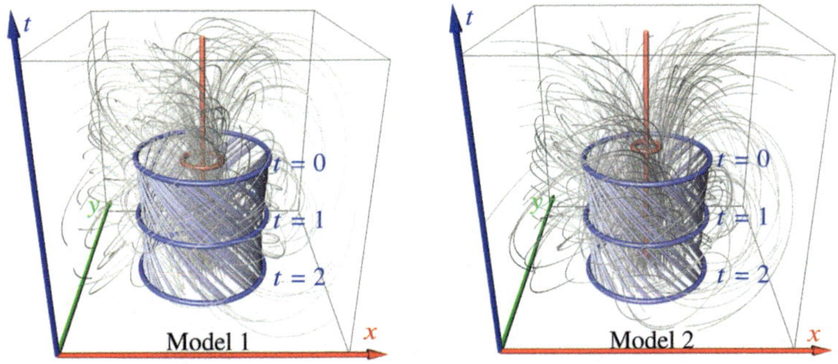

Fig. 6 Bent second-order inertial vortex corelines in the VORTEX RING data set for Model 1 (left, $r = 0.1$) and Model 2 (right, $R = 0.8$ and $St = 0.1$) using $\mathbf{g} = (0, 0, -1)$. The second-order corelines (•) are shown for three different time steps, and the first-order corelines (•) for $t = 0$. Note that inertial pathlines (•) remain near the vortex over time, while inertial pathlines (•) from $t = 0$ show the rotating motion

along the coreline is denoted by s and the vortex translates with speed a along the z-axis.

$$\mathbf{u}(x, y, z, t) = \begin{pmatrix} -x \cdot (z + a \cdot t) - s \cdot y \\ -y \cdot (z + a \cdot t) + s \cdot x \\ x^2 + y^2 - 1 - a \end{pmatrix}. \tag{45}$$

In our examples, we set $a = s = 1$. We consider the flow in the spatial domain $[-2, 2]^3$. In this flow, the vortex coreline is bent and thus the first-order methods fail to detect the correct coreline (•), as shown in Fig. 6. For the given model parameterization, the inertial particles of Model 1 exhibit more inertia than the particles in Model 2, leading them onto wider paths. The second-order vortex corelines (•) of both models are similar to each other. Note that the first-order criterion gives a wrong solution, since the coreline is not straight, which shows the necessity of our second-order vortex coreline criterion. Particles seeded on a vortex coreline are expected to remain close to the coreline over time, which is the case for our second-order condition.

Helicopter. Our last numerical example stems from brown-out engineering [1, 2], which studies the uplift of dust and sand when helicopters or airplanes approach the ground. The simulation shown in Fig. 7 contains a model rotor spinning 75 Hz. The strong vertical airflow pushes down onto the sediment bed, as it carries tip vortices from the rotor into the domain. Due to numerical dissipation, these vortices are not well preserved in the simulation data. Very well distinguishable, however, is the large vortex ring that forms around the helicopter. The first-order and second-order corelines agree for this ring, since the velocity component along the vortex coreline is zero. There, particles are not moving along the vortex coreline, but stay stationary on it. Studying this intrinsically curved vortex is important, since it is a strong driver of sediment uplift, which causes mechanical wear of the blades and, more importantly, a

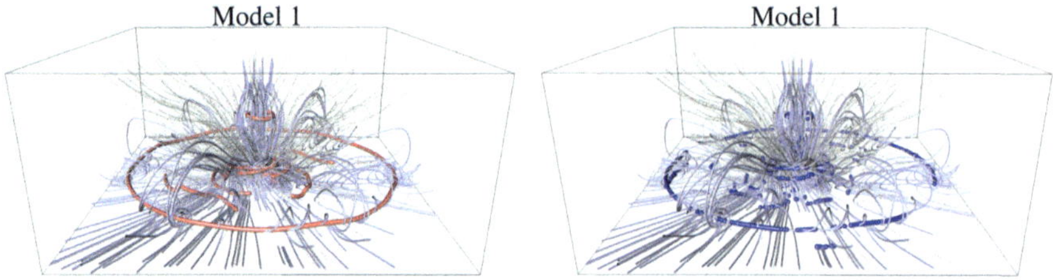

Fig. 7 First-order (•) (left) and second-order (•) (right) inertial vortex corelines in the HELICOPTER data set for Model 1 using $\mathbf{g} = \mathbf{0}$ and $r = 0.001$. Near the ground, a large vortex ring forms around the hovering helicopter. The trajectories show the paths of sand particles. Differences between the methods can be seen below the rotor

limited view that frequently causes accidents. Differences between the two methods can be seen in the flow regime below the rotor blades, where the vortex coreline spirals outwards. These vortices eventually reach the sediment bed, causing uplift. Further, we can see that the second-order lines are more spurious than the first-order lines, which is a consequence of the numerical challenges of higher-order methods.

5.4 Memory Consumption and Performance

For all performance measurements, we used an Intel i7-6700K CPU with 32 GB RAM. All presented feature definitions are local and their extraction can thus easily be parallelized. In unsteady 2D flows, we only have to compute a 2D vector field in which the critical points are searched. This sums up to $2N$ double variables, where N is the number of grid points. Note that the time slices can be processed sequentially, which is why we only need to store a single time slice at a time. In 3D, we store for the first-order method two 3D vector fields, i.e., $6N$ double variables. For the 3D second-order methods, we store for each grid point 4 vector fields (a, b, c, d), 3 matrix fields (B, C, D) and 1 tensor field. This accumulates to $66N$ double variables. In general, the extraction performance is linear in the number of voxels that are tested for parallel vectors. The number of tested voxels depends on the amount and extent of swirling motion in the domain. The computation time of the quantities resulting in the vortex criterion and of the actual numerical extraction of the corelines are listed in Table 1. While the 2D and first-order 3D extraction can be carried out in the order of seconds (similar to the traditional massless case), the second-order 6D extraction is computationally still expensive, taking multiple hours on numerical data. Further speed-up could be achieved by pruning the search space, e.g., by considering only visible voxels in the domain or by specifying a region-of-interest, which have both not been implemented yet.

Table 1 Computation time in seconds for the vector fields (VF) from Eq. (20) for first order space-time 2D, Eq. (25) for 3D and Eq. (32) for second order and corelines extraction with parallel vectors (PV) for the two models (M).

		Grid	Memory	Order	M	VF	PV
MOVING CENTER	2D	$128 \times 128 \times 128$	256 KB	1st	1	0.11	0.30
					2	9.12	13.27
CYLINDER	2D	$640 \times 80 \times 1501$	800 KB	1st	1	13.74	2.61
					2	80.67	2.73
SQUARED CYLINDER	3D	$192 \times 64 \times 48$	27 MB	1st	1	0.55	6.74
					2	3.82	14.11
DELTA WING	3D	$250 \times 125 \times 100$	35.2 MB	1st	1	2.47	0.07
					2	16.36	0.77
VORTEX RING	3D	$128 \times 128 \times 128$	96 MB	1st	1	0.59	0.30
					2	0.74	0.26
			1.03 GB	2nd	1	0.48	0.85 h
					2	0.82	1.1 h
HELICOPTER	3D	$128 \times 256 \times 256$	384 MB	1st	1	8.25	1.37
			4.12 GB	2nd	1	11.18	14 h

5.5 *Discussion*

Parameters. A benefit of our Galilean-invariant method, compared to the objective approach of Günther and Theisel [7], is that our reference frame optimization is parameter-free. While the objective method required the specification of a neighborhood region to regularize a linear system, the Galilean invariant approach has a local analytic solution.

Temporal Coherence. Our coreline definitions are local and do not involve any spatial or temporal smoothing kernels. Consequentially, temporal coherence is not directly enforced, which can result in temporal flickering. This is a general limitation of local methods [6, 27]. Obtaining smoother results is typically achieved through a pre-processing or post-processing step, which are orthogonal problems to the feature definition.

Computation Time. By using an iterative Newton refinement, the computation time of the 6D parallel vectors extraction reduced by about factor 40 compared to the Bézier-based subdivision in previous work [7], but it can still see improvements. An implicit or view-dependent computation of the line structures or a smarter selection of numerical parameters might be paths to faster updates.

Reference Frame Invariance. The shape of a coreline and its motion are two independent phenomena. In fact, in the VORTEX RING data set in Fig. 6, the objective approach [7] and the first-order Galilean-invariant method [6] will both give an identical (and wrong) result, since the vortex ring is not straight and performs an equal-speed translation. In this paper, we introduced a criterion to extract more general inertial coreline shapes. For this, we concentrated on vortex corelines that perform Galilean transformations. To handle more arbitrary motions in the future, we would like to apply the second-order criterion in reference frames that are objective.

Limit Case for Tracer Particles. A natural question that arises for feature definitions of inertial particles is whether they are consistent with the massless case. In the Appendix 2, we show that our inertial first-order and second-order criteria approach the massless case in the limit for $\kappa \rightarrow 0$, as desired.

6 Conclusion

In this paper, we developed vortex coreline extractors for inertial particles for two particle models. Based on the selection of an optimal Galilean reference frame, in which the flow field becomes steady, we introduced generalized coreline criteria that include both particle models as special cases. After covering the 2D case, we discussed the first-order and second-order inertial coreline extraction in 3D, which required a 3D or 6D parallel vectors extraction. For the latter, we combined a Bézier-based subdivision approach with a subsequent iterative Newton refinement. Our work connects to reference frame optimization, shedding light on the formal mathematical properties of previous work [6] that derived a criterion geometrically.

Acknowledgements The CYLINDER flow was numerically simulated with Gerris Flow solver [48], The DELTA WING vector field was kindly provided by Markus Rütten and the SQUARE CYLINDER flow was simulated by Camarri et al. [49] and resampled by Tino Weinkauf. The HELICOPTER flow was simulated by Benjamin Kutz [3]. For all visualizations, we used Amira [50]. This work was supported by the Swiss National Science Foundation (SNSF) Ambizione grant no. PZ00P2_180114 and by ETH Research Grant ETH-07 18-1.

Appendix 1 - Derivation of First-Order 3D Criterion

Next, we show how the 6D parallel vectors condition of the first-order case:

$$\tilde{\mathbf{v}}^* \parallel \tilde{\mathbf{J}}^*\tilde{\mathbf{v}}^* \quad \Rightarrow \quad \begin{pmatrix}\mathbf{v}+\mathbf{d}\\ \mathbf{k}-\frac{\mathbf{v}}{\kappa}\end{pmatrix} \parallel \begin{pmatrix}\mathbf{k}-\frac{\mathbf{v}}{\kappa}\\ \nabla\mathbf{k}(\mathbf{v}+\mathbf{d}) - \frac{\mathbf{k}-\frac{\mathbf{v}}{\kappa}}{\kappa}\end{pmatrix}. \tag{46}$$

can be simplified to a 3D criterion. First, we look at the position subspace. Multiplying with κ and adding $\mathbf{v}+\mathbf{d}$ to the right hand side gives:

$$\mathbf{v}+\mathbf{d} \parallel \mathbf{k}-\frac{\mathbf{v}}{\kappa} \quad \Leftrightarrow \quad \mathbf{v}+\mathbf{d} \parallel \kappa\mathbf{k}-\mathbf{v} \tag{47}$$

$$\Leftrightarrow \quad \mathbf{v}+\mathbf{d} \parallel \kappa\mathbf{k}+\mathbf{d}, \tag{48}$$

Equating Eq. (47) and Eq. (48) gives:

$$\mathbf{k}-\frac{\mathbf{v}}{\kappa} \parallel \kappa\mathbf{k}+\mathbf{d}. \tag{49}$$

Considering the velocity subspace of Eq. (46):

$$\mathbf{k}-\frac{\mathbf{v}}{\kappa} \parallel \nabla\mathbf{k}\,(\mathbf{v}+\mathbf{d})-\frac{\mathbf{k}-\frac{\mathbf{v}}{\kappa}}{\kappa}, \tag{50}$$

and multiplying the right hand side with κ gives:

$$\mathbf{k}-\frac{\mathbf{v}}{\kappa} \parallel \kappa\nabla\mathbf{k}\,(\mathbf{v}+\mathbf{d})-\mathbf{k}+\frac{\mathbf{v}}{\kappa}\,. \tag{51}$$

Adding the left hand side to the right hand side and dividing by κ:

$$\mathbf{k}-\frac{\mathbf{v}}{\kappa} \parallel \nabla\mathbf{k}\,(\mathbf{v}+\mathbf{d}). \tag{52}$$

Finally, substituting Eq. (49) on the left hand side of Eq. (52) and inserting Eq. (48) on the right hand side of Eq. (52) gives Eq. (25):

$$\kappa\mathbf{k}+\mathbf{d} \parallel \nabla\mathbf{k}\,(\kappa\mathbf{k}+\mathbf{d}), \tag{53}$$

which is a 3D condition, independent of the particle velocity $\mathbf{v}$.

Appendix 2 - Tracer Particles as Limit Case

Next, we show that our inertial first-order and second-order criteria approach the massless case in the limit. The proofs of Model 1 and 2 are analogue. For brevity, we show the derivation for Model 1.

Inertial Motion. First, the motion of inertial particles is consistent with tracer particles for $r \to 0$, as shown by Günther and Theisel [6]: Rearranging the velocity subspace of $\tilde{\mathbf{v}}$ in Eq. (7) for $\mathbf{v}$ and substituting in the position subspace of $\tilde{\mathbf{v}}$ gives with Eq. (9):

$$\lim_{r\to 0}\frac{\mathrm{d}\mathbf{x}}{\mathrm{d}t}=\mathbf{v}=\mathbf{u}(\mathbf{x},t)\underbrace{-r\frac{\mathrm{d}\mathbf{v}}{\mathrm{d}t}+r\mathbf{g}}_{0}. \tag{54}$$

Vortex Centers in 2D. The motion of inertial particles is described with Eq. (9). We consider the limit $r \to 0$ for tracer particles:

$$\kappa = r \qquad \kappa\mathbf{k} = \mathbf{u}(\mathbf{x}, t) + r\mathbf{g} \tag{55}$$

$$\lim_{r\to 0} \kappa\mathbf{k} = \mathbf{u}(\mathbf{x}, t). \tag{56}$$

With $\mathbf{d} = \mathbf{J}^{-1}\mathbf{u}_t = -\mathbf{f}$ from Eq. (15), we insert the limit in Eq. (56) into the general 2D vortex center condition in Eq. (17):

$$\kappa\mathbf{k} + \mathbf{d} = \mathbf{0} \qquad \overset{r\to 0}{\Rightarrow} \qquad \mathbf{u}(\mathbf{x}, t) - \mathbf{f} = \mathbf{0}, \tag{57}$$

which is the Galilean invariant 2D vortex coreline criterion for massless particles by Weinkauf et al. [27], cf. Eq. (55) in [23].

First-Order Corelines in 3D. For the first-order vortex corelines in 3D, we insert the limit in Eq. (56) into the general first-order vortex coreline condition in Eq. (25):

$$\kappa\mathbf{k} + \mathbf{d} \parallel \nabla\mathbf{k}\,(\kappa\mathbf{k} + \mathbf{d}) \qquad \overset{r\to 0}{\Rightarrow} \qquad \mathbf{u} - \mathbf{f} \parallel \frac{\mathbf{J}}{r}\,(\mathbf{u} - \mathbf{f}) \tag{58}$$

$$\mathbf{u} - \mathbf{f} \parallel \mathbf{J}\,(\mathbf{u} - \mathbf{f}), \tag{59}$$

which is the Galilean invariant first-order 3D vortex coreline criterion for massless particles by Weinkauf et al. [27], cf. Eq. (53) in [23].

Second-Order Corelines in 3D. We consider the transformed velocity $\tilde{\mathbf{v}}^*$ in Eq. (13) and the rate of acceleration $\tilde{\mathbf{b}}^*$:

$$\tilde{\mathbf{b}}^* = \begin{pmatrix} \nabla\mathbf{k}(\mathbf{v} + \mathbf{d}) + \frac{\mathbf{v}}{\kappa^2} - \frac{\mathbf{k}}{\kappa} \\ \nabla(\nabla\mathbf{k})(\mathbf{v} + \mathbf{d})(\mathbf{v} + \mathbf{d}) - \frac{\nabla\mathbf{k}}{\kappa}(\mathbf{v} + \mathbf{d}) + \nabla\mathbf{k}(\mathbf{k} - \frac{\mathbf{v}}{\kappa}) - \frac{\mathbf{v}}{\kappa^3} + \frac{\mathbf{k}}{\kappa^2} \end{pmatrix} \tag{60}$$

in the optimal reference frame, i.e., $\mathbf{d}$ is the translation rate:

$$\tilde{\mathbf{v}}^* = \begin{pmatrix} \mathbf{v} + \mathbf{d} \\ \mathbf{a}^* \end{pmatrix}, \qquad \tilde{\mathbf{b}}^* = \begin{pmatrix} \mathbf{b}^* \\ \mathbf{c}^* \end{pmatrix}. \tag{61}$$

We rearrange the velocity subspace in Eq. (60), denoted as $\mathbf{c}^*$ with

$$\mathbf{c}^* = \nabla(\nabla\mathbf{k})(\mathbf{v} + \mathbf{d})(\mathbf{v} + \mathbf{d}) - \frac{\nabla\mathbf{k}}{\kappa}(\mathbf{v} + \mathbf{d}) + \nabla\mathbf{k}(\mathbf{k} - \frac{\mathbf{v}}{\kappa}) - \frac{\mathbf{v}}{\kappa^3} + \frac{\mathbf{k}}{\kappa^2}$$

and with $\mathbf{a}^* = \mathbf{k} - \frac{\mathbf{v}}{\kappa} = \mathbf{J}(\mathbf{u} - \mathbf{f})$ [23] and $\mathbf{b}^*$ in Eq. (60) into

$$\mathbf{c}^* = \nabla(\nabla\mathbf{k})(\mathbf{v} + \mathbf{d})(\mathbf{v} + \mathbf{d}) + \nabla\mathbf{k}\,(\mathbf{J}(\mathbf{u} - \mathbf{f})) - \frac{\mathbf{b}^*}{\kappa}. \tag{62}$$

Inserting Model 1 with Eqs. (56) and (57) gives for $r \to 0$:

$$\mathbf{c}^* = \frac{1}{r}\nabla\mathbf{J}(\mathbf{u}-\mathbf{f})(\mathbf{u}-\mathbf{f}) + \frac{\mathbf{J}}{r}\left(\mathbf{J}(\mathbf{u}-\mathbf{f})\right) - \frac{\mathbf{b}^*}{r} \tag{63}$$

$$\Leftrightarrow\ r \cdot \mathbf{c}^* = \nabla\mathbf{J}(\mathbf{u}-\mathbf{f})(\mathbf{u}-\mathbf{f}) + \mathbf{J}\left(\mathbf{J}(\mathbf{u}-\mathbf{f})\right) - \mathbf{b}^* = \mathbf{0}. \tag{64}$$

Rearranging for the rate of acceleration $\mathbf{b}^*$ gives the material derivative of the steady acceleration $\frac{D}{Dt}(\mathbf{J}\mathbf{u}^*)$ in the optimal frame:

$$\lim_{r\to 0} \mathbf{b}^* = \left(\nabla\mathbf{J}\mathbf{u}^* + \mathbf{J}\mathbf{J}\right)\mathbf{u}^* \quad \text{with} \quad \mathbf{u}^* = \mathbf{u} - \mathbf{f}. \tag{65}$$

Thus, the position subspace simplifies to $\mathbf{u}^* \parallel \mathbf{b}^*$, which is equivalent to the criterion of Roth and Peikert [9] in the optimal steady reference frame. Thus, all proposed inertial vortex criteria are consistent with the massless case for $r \to 0$.

References

1. Syal, M., Govindarajan, B., Leishman, J.G.: Mesoscale sediment tracking methodology to analyze brownout cloud developments. In: 66th Annual Forum of Proceedings of the American Helicopter Society (2010)
2. Sydney, A., Baharani, A., Leishman, J.G.: Understanding brownout using near-wall dual-phase flow measurements. In: 67th Annual Forum of Proceedings of the American Helicopter Society, Virginia Beach, VA, May 2011
3. Kutz, B.M., Gunther, T., Rumpf, A., Kuhn, A.: Numerical examination of a model rotor in brownout conditions. In: Proceedings of the American Helicopter Society, no. AHS2014-000343 (2014)
4. Karl, D.M.: A sea of change: biogeochemical variability in the North Pacific Subtropical Gyre. Ecosystems **2**(3), 181–214 (1999)
5. Bordas, R.: Optical measurements in disperse two-phase flows: application to rain formation in cumulus clouds. Ph.D. thesis, University of Magdeburg (2011)
6. Günther, T., Theisel, H.: Vortex cores of inertial particles. IEEE Trans. Vis. Comput. Graph. **20**(12), 2535–2544 (2014). (Proceedings of the IEEE SciVis)
7. Günther, T., Theisel, H.: Objective vortex corelines of finite-sized objects in fluid flows. IEEE Trans. Vis. Comput. Graph. **25**(1), 956–966 (2019). (Proceedings of the IEEE Scientific Visualization 2018)
8. Sujudi, D., Haimes, R.: Identification of swirling flow in 3D vector fields. Technical report, Department of Aeronautics and Astronautics, MIT (1995). AIAA paper: 95-1715
9. Roth, M., Peikert, R.: A higher-order method for finding vortex core lines. In: Proceedings of the IEEE Visualization, pp. 143–150 (1998)
10. Peikert, R., Roth, M.: The "parallel vectors" operator–a vector field visualization primitive. In: Proceedings of the IEEE Visualization, pp. 263–270 (1999)
11. Lugt, H.J.: The dilemma of defining a vortex. In: Recent Developments in Theoretical and Experimental Fluid Mechanics, pp. 309–321. Springer (1979)
12. Robinson, S.K.: Coherent motions in the turbulent boundary layer. Ann. Rev. Fluid Mech. **23**(1), 601–639 (1991)
13. Crowe, C., Sommerfield, M., Tsuji, Y.: Multiphase Flows with Droplets and Particles. CRC Press, Boca Raton (1998)
14. Haller, G., Sapsis, T.: Where do inertial particles go in fluid flows? Physica D **237**, 573–583 (2008)
15. Sudharsan, M., Brunton, S.L., Riley, J.J.: Lagrangian coherent structures and inertial particle dynamics. ArXiv e-prints 1512.05733 (2015)

16. Wan, Z.Y., Sapsis, T.P.: Machine learning the kinematics of spherical particles in fluid flows. J. Fluid Mech. **857**, R2 (2018). https://doi.org/10.1017/jfm.2018.797
17. Benzi, R., Biferale, L., Calzavarini, E., Lohse, D., Toschi, F.: Velocity-gradient statistics along particle trajectories in turbulent flows: the refined similarity hypothesis in the Lagrangian frame. Phys. Rev. E **80**, 066318 (2009)
18. Cartwright, J.H.E., Feudel, U., Karolyi, G., Moura, A., Piro, O., Tel, T.: Dynamics of finite-size particles in chaotic fluid flows. In: Thiel, M., Kurths, J., Romano, M.C., Károlyi, G., Moura, A. (eds.) Nonlinear Dynamics and Chaos: Advance and Perspectives. Understanding Complex Systems, pp. 51–87. Springer, Heidelberg (2010)
19. Bec, J., Biferale, L., Cencini, M., Lanotte, A.S., Toschi, F.: Spatial and velocity statistics of inertial particles in turbulent flows. J. Phys: Conf. Ser. **333**(1), 012003 (2011)
20. Benczik, I.J., Toroczkai, Z., Tel, T.: Selective sensitivity of open chaotic flows on inertial tracer advection: catching particles with a stick. Phys. Rev. Lett. **89**, 164501 (2002)
21. Babiano, A., Cartwright, J.H.E., Piro, O., Provenzale, A.: Dynamics of a small neutrally buoyant sphere in a fluid and targeting in Hamiltonian systems. Phys. Rev. Lett. **84**, 5764–5767 (2000)
22. Vilela, R.D., de Moura, A.P.S., Grebogi, C.: Finite-size effects on open chaotic advection. Phys. Rev. E **73**, 026302 (2006)
23. Günther, T., Theisel, H.: The state of the art in vortex extraction. Comput. Graph. Forum **37**(6), 149–173 (2018)
24. Globus, A., Levit, C., Lasinski, T.: A tool for visualizing the topology of three dimensional vector fields. In: Proceedings of the IEEE Visualization, pp. 33–40 (1991)
25. Sahner, J., Weinkauf, T., Hege, H.-C.: Galilean invariant extraction and iconic representation of vortex core lines. In: Proceedings of the Eurographics/IEEE VGTC Symposium on Visualization (EuroVis), pp. 151–160 (2005)
26. Fuchs, R., Peikert, R., Hauser, H., Sadlo, F., Muigg, P.: Parallel vectors criteria for unsteady flow vortices. IEEE Trans. Vis. Comput. Graph. **14**(3), 615–626 (2008)
27. Weinkauf, T., Sahner, J., Theisel, H., Hege, H.-C.: Cores of swirling particle motion in unsteady flows. IEEE Trans. Vis. Comput. Graph. **13**(6), 1759–1766 (2007). (Proceedings of the Visualization)
28. Theisel, H., Seidel, H.-P.: Feature flow fields. In: Proceedings of the Symposium on Data Visualisation, pp. 141–148 (2003)
29. Günther, T., Schulze, M., Theisel, H.: Rotation invariant vortices for flow visualization. IEEE Trans. Vis. Comput. Graph. **22**(1), 817–826 (2016). (Proceedings of the IEEE SciVis 2015)
30. Günther, T., Gross, M., Theisel, H.: Generic objective vortices for flow visualization. ACM Trans. Graph. **36**(4), 141:1–141:11 (2017). (Proceedings of the SIGGRAPH)
31. Hadwiger, M., Mlejnek, M., Theuÿl, T., Rautek, P.: Time-dependent flow seen through approximate observer killing fields. IEEE Trans. Vis. Comput. Graph. **25**(1), 1257–1266 (2019). (Proceedings of IEEE Scientific Visualization 2018)
32. Chong, M.S., Perry, A.E., Cantwell, B.J.: A general classification of three-dimensional flow fields. Phys. Fluids A **2**(5), 765–777 (1990)
33. Rojo, I.B., Günther, T.: Vector field topology of time-dependent flows in a steady reference frame. IEEE Trans. Vis. Comput. Graph. **26**(1), 280–290 (2019). (Proceedings of the IEEE Scientific Visualization 2019)
34. Günther, T., Theisel, H.: Hyper-objective vortices. IEEE Trans. Vis. Comput. Graph. **26**(3), 1532–1547 (2020)
35. Wiebel, A.: Feature detection in vector fields using the Helmholtz-Hodge decomposition. Diploma thesis, University of Kaiserslautern (2004)
36. Wiebel, A., Garth, C., Scheuermann, G.: Computation of localized flow for steady and unsteady vector fields and its applications. IEEE Trans. Vis. Comput. Graph. **13**(4), 641 (2007)
37. Bhatia, H., Pascucci, V., Kirby, R.M., Bremer, P.-T.: Extracting features from time-dependent vector fields using internal reference frames. Comput. Graph. Forum **33**(3), 21–30 (2014). (Proceedings of the EuroVis)
38. Bujack, R., Hlawitschka, M., Joy, K.I.: Topology-inspired Galilean invariant vector field analysis. In: IEEE Pacific Visualization Symposium, pp. 72–79, April 2016

39. Kim, B., Günther, T.: Robust reference frame extraction from unsteady 2D vector fields with convolutional neural networks. Comput. Graph. Forum **38**(3), 285–295 (2019). (Proceedings of the EuroVis)
40. Günther, T., Gross, M.: Flow-induced inertial steady vector field topology. Comput. Graph. Forum **36**(2), 143–152 (2017). (Proceedings of the Eurographics)
41. Helman, J.L., Hesselink, L.: Representation and display of vector field topology in fluid flow data sets. Computer **22**(8), 27–36 (1989)
42. Rockwood, A., Heaton, K., Davis, T.: Real-time rendering of trimmed surfaces. In: ACM SIGGRAPH Computer Graphics, vol. 23, pp. 107–116. ACM (1989)
43. Hoschek, J., Lasser, D.: Fundamentals of computer aided geometric design. AK Peters (1993)
44. Kelley, C.T.: Iterative methods for linear and nonlinear equations. Front. Appl. Math. **16**, 575–601 (1995)
45. Van Gelder, A., Pang, A.: Using PVsolve to analyze and locate positions of parallel vectors. IEEE Trans. Vis. Comput. Graph. **15**(4), 682–695 (2009)
46. Hofmann, L., Sadlo, F.: The dependent vectors operator. Comput. Graph. Forum **38**(3), 261–272 (2019). https://doi.org/10.1111/cgf.13687. (Proceedings of the EuroVis)
47. Rojo, B.I., Gross, M., Gunther, T.: Visualizing the phase space of heterogeneous inertial particles in 2D flows. Comput. Graph. Forum **37**(3), 289–300 (2018). (Proceedings of the EuroVis)
48. Popinet, S.: Free computational fluid dynamics. Cluster World **2**, 6 (2004)
49. Camarri, S., Salvetti, M.-V., Buffoni, M., Iollo, A.: Simulation of the three-dimensional flow around a square cylinder between parallel walls at moderate Reynolds numbers. In: XVII Congresso di Meccanica Teorica ed Applicata (2005)
50. Stalling, D., Westerhoff, M., Hege, H.-C.: Amira: a highly interactive system for visual data analysis. In: The Visualization Handbook, pp. 749–767. Elsevier (2005)
51. Oster, T., Rossl, C., Theisel, H.: Core lines in 3D second-order tensor fields. Comput. Graph. Forum **37**(3), 327–337 (2018). (Proceedings of the EuroVis)

Implicit Visualization of 2D Vector Field Topology for Periodic Orbit Detection

Alexander Straub, Grzegorz K. Karch, Filip Sadlo, and Thomas Ertl

Abstract We present implicit visualization of 2D vector field topology, and show its utility for validating and guiding approaches for periodic orbit extraction. Instead of following the traditional approach by explicit extraction of the topological skeleton, we investigate its implicit visualization by approaches that label the regions that are separated by the skeleton. While such approaches perform well for gradient fields, they fail, in particular, to visualize periodic orbits. This motivates us to complement the label-based approach with a closely related distance-based metric. We show that our approach is able to reveal periodic orbits, also in configurations in which the state-of-the-art techniques for periodic orbit extraction fail, and demonstrate their utility for interactive extraction of all periodic orbits of a 2D vector field.

1 Introduction

There is a wide range of problems in science and engineering that involve vector fields—and topological analysis is an effective means for their analysis. It is in particular the reduction of the complex vector field to the so-called topological skeleton, which has proven very useful for obtaining qualitative insight into the overall structure of such fields. The topological skeleton consists of the critical points (isolated

A. Straub (✉) · G. K. Karch · T. Ertl
University of Stuttgart, Stuttgart, Germany
e-mail: alexander.straub@visus.uni-stuttgart.de

G. K. Karch
e-mail: gkarch@nvidia.com

T. Ertl
e-mail: thomas.ertl@vis.uni-stuttgart.de

F. Sadlo
Heidelberg University, Heidelberg, Germany
e-mail: sadlo@uni-heidelberg.de

I. Hotz et al. (eds.), *Topological Methods in Data Analysis and Visualization VI*, Mathematics and Visualization, https://doi.org/10.1007/978-3-030-83500-2_9

zeros of the vector field), the separatrices (manifolds of streamlines that converge to saddle-type critical points in forward or reverse time), as well as periodic orbits (isolated closed streamlines) and their manifolds (streamlines converging to saddle-type periodic orbits). Beyond that, there are also more exotic structures, such as invariant tori, Cantori, KAM tori [15], and strange attractors. However, whereas the extraction of critical points in discretized vector fields is trivial, and the only substantial difficulty with the extraction of separatrices is to choose sufficiently long integration time, e.g., in case of slowly converging (spiraling) streamlines, the detection and extraction of periodic orbits is more involved, and the extraction and visualization of the different types of tori and strange attractors is still in its infancy.

In this paper, we address part of these issues in the traditional visualization of these structures, in particular the periodic orbits. Instead of aiming at their explicit extraction and depiction, we investigate approaches for implicit visualization of the topological skeleton by means of derived scalar fields. Although our approach is generic, we exemplify it here only for 2D vector fields, and would like to address its application to 3D flows (which can exhibit the aforementioned tori and strange attractors) as future work.

Our contributions include:

- a set of scalar fields that enable implicit visualization of vector field topology,
- refinement and progressive computation of these fields, and
- combination of our approach with existing periodic orbit extraction techniques.

2 Related Work

The concept of vector field topology was introduced in flow visualization by Helman and Hesselink for 2D [8] and later for 3D vector fields [9]. Besides critical points, Helman and Hesselink account for solid boundaries by including attachment and separation points in 2D fields, and attachment and separation lines in 3D fields. Nevertheless, they did not account for periodic orbits. Asimov [1] provides a good introduction to the topic and gives concise definitions for many topological concepts, including periodic orbits and the Poincaré maps used for their analysis. Wischgoll and Scheuermann [22, 23] present approaches for determining if a streamline converges to a periodic orbit, and, based on this, they propose a simple seeding scheme for extracting periodic orbits. As the authors state and one can show with our approach, this simple seeding strategy can miss periodic orbits, i.e., exhibit false negatives. Alternative methods for the extraction of periodic orbits were introduced by Chen et al. [3] using Morse decomposition, and by Theisel et al. [19] intersecting stream surfaces in a derived 3D domain. While both methods are generally able to detect all periodic orbits and do not rely on seeding strategies, they tend to suffer, compared to our technique, from numerical problems and performance issues.

Later contributions to topological analysis of vector fields include saddle connectors, i.e., the intersection ribbons of forward and reverse 2-manifold separatrices in

3D fields by Theisel et al. [18], and boundary switch curves [21], i.e., points/curves on open domain boundaries separating inflow from outflow, which also give rise to separating manifolds. While our approach can reveal dynamics related to open domain boundaries, we focus on fields without open boundaries.

Almost all previous works on the visualization of vector field topology carry out explicit extraction of the topological skeleton. They typically start with extracting the critical constructs, such as critical points and periodic orbits, followed by extraction of the separating manifolds by seeding streamlines at those critical constructs, which exhibit saddle-type flow behavior. Exceptions from this overall approach include the work by Park et al. [11], which seeds forward and reverse streamlines and counts the number of intersections per cell on a predefined grid, resulting in density fields. These fields reveal the attracting and repelling structures, but cannot capture remote parts of separatrices, nor the regions of similar flow behavior. A similar approach was presented by Zhang et al. [24, 25], who construct a so-called Φ-field from the angle differences along integral curves. The derived gradient field $\|\nabla\Phi\|$ can then be used to visualize repelling structures, similar to our distance gradient field. While their approach works especially well for regions where streamlines diverge, e.g., near saddle points, it fails at detecting nested periodic orbits. This is because streamlines on either side of a periodic orbit diverge only slightly, while rotating in the same direction. Another exception, which captures regions of similar streamline convergence in an implicit manner, is the approach by Shi et al. [13, 14], which analyzes the dynamics in periodic 2D time-dependent vector fields. The periodicity of such flows, and the fact that pathlines represent streamlines in space-time representation of a time-dependent vector field, enables the authors to reveal the basins of attraction in the resulting Poincaré maps by dense sampling of the spatial domain, integrating forward and reverse space-time streamlines for each of the samples, and identifying the critical point in the Poincaré map that a streamline converges to. This approach is very similar to the basic idea of our approach, but, e.g., cannot reveal periodic orbits. We would like to refer to our discussion and comparison to other methods in Sect. 4 for further details.

Our approach also shares similarities with the implicit extraction of stream surfaces by van Wijk [20], and its extension to other integral surface types by Stöter et al. [16]. Van Wijk also generates a derived scalar field by determining where densely seeded streamlines go, but whereas he determines their intersection with planes, e.g., at inflow and outflow regions, we determine their convergence to critical points. Finally, less closely related works that investigate convergence of streamlines, are the explicit ones by Friederici et al. [4, 5], which augment separatrices with respect to their separating property, including convergence. Nevertheless, these works do not address visualization of regions of similar streamline behavior, as we do.

3 Motivation

Traditionally, vector field topology is visualized by means of explicit extraction and depiction of the topological skeleton. The extraction of critical points and integration of the separatrices therefrom is accurate, efficient, and comparably simple to implement. However, it is (I) often nontrivial to identify the regions of similar streamline behavior from the skeleton alone, in particular when the flow exhibits rotational motion, including densely spiraling and intertwined separatrices (Fig. 1b). A further difficulty in this context is (II) that it is hard to judge if the integration duration for computing the separatrices has been chosen long enough, i.e., if the streamlines converged sufficiently to the connecting critical points or periodic orbits, and thus the respective regions are sufficiently separated by the separatrices. Notice, that it would take infinite integration time to make the separatrices connect and fully separate the regions, and that the streamlines can get arbitrarily close to (and spend arbitrarily long integration time at), e.g., saddle points before they reach the alpha/omega set, e.g., a sink. This circumstance also impedes the extraction of the regions by detection and meshing of closed loops in the topological skeleton. Last but not least, extracting all periodic orbits is computationally and algorithmically demanding (III). This becomes apparent from the limitations of the state-of-the-art methods by Wischgoll and Scheuermann [22, 23], Theisel et al. [19], and Chen et al. [3], which we discuss in Sect. 6.4. From these methods, Wischgoll and Scheuermann provide a criterion to determine if a given streamline converges to a periodic orbit. Although the seeding strategy used in their approach can suffer from false negatives, their criterion lends itself for interactive exploration of periodic orbits. Since our approach, on the other hand, suffers only from false positives, we use it for guiding the interactive exploration and thus extraction of *all* periodic orbits of a vector field. Hence, we also use it for evaluation of the techniques by Wischgoll and Scheuermann, Theisel et al., and Chen et al. In the following, we focus primarily on the comparison with Wischgoll and Scheuermann's approach, and defer the discussion and evaluation of the other approaches to Sect. 6.4.

4 Method

We address issues (I)–(III) from Sect. 3 by means of derived scalar fields. As introduced in Sect. 2, Shi et al. [13] obtain a scalar field, which encodes, for each point in space, the convergence structure of the streamline started at that point. If the forward trajectory converges to a sink, it encodes the ID of that sink. Otherwise, the field stores if the respective trajectory left the domain or did not converge within the given integration time. The same procedure is also carried out in reverse time direction, i.e., the streamline is integrated in backward time and its convergence to sources is determined. Thus, a forward and a reverse scalar field are obtained. We maintain two similar scalar fields, which we denote *label fields*. However, in contrast to Shi et al.,

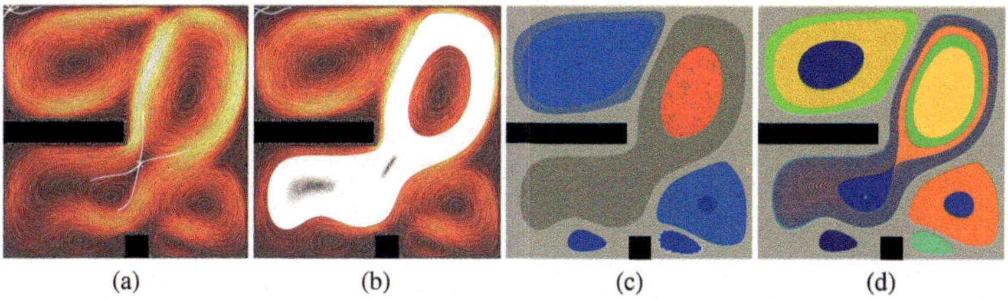

Fig. 1 Buoyant Flow dataset. **a** Short integration of separatrices from saddle-type critical point (cross) only reveals local structure. **b** Long integration, on the other hand, results in visual clutter because the separatrices converge to periodic orbits in this case. **c** Seeding a streamline at each point of the domain and assigning to it the label of the critical point that the streamline's end is closest to [13, 14] leads to insignificant visualization (noise). **d** Our approach, which determines the label of the critical point closest to the *entire* streamline, reveals much more topological structure

we address *steady* (rotational) flow. Our first step is to additionally include critical points of type attracting focus and repelling focus in the convergence analysis. More precisely, our *set of convergence structures* C, for which we determine convergence, consists of all critical points, including saddles. And we assign an identifier $\iota_{\boldsymbol{\zeta}}$ (i.e., a label) to each convergence structure $\boldsymbol{\zeta} \in C$.

To the best of our knowledge, Shi et al. detect convergence by testing the Euclidean distance between the streamline's endpoint and the critical points against a user-defined threshold. While this works well in irrotational flow (see, e.g., Fig. 2), it is generally not applicable for rotational flow (see Fig. 1c). The reason for the resulting erroneous (noisy) regions of this approach is that the streamlines may converge to a periodic orbit, in which case their endpoints are distributed along the orbit, with the result that they are typically nearest to different critical points.

This motivates us to determine the critical point closest to the *entire* streamline, instead of just the endpoint (which leads to more consistent results in such configurations, see Fig. 1d). That is, we measure the distance between a convergence structure and the point on the streamline closest to it. In our prototype, we achieve this by evaluating, at each integration step of a streamline, the distance from the current point to all convergence structures $\boldsymbol{\zeta} \in C$, and maintaining the identifier $\iota_{\boldsymbol{\zeta}}$ and distance $\delta_{\boldsymbol{\zeta}}$ of the closest convergence structure. For a 2D vector field $\mathbf{u}(\mathbf{x}) \in \mathbb{R}^2$ with $\mathbf{x} \in \Omega \subseteq \mathbb{R}^2$, this provides us with a respective forward-integrated label field

$$l^{+}(\mathbf{x}) := \underset{\iota_{\boldsymbol{\zeta}}}{\arg\min} \, \|\boldsymbol{\xi}_{\mathbf{x}}(s) - \boldsymbol{\zeta}\| \,, \quad \forall s \in [0, \tau], \quad \forall \boldsymbol{\zeta} \in C, \tag{1}$$

with

$$\boldsymbol{\xi}_{\mathbf{x}_0}(s) := \mathbf{x}_0 + \int_0^s \mathbf{u}\left(\boldsymbol{\xi}_{\mathbf{x}_0}(\varsigma)\right) d\varsigma \tag{2}$$

being the point on the streamline $\boldsymbol{\xi}_{\mathbf{x}_0}(\cdot)$, seeded at point $\mathbf{x}_0$ and integrated for time s, and τ being the total integration time of the label field. If the streamline leaves

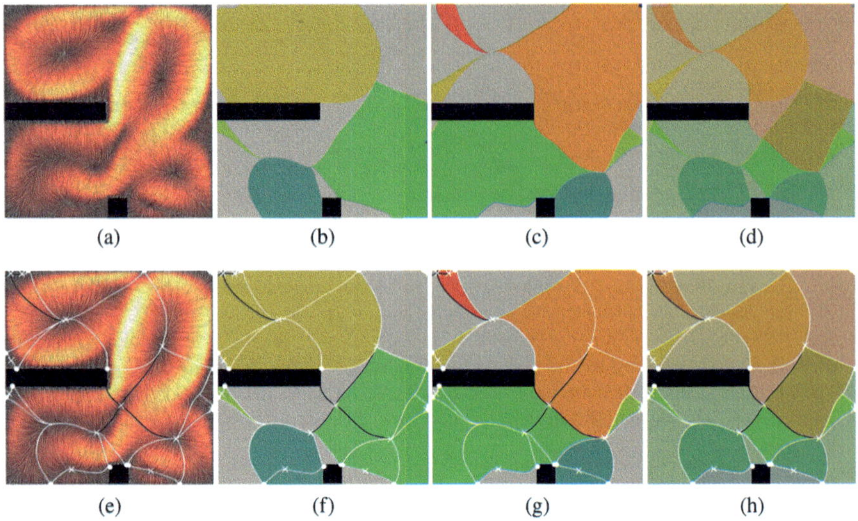

Fig. 2 Rotated Flow dataset, with convergence structures consisting of critical points (crosses). **a** Line integral convolution [2], with higher speed by brighter color. **b** Forward-integrated label field $l^+(\mathbf{x})$. **c** Reverse-integrated label field $l^-(\mathbf{x})$. **d** Composition of (b) and (c) provides implicit visualization of vector field topology. **e–h** Same as top row, but with superimposed topological skeleton for validation, with separatrices seeded at saddles (black lines) and boundary switch points (dots). Distance fields $d^+(\mathbf{x})$ and $d^-(\mathbf{x})$ are not shown, as they are very close to zero because all streamlines converge to critical points (except those in the small top left region)

the domain or stops during integration, we store an "invalid" label in the label field, which we map to gray color in our results. Notice, that we use adaptive Runge–Kutta–Fehlberg 4(5) integration, which may lead to such a stop if the step size becomes too small. Analogously, we obtain a reverse-integrated label field

$$l^-(\mathbf{x}) := \underset{l_\zeta}{\arg\min} \, \|\boldsymbol{\xi}_\mathbf{x}(s) - \boldsymbol{\zeta}\|, \quad \forall s \in [-\tau, 0], \quad \forall \boldsymbol{\zeta} \in C. \tag{3}$$

By assigning each label a unique color, we obtain a respective visual representation of l^+ (Fig. 2b) and l^- (Fig. 2c), and by compositing both representations, we obtain our label-based implicit visualization of vector field topology (Fig. 2d). In irrotational flow, our composed visualization is consistent with the topological skeleton (see Fig. 2h, in which case the topological skeleton includes separatrices of saddle points and boundary switch points). In case of open domain boundaries, however, configurations (with critical points outside of the data domain) can be constructed, in which the label fields cannot reveal such separatrices of boundary switch points. This is in analogy to the approach due to Shi et al. [13].

In rotational flow, with C consisting of all critical points, our forward and reverse label fields $l^+(\mathbf{x})$ and $l^-(\mathbf{x})$ are able to reveal some topological structure, but at the same time they suffer from false positives and false negatives. For example,

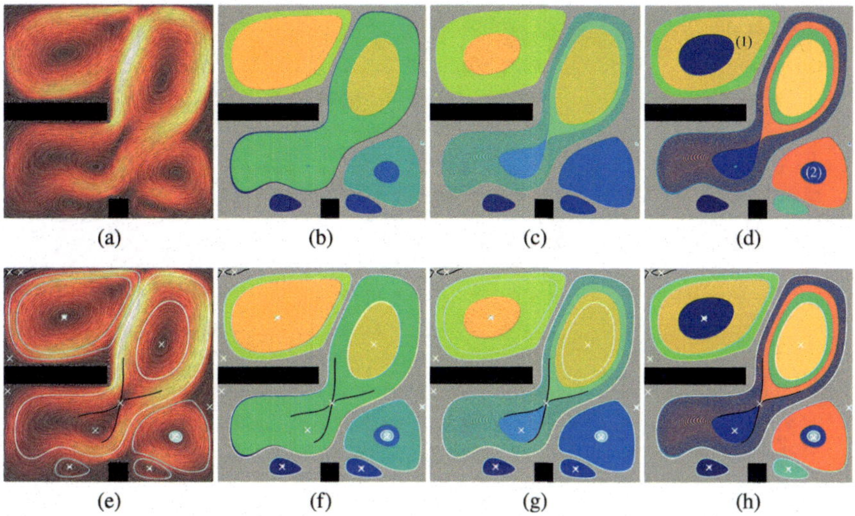

Fig. 3 Buoyant Flow dataset, with convergence structures consisting of critical points (crosses). **a** Line integral convolution, with higher speed by brighter color. **b** Forward-integrated label field $l^+(\mathbf{x})$. **c** Reverse-integrated label field $l^-(\mathbf{x})$. **d** Composition of (b) and (c) reveals some topological structure, but exhibits false positives and false negatives. **e–h** Same as top row, but with superimposed topological skeleton for validation, with separatrices seeded at saddles (black lines), periodic orbits extracted according to Wischgoll and Scheuermann [22] (white), and remaining periodic orbits (turquoise) seeded manually with the help of our distance fields. See Figs. 5 and 6 for more details

the boundary (1) in Fig. 3d does not represent a periodic orbit, as can be seen by validation with the explicitly extracted topological skeleton (Fig. 3h). On the other hand, the periodic orbits at (2) are not captured by the label fields. Other issues, which are also present in explicit visualization of vector field topology, include too short integration time τ and incomplete C. For example, tightly spiraling separatrices, as present in Fig. 1b, make it difficult to determine sufficiently large τ. On the other hand, as discussed above, there is no technique available that robustly provides all periodic orbits of a vector field, and thus C is potentially incomplete with respect to periodic orbits. Notice that in our approach, which measures the distance of a convergence structure to the entire streamline instead of only its endpoint, one could add a single point located on each known periodic orbit to the convergence set C to visualize the periodic orbits using the label fields. However, in our example, only the white orbits in Fig. 3e have been automatically determined by Wischgoll and Scheuermann's [22] approach.

These observations motivate us to complement our approach with respective *distance fields*. We store the distance δ_ζ during forward integration in the forward-integrated distance field

$$d^+(\mathbf{x}) := \min \|\boldsymbol{\xi}_{\mathbf{x}}(s) - \boldsymbol{\zeta}\|, \quad \forall s \in [0, \tau], \quad \forall \boldsymbol{\zeta} \in C, \tag{4}$$

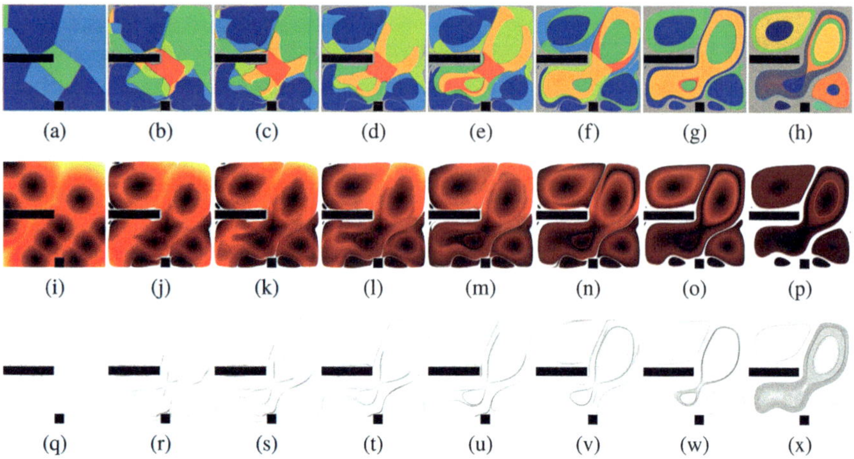

Fig. 4 Buoyant Flow dataset, with increasing integration time τ from left to right. **a**–**h** Composite label field, **i**–**p** combined distance field, and **q**–**x** combined distance field gradient

and analogously, during reverse integration, in the reverse-integrated distance field

$$d^-(\mathbf{x}) := \min \|\boldsymbol{\xi}_{\mathbf{x}}(s) - \boldsymbol{\zeta}\|, \quad \forall s \in [-\tau, 0], \quad \forall \boldsymbol{\zeta} \in C. \tag{5}$$

It is the purpose of $d^+(\mathbf{x})$ and $d^-(\mathbf{x})$ to help with all abovementioned issues. If a forward/reverse streamline converges sufficiently close to a convergence structure, the respective entry in $d^+(\mathbf{x})$ or $d^-(\mathbf{x})$ is very close to zero. On the other hand, if C is incomplete, e.g., because (a point on) a periodic orbit has not been added to C, the streamlines that converge to that structure will obtain labels of nearby convergence structures, e.g., of critical points, and the resulting entries in the distance fields will be accordingly large. That is, the distance fields serve as an uncertainty measure of our implicit topological visualization, i.e., regions with non-negligible values indicate issues with integration or missing convergence structures.

Figure 4 exemplifies this. Figure 4a–h shows the (composed) evolution of $l^+(\mathbf{x})$ and $l^-(\mathbf{x})$. For small τ (Fig. 4a), the label field reflects the Voronoi diagram of the convergence structures, while their structure develops with increasing τ (Fig. 4b–h). Figure 4i–p shows the evolution of $\|\left(d^+(\mathbf{x}), d^-(\mathbf{x})\right)^\top\|$. On the one hand, one can see that this field is monotonically decreasing with increasing τ, i.e., τ needs to be increased as long as the distance fields are decreasing. On the other hand, one can see that, although the values decrease with increasing τ, larger values remain along the periodic orbits, as discussed above. This motivated us to introduce the *gradient of the distance fields*, i.e., $\nabla d^+(\mathbf{x})$ and $\nabla d^-(\mathbf{x})$. Figure 4q–x shows the combined gradient $\|\left(\nabla d^+(\mathbf{x}), \nabla d^-(\mathbf{x})\right)^\top\|$. It can be seen that, as τ increases, the gradient field increasingly reveals the periodic orbits, which are apparent as sharp lines of high values (mapped here to darker levels of gray).

Fig. 5 Buoyant Flow dataset. **a–c** Label fields (same as Fig. 3b–d), **d–f** respective distance fields, and **g–i** respective distance field gradient. **a**, **d**, **g** Forward integration, **b**, **e**, **h** reverse integration, and **c**, **f**, **i** their composed representation

A more detailed look at this dataset in terms of the distance fields and their gradient is provided in Fig. 5. The dark sharp lines in, e.g., Fig. 5i represent candidates for periodic orbits. Figure 6 shows a validation with the explicitly extracted topological skeleton. While some of the candidates have already been detected by Wischgoll and Scheuermann's approach (white in Fig. 3e), we tested the remaining candidates by interactive seeding of streamlines in their vicinity, and using Wischgoll and Scheuermann's convergence criterion to determine if the streamlines converge to periodic orbits. The orbits that we determined in this interactive approach are colored

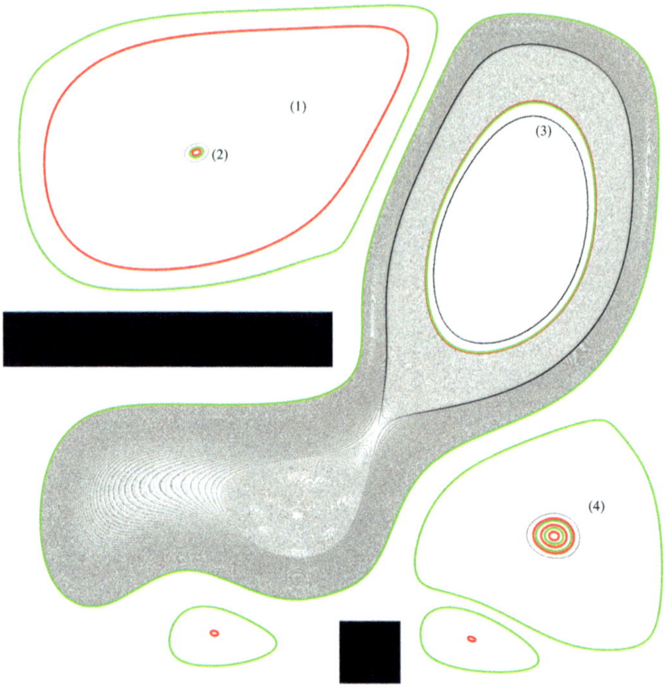

Fig. 6 Buoyant Flow dataset. Composite forward and reverse distance field gradient (gray), with overlaid periodic orbits (attracting green, repelling red)

turquoise in Fig. 3e. All extracted periodic orbits are depicted by color-coded curves in Fig. 6. We observe two main features in this gradient field: boundaries of almost uniform regions, which do not represent candidates ((1) and (4)), and sharp ridges (dark lines) which do represent candidates for periodic orbits. Nevertheless, some of the dark lines (e.g., (2) and (3)) represent false positives of our approach, i.e., candidates that did not lead to a detected orbit. These false positives turned out to be caused by insufficiently accurate streamline integration. For interactive investigation using streamlines and Wischgoll and Scheuermann's convergence criterion, we had to use an integration step size in the order of one-thousandth of the data grid's cell size to be able to reveal that there were no periodic orbits at these locations. However, for the computation of our label fields and distance fields, we could not afford such highly accurate integration due to its high computational cost, and thus instead used adaptive step size control, which caused these false positives.

Overall, we have shown the utility of our approach for: (I) validating automatic periodic orbit extraction techniques, (II) supporting the user in interactive seeding of streamlines for periodic orbit extraction, and (III) showing that the extraction scheme proposed by Wischgoll and Scheuermann [22], which seeds streamlines for periodic orbit extraction only at the critical points, cannot extract nested periodic orbits.

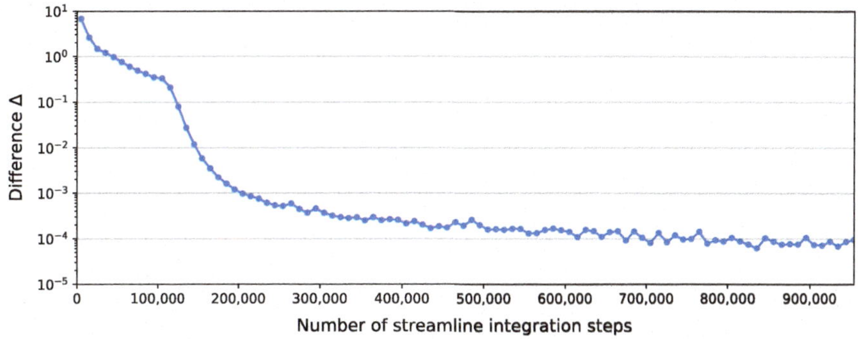

Fig. 7 Monitoring of improvement Δ w.r.t. increasing integration time τ. In this case, one can conclude that approximately 10^6 integration steps suffice. Please note the logarithmic scale

A further issue with a naive application of our approach is that a high resolution of the label fields and distance fields is required for pleasant and detailed representations. However, since each sample of these fields requires the integration of a computationally costly streamline, and since the label fields exhibit regions with uniform label, the sampling of our fields lends itself well for adaptive refinement, which is presented in Sect. 5.2. Finally, to support appropriate choice of the integration time τ, we employ progressive increase of τ, combined with monitoring of the change of the resulting visual representation (Sect. 5.1).

Overall, our work can be considered an extension of the work by Shi et al. [13], with the following main differences:

- we address steady rotational flow and include spiral critical points,
- we determine convergence without a threshold,
- we provide distance fields as a measure for visualization uncertainty, and
- we employ adaptive refinement and progressive computation for efficiency.

5 Algorithm

Algorithm 1 provides an overview of our approach. The procedure starts with initial seeds $\mathcal{X}_0$, consisting of the original nodes of the data grid, and the set of convergence structures C. After integration of a streamline (Sect. 5.1) at each of these seeds, their distances to the convergence structures are determined, and the label fields and distance fields updated accordingly. Then, as second step, the grid is refined (Sect. 5.2), and streamlines at the resulting new seeds $\mathcal{X}^{new}$ are integrated. This step is repeated until the grid is refined according to prescribed thresholds.

Algorithm 1. Outline

Input: Vector field $\mathbf{u}(\mathbf{x})$, initial seed $\mathbf{x}_0 \in \mathcal{X}_0$, convergence structures $\boldsymbol{\zeta} \in C$ with labels $\iota_{\boldsymbol{\zeta}}$
Parameters: Integration time N, distance difference threshold γ, and edge length threshold η
Output: Forward and reverse labels l_p^+, l_p^-, distances d_p^+, d_p^-, and streamline endpoints $\mathbf{x}_p^+, \mathbf{x}_p^-$

/ Function for updating labels and distances */*
function UPDATE_LABELS_AND_DISTANCES($\mathbf{x}, l, d, C$)
  for each point $\boldsymbol{\zeta} \in$ convergence structures C **do**
    if $\|\mathbf{x} - \boldsymbol{\zeta}\| < d$ **then**
      $\{l, d\} \leftarrow \{\iota_{\boldsymbol{\zeta}}, \|\mathbf{x} - \boldsymbol{\zeta}\|\}$
    end if
  end for
  return $\{l, d\}$
end function

/ Use previously computed sample positions, labels and distances, or initialize values */*
if not first execution **then**
  $\left(\{\mathbf{x}_0, \mathbf{x}^+, \mathbf{x}^-, l^+, l^-, d^+, d^-\}_p \in \mathcal{Y}\right) \leftarrow$ load previous results
  $n \leftarrow$ previous stream line integration time
else
  $\left(\{\mathbf{x}_0, \{\mathbf{x}, l, d\}^{\{+,-\}}\}_p \in \mathcal{Y}\right) \leftarrow \{\mathbf{x}_0, \{\mathbf{x}_0, -1, \infty\}\}$
  $n \leftarrow 0$
end if

/ Integrate streamlines, and update labels and distances */*
for each $\{\mathbf{x}_0, \{\mathbf{x}, l, d\}^{\{+,-\}}\}_p \in$ loaded/initialized streamlines $\mathcal{Y}$ **do**
  for each $\sigma \in \{+, -\}$ **do**
    for n to N **do**
      $\mathbf{x}_p^\sigma \leftarrow$ advect $\mathbf{x}_p^\sigma$ with Runge–Kutta on $\sigma\mathbf{u}(\mathbf{x})$
      $l_p^\sigma, d_p^\sigma \leftarrow$ UPDATE_LABELS_AND_DISTANCES($\mathbf{x}_p^\sigma, l_p^\sigma, d_p^\sigma, C$)
    end for
  end for
end for

/ Refine grid, re-triangulate, and update labels and distances */*
$t \leftarrow$ Delaunay triangulation of input seeds $\mathcal{X}_0$
repeat
  $t, \mathcal{X}^{new} \leftarrow$ refine triangulation t with parameters γ and η
  for each point $\mathbf{x} \in$ new sample positions $\mathcal{X}^{new}$ **do**
    $\{\mathbf{x}_0, \{\mathbf{x}, l, d\}^{\{+,-\}}\}_p \leftarrow \{\mathbf{x}, \{\mathbf{x}, -1, \infty\}\}$
    for each $\sigma \in \{+, -\}$ **do**
      for 0 to N **do**
        $\mathbf{x}_p^\sigma \leftarrow$ advect $\mathbf{x}_p^\sigma$ with Runge–Kutta on $\sigma\mathbf{u}(\mathbf{x})$
        $l_p^\sigma, d_p^\sigma \leftarrow$ UPDATE_LABELS_AND_DISTANCES($\mathbf{x}_p^\sigma, l_p^\sigma, d_p^\sigma, C$)
      end for
    end for
    $\mathcal{Y} \leftarrow \mathcal{Y} \cup \{\mathbf{x}_0, \{\mathbf{x}, l, d\}^{\{+,-\}}\}_p$
  end for
until $\mathcal{X}^{new} = \emptyset$

/ Return results, which can be used to initialize further computation */*
return all $\{\mathbf{x}_0, \{\mathbf{x}, l, d\}^{\{+,-\}}\}_p \in \mathcal{Y}$

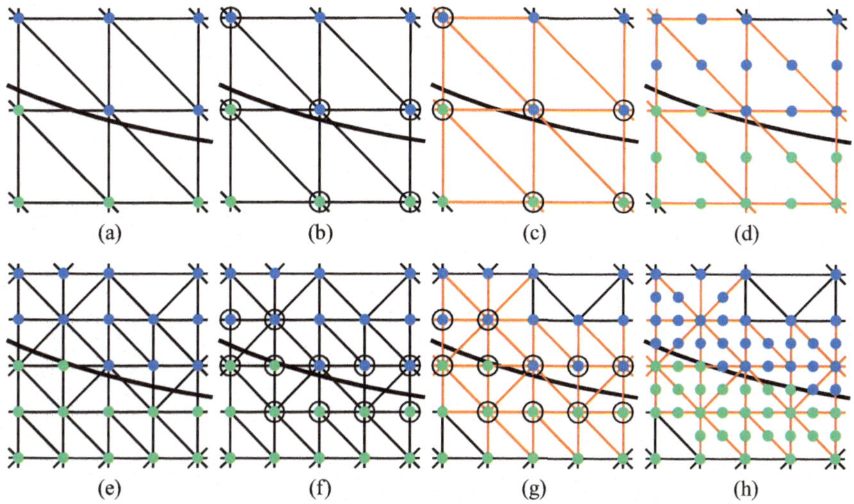

Fig. 8 Refinement steps. Separatrix (black curve) between two topological regions (green and blue). Starting with the initial triangulation (a), nodes with different neighbor labels are marked (circles) (b). In the second stage, all edges with a marked node are selected (orange) (c) and midpoints of these edges added (d). Existing and new nodes triangulated (e). Procedure is iterated (f)–(h)

5.1 Integration

As discussed in Sect. 4, $d^+(\mathbf{x})$ and $d^-(\mathbf{x})$ decrease monotonically as integration time τ is increased. Therefore, we can use them for estimating if τ has been chosen sufficiently large. We compute the streamlines progressively, i.e., after all streamlines have been advanced by n integration steps, we compute ϵ_{i+1}, which is the average of $d^+(\mathbf{x})$ and $d^-(\mathbf{x})$, and subtract it from the ϵ_i that resulted before advancing the streamlines. This provides us with $\Delta_{i+1} := \epsilon_{i+1} - \epsilon_i$, measuring the "improvement" of our fields, which we can monitor by plotting (Fig. 7). This plotting supports the user in finding an appropriate τ, i.e., in deciding when to stop the process. Δ could also be tested automatically with respect to a user-defined threshold, in particular to switch between integration phases and refinement phases (see below). We did, however, not implement such switching and refine only after τ has been determined.

5.2 Refinement

As motivated above, we employ adaptive refinement to efficiently obtain high-resolution label fields and (gradient of) distance fields. We refine the label fields and distance fields together, i.e., they are computed on the same refined grid. The refinement is steered by a distance threshold γ and an edge length threshold η. While

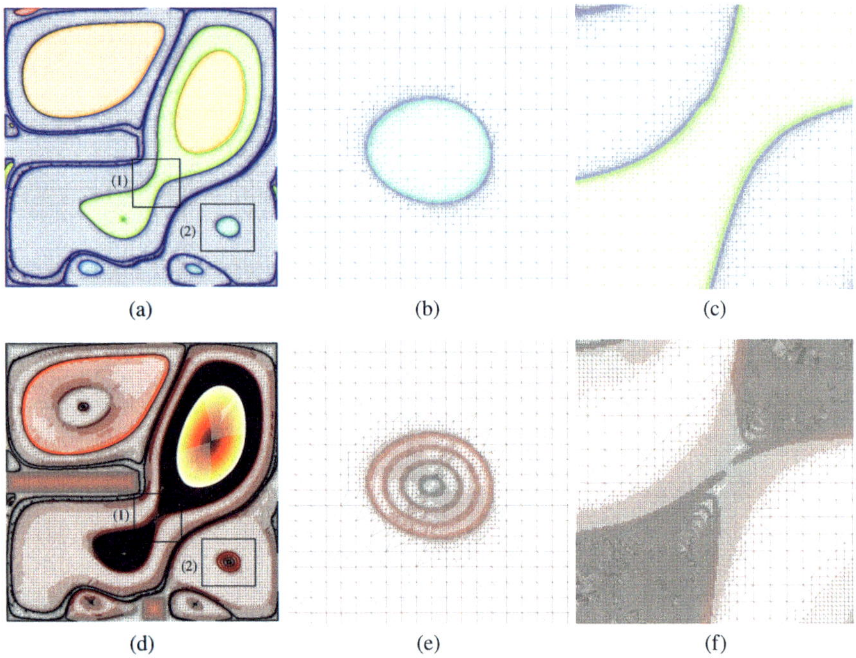

Fig. 9 Exemplary refinement for the Buoyant Flow dataset in wireframe representation. **a**–**c** Refinement at label boundaries, and **d**–**f** in regions of high gradient in the respective distance field. Refinement reveals **c**, **f** manifolds around saddles (1) and **b**, **e** periodic orbits (2)

the former is used to detect and subdivide edges that connect nodes whose distance values differ by more than the threshold γ, the latter is used to prevent over-refinement by setting a minimum edge length. Additionally, edges are refined that connect nodes with different values in the label fields. Note, that it suffices that one of the criteria is met for either forward or reverse integration.

In order to capture all relevant regions, first all nodes are marked for refinement that share an edge that meets one of the above criteria. Then, all edges connected to a marked node are subdivided by placing a new seed at their mid. Together, the nodes from the previous iteration and the new seeds are triangulated using Delaunay triangulation. Figure 8 illustrates the procedure for the label criterion, and Fig. 9 visualizes a possible resulting grid for the Buoyant Flow dataset. Here, one can observe that the label criterion leads to a refinement at the boundaries between different regions, which are defined by the label field. The distance criterion, on the other hand, induces refinement in the neighborhood of periodic orbits and streamlines passing near the saddle.

5.3 Implementation

Because very long streamlines of high accuracy are needed, our prototype employs GPU computation using CUDA [10] for parallel streamline integration. Delaunay triangulation of the seed points is carried out using the Computational Geometry Algorithms Library (CGAL) [17] after each refinement. Our prototype is further designed to be highly flexible and adaptive, allowing the user to change parameters at runtime. Additionally, intermediate results are visualized directly for monitoring, and they can be saved to file to later resume the computation. Our implementation, as part of the *flowvis* plugin for MegaMol [6, 7], is published on GitHub at https://github.com/UniStuttgart-VISUS/implicit-topology.

6 Results

We exemplify the utility of our approach using three datasets originating from a time-dependent convection-driven 2D flow simulation of air in a closed container [12]. Due to the compressibility of air, it exhibits non-vanishing divergence. The Buoyant Flow dataset (Sect. 6.2), represents the time step at 12.524952 s (physical time), in a phase where the flow is well developed and highly time-dependent. The Rotated Flow dataset (Sect. 6.1), on the other hand, has been obtained from this vector field by rotating each vector by 90° in clockwise direction. Our third dataset, the Buoyant Flow II (Sect. 6.3), represents the time step at 1.849999 s, where the flow is still developing and exhibits less tightly spiraling streamlines. The dataset is available at https://dx.doi.org/10.11588/data/LFH1LI.

6.1 Rotated Flow

As this vector field is dominated by irrotational dynamics, all seeded streamlines eventually converge to a source or a sink, or leave the domain in either forward or reverse direction. Therefore, the computed distance fields approach zero for sufficiently long integration time τ. Thus, we limit the visualization to the resulting label fields, which are shown in Fig. 2b for forward, in Fig. 2c for reverse integration, and Fig. 2d the composed label fields. As one can observe, our method is capable of implicitly visualizing manifolds of boundary switch points, although they have not been added to C. In this special case, the results of our approach match the ones using the method by Shi et al. [13] for this steady field.

For this example, the implicit visualization of the topology is computationally cheap. With merely 2,000 streamline integration steps per seed, all topological features have been revealed. As only the labels are of interest here, it suffices to refine

Table 1 Computation time and number of seeded streamlines for the Buoyant Flow (I) and Buoyant Flow II (II) datasets. The number of integration steps per streamline is 10^6, with refinement parameters $\gamma = 2.5 \times 10^{-4}$ and $\eta = 1.2 \times 10^{-4}$. Note that the error threshold for the adaptive step size control is 3×10^{-6} for the Buoyant Flow, and 7×10^{-6} for Buoyant Flow II dataset, which explains the better performance of the latter. Times are given in either seconds or minutes

Refinement step		Grid refinement [s]		Seeded streamlines		Integration [min]	
I	II	I	II	I	II	I	II
-	-	-	-	10,201	10,201	0.63	0.69
1	1	<1	<1	23,549	20,983	1.35	1.22
2	2	<1	<1	84,938	68,508	4.40	3.27
3	3	<1	<1	270,976	156,929	13.40	7.61
4	4	3	<1	734,185	263,953	34.67	12.89
5	5	3	<1	30,142	4,420	1.70	0.46
6	6	3	<1	6,252	830	0.43	0.36
7	7	3	<1	2,064	344	0.38	0.34
8–57	8–18	111	9	2,367	797	4.79	2.51
57	18	124	13	1,164,674	516,764	61.76	29.35

the grid at label boundaries. Therefore, the cost of the refinement and integration of respective streamlines is very low, and the computation completed in under a minute.

6.2 *Buoyant Flow*

Besides the structures already discussed above, one can also observe large gradients near saddles. Looking at the gradient field for forward (Fig. 5g) and reverse (Fig. 5h) integration, one can identify the respective separatrices.

Contrary to the computation for the Rotated Flow dataset, longer streamline computation is necessary in this example. This is because there are many regions in which streamlines converge only slowly to critical points or periodic orbits. Here, 10^6 streamline integration steps were necessary. Further, much more cells had to be refined, as—additionally to label boundaries—refinement had to be also performed with respect to the distance fields. For this, the parameters were set to $\gamma = 2.5 \times 10^{-4}$ and $\eta = 1.2 \times 10^{-4}$. This way, more than one million additional seed points were generated. Although our method employs parallel streamline computation on the GPU, this step still consumes most of the time. The complete computation was conducted in little more than an hour. Details are provided in Table 1.

Fig. 10 Buoyant Flow II dataset. **a**–**c** Label fields, **d**–**f** distance fields, and **g**–**i** distance gradient fields. **a**, **d**, **g** Forward, **b**, **e**, **h** reverse integration, and **c**, **f**, **i** their composition

6.3 Buoyant Flow II

As introduced, this example consists of an early time step of the same simulation, exhibiting less tightly spiraling streamlines. Here, the same refinement parameters $\gamma = 2.5 \times 10^{-4}$ and $\eta = 1.2 \times 10^{-4}$ were used as for the Buoyant Flow dataset. Performance details are again given in Table 1. Figure 10 shows the label fields, distance fields, and distance gradient fields. Compared to the Buoyant Flow dataset, this example exhibits less "gray area", i.e., less area whose streamlines stopped at the domain boundary. Figure 11 shows the distance gradient field together with the

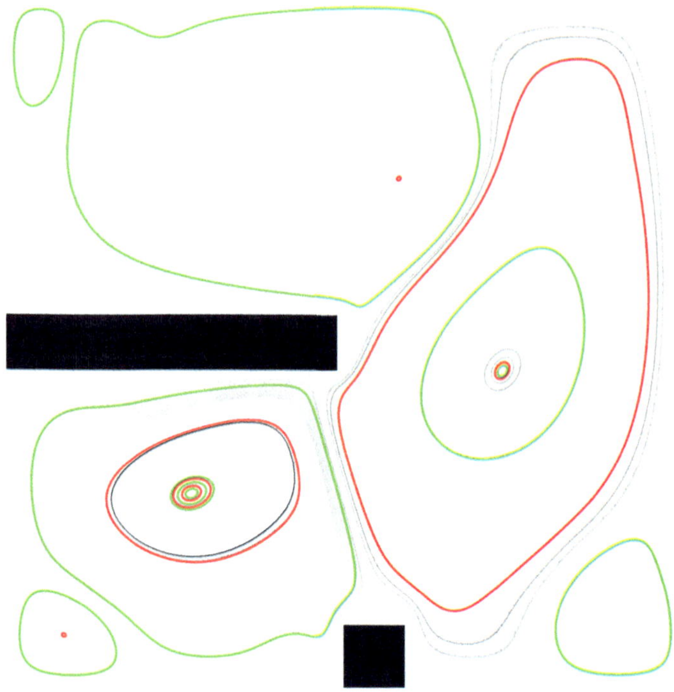

Fig. 11 Buoyant Flow II dataset. Composite forward and reverse distance field gradient (gray), with overlaid periodic orbits (attracting green, repelling red)

topological skeleton. Again, the distance gradient field guided us in our interactive seeding to the extraction of periodic orbits that have not been captured by Wischgoll and Scheuermann's approach, and again in this case we experienced false positives due to inaccuracies introduced by the adaptive step size control during integration.

6.4 Discussion

Although it is rather difficult to prove without a ground truth (robust existing extraction technique), we discuss why our technique detects all periodic orbits. In the neighborhood of a periodic orbit, three types of streamline behavior are possible:

- streamlines seeded on either side converge to a critical point,
- streamlines seeded on either side converge to another periodic orbit, or
- streamlines seeded on one side converge to a critical point, whereas the ones seeded on the other side converge to a periodic orbit.

In the first case, already the label field captures the periodic orbit, as it coincides with the region boundary. For the second case, it is theoretically possible to miss the

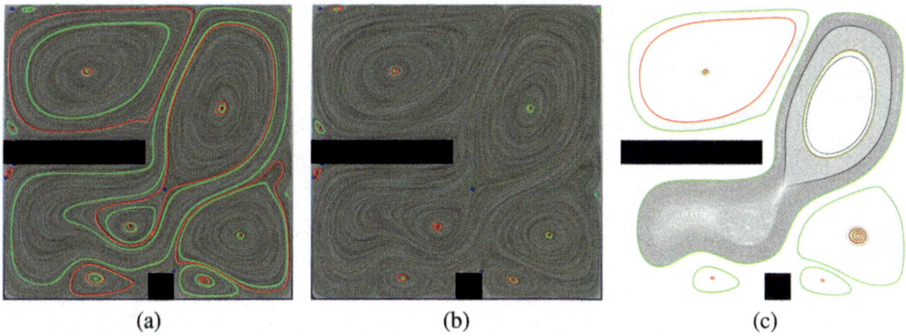

Fig. 12 Periodic orbit extraction using Morse decomposition with 2nd-order Runge–Kutta (a), and 4th-order Runge–Kutta (b) integration. The images are obtained using the reference implementation by Chen et al. [3]. (c) Our validated results for comparison. The method by Chen et al. finds many of the periodic orbits. However, for this dataset, there are also false positives and false negatives

periodic orbit if and only if the other periodic orbits, which both seeded streamlines converge to, have the same minimum distance to a critical point. Then, streamlines seeded on both sides would exhibit the same minimum distance value, and the distance gradient would vanish. As such configuration would be degenerate, we can conclude that also in this case, the discontinuity in the discretized distance field will reveal the periodic orbit. In the last case, the minimum distance on one side is negligible because it converges to a critical point, while on the other side it is non-zero (approaching a periodic orbit) and thus the discretized distance gradient reveals the periodic orbit.

In the following, we compare our method to previous techniques, which take completely different approaches to find periodic orbits. Chen et al. [3] employ Morse decomposition, building a Morse connection graph and its refinement, an entity connection graph. This way, they find both attracting and repelling periodic orbits in 2D steady vector fields and in surface flows. Like our approach, this method is able to detect nested periodic orbits. However, numerical instability can lead to false positive and false negative detection. This is illustrated in Fig. 12, where different integration schemes lead to different sets of detected closed streamlines.

Theisel et al. [19], on the other hand, extend the 2D static vector field $\mathbf{u} = (u_x\ u_y)^\top$ to a 3D static vector field $\mathbf{u}' = (u_x\ u_y\ 0)^\top$. There, they integrate stream surfaces in forward and reverse time, starting from seeding lines that connect critical points. Intersections between these forward and reverse stream surfaces, started from the same seeding line, coincide with periodic orbits. As their method for choosing seeding lines guarantees that every periodic orbit can be found, it does not suffer from the same issue as the seeding scheme by Wischgoll and Scheuermann. However, the extension to 3D space introduces an additional parameter for controlling the spatial resolution of the stream surface mesh. Especially for (nested) periodic orbits, to which streamlines converge very slowly, a fine spatial and temporal resolution is necessary to detect them all. At the same time, periodic orbits can lie halfway between two critical points, thus requiring large integration time. This is, for example, the case in

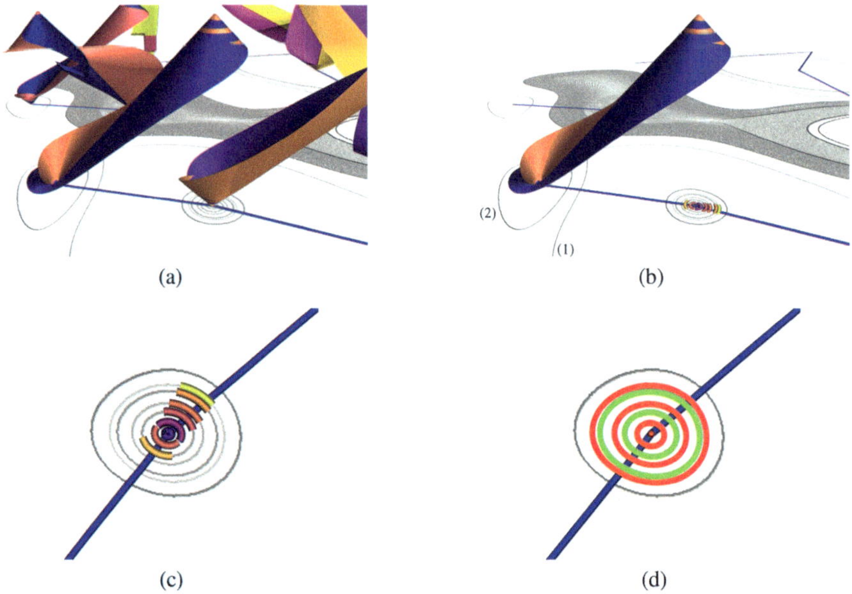

Fig. 13 Stream-surface-based periodic orbit extraction according to Theisel et al. [19], showing **a** all stream surfaces (with color-coded forward and reverse integration), and **b** the extracted periodic orbits for an exemplary region. The background shows our distance gradient field for reference. Comparing the results of the method by Theisel et al. **c** with our method **d**, one can observe that both find the correct nested periodic orbits close to the critical point. However, using Theisel et al.'s approach, we were not able to find the periodic orbits farther away from the critical points, annotated (1) and (2) in **b**. Additionally, their technique can suffer from multiple extractions **c**

fields such as the Buoyant Flow dataset, and can be observed in Fig. 13: all nested periodic orbits close to the critical point are found, however, the integration time was not large enough to find the one farther away. Additionally, as stated by Theisel et al., it is possible that the stream surfaces intersect twice for the same periodic orbit. However, due to numerical instabilities, many more intersection points (156) were found in our case. Computing their distance to each other and removing close ones from the results is an option, but in the presence of very close nested periodic orbits, this could lead to the elimination of valid ones, thus leading to false negatives.

Wischgoll and Scheuermann [22], as discussed throughout our paper, suffer from a lack of a seeding strategy that provides all periodic orbits. This issue can be mitigated by our interactive seeding approach. Another, naive possibility would be to apply their method with a very high seeding density. While this would likely solve the seeding problem and detect all periodic orbits within a dataset for sufficiently small integration steps, it would also be very costly. Because of the required fine temporal resolution, a large number of integration steps would be needed. As our method can make use of dynamic step size control to approximate the convergence of the vector field, our approach is much faster, reducing the costly seeding at the drawback of possible false positives.

7 Conclusion

We presented implicit visualization of 2D vector field topology by means of label fields and distance fields, and demonstrated its utility for evaluating approaches for periodic orbit extraction. We have shown that it is sufficient to include only critical points in the computation of these fields. Since the distance fields indicate topological structures that have not been included in the computation, they readily reveal periodic orbits. This enabled us to demonstrate that state-of-the-art approaches for periodic orbit extraction can miss orbits, and at the same time guided us to the missing ones. As future work, we plan to extend our approach to 3D vector fields, and to more advanced topological structures, such as strange attractors. Additionally, we further would like to investigate the influence of the input convergence structures on the extracted topology by selectively removing structures from the input set C.

Acknowledgements We kindly thank Guoning Chen and Eugene Zhang for their help, providing us with the implementation of their Morse decomposition for periodic orbit detection, as well as the discussion of their work. This work was partially funded by Deutsche Forschungsgemeinschaft (DFG) as part of the Cluster of Excellence EXC 2075 "SimTech" (390740016), Transregional Collaborative Research Center SFB/Transregio 75 (84292822), and International Research Training Group GRK 2160 "DROPIT" (270852890) at the University of Stuttgart.

References

1. Asimov, D.: Notes on the topology of vector fields and flows. Technical Report RNR-93-003, NASA Ames Research Center (1993)
2. Cabral, B., Leedom, L.C.: Imaging vector fields using line integral convolution. Proc. SIGGRAPH **93**, 263–270 (1993)
3. Chen, G., Mischaikow, K., Laramee, R.S., Pilarczyk, P., Zhang, E.: Vector field editing and periodic orbit extraction using Morse decomposition. IEEE Trans. Vis. Comput. Graph. **13**(4), 769–785 (2007)
4. Friederici, A., Günther, T., Rössl, C., Theisel, H.: Finite time steady vector field topology – Theoretical foundation and 3D case. Vis. Model. Vis. VMV **2017**, 95–102 (2017)
5. Friederici, A., Rössl, C., Theisel, H.: Finite time steady 2D vector field topology. In: Topological Methods in Data Analysis and Visualization IV, pp. 253–266 (2017)
6. Gralka, P., et al.: MegaMol – A comprehensive prototyping framework for visualizations. Eur. Phys. J. Special Top. **227**(14), 1817–1829 (2019)
7. Grottel, S., Krone, M., Müller, C., Reina, G., Ertl, T.: MegaMol – A prototyping framework for particle-based visualization. IEEE Trans. Vis. Comput. Graph. **21**(2), 201–214 (2015)
8. Helman, J., Hesselink, L.: Representation and display of vector field topology in fluid flow data sets. Computer **22**(8), 27–36 (1989)
9. Helman, J., Hesselink, L.: Visualizing vector field topology in fluid flows. IEEE Comput. Graph. Appl. **11**(3), 36–46 (1991)
10. Nickolls, J., Buck, I., Garland, M., Skadron, K.: Scalable parallel programming with CUDA. Queue **6**(2), 40–53 (2008)
11. Park, S.W., Yu, H., Hotz, I., Kreylos, O., Linsen, L., Hamann, B.: Structure-accentuating dense flow visualization. In: Joint Eurographics – IEEE VGTC Symposium on Visualization, EuroVis 2006, pp. 163–170 (2006)
12. Sadlo, F.: Buoyant 2D airflow with two obstacles [Data] (2020)

13. Shi, K., Theisel, H., Weinkauf, T., Hauser, H., Hege, H.-C., Seidel, H.-P.: Path line oriented topology for periodic 2D time-dependent vector fields. In: Joint Eurographics – IEEE VGTC Symposium on Visualization, EuroVis 2006, pp. 139–146 (2006)
14. Shi, K., Theisel, H., Weinkauf, T., Hauser, H., Hege, H.-C., Seidel, H.-P.: Extracting separation surfaces of path line oriented topology in periodic 2D time-dependent vector fields. J. WSCG **15**(1–3), 75–82 (2007)
15. Sotiropoulos, F., Ventikos, Y., Lackey, T.C.: Chaotic advection in three-dimensional stationary vortex-breakdown bubbles: Šil'nikov's chaos and the devil's staircase. J. Fluid Mech. **444**, 257–297 (2001)
16. Stöter, T., Weinkauf, T., Seidel, H.-P., Theisel, H.: Implicit integral surfaces. Vis. Model. Vis. VMV **2012**, 127–134 (2012)
17. The CGAL Project. CGAL User and Reference Manual 5.0. CGAL Editorial Board (2019)
18. Theisel, H., Weinkauf, T., Hege, H.-C., Seidel, H.-P.: Saddle connectors – An approach to visualizing the topological skeleton of complex 3D vector fields. In: Proceedings of IEEE Visualization 2003, pp. 225–232 (2003)
19. Theisel, H., Weinkauf, T., Hege, H.-C., Seidel, H.-P.: Grid-independent detection of closed stream lines in 2D vector fields. Vis. Model. Vis. VMV **2004**, 421–428 (2004)
20. van Wijk, J.J.: Implicit stream surfaces. In: Proceedings of IEEE Visualization 1993, pp. 245–252 (1993)
21. Weinkauf, T., Theisel, H., Hege, H.-C., Seidel, H.-P.: Boundary switch connectors for topological visualization of complex 3D vector fields. In: Joint Eurographics - IEEE TCVG Symposium on Visualization, VisSym 2004, pp. 183–192 (2004)
22. Wischgoll, T., Scheuermann, G.: Detection and visualization of closed streamlines in planar flows. IEEE Trans. Vis. Comput. Graph. **7**(2), 165–172 (2001)
23. Wischgoll, T., Scheuermann, G.: Locating closed streamlines in 3D vector fields. In: Joint Eurographics – IEEE TCVG Symposium on Visualization, VisSym 2002, pp. 227–232 (2002)
24. Zhang, L., Chen, G., Laramee, R.S., Thompson, D., Sescu, A.: Flow visualization based on a derived rotation field. Vis. Data Anal. **2016**, 1–10 (2016)
25. Zhang, L., Nguyen, D., Thompson, D., Laramee, R.S., Chen, G.: Enhanced vector field visualization via Lagrangian accumulation. Comput. Graph. **70**, 224–234 (2018)

Visually Evaluating the Topological Equivalence of Bounded Bivariate Fields

Daisuke Sakurai and Takahiro Yamamoto

Abstract We apply visualization to evaluating a new topological equivalence relation, which we call the *topological $\mathcal{B}_+$-equivalence*. It has been used in our separate, yet ongoing, study in mathematics. The equivalence is a building block for the topological study of maps of bounded manifolds into the plane (aka *bounded bivariate fields*). In that study, we have introduced a few invariants that approximate the equivalence, which is hard to treat directly. In this chapter dedicated to the visualization community, we show that visualizing the Reeb space gives us a near-instant way of evaluating the invariants. The process has traditionally required an unpredictable amount of time due to manual analysis of high-order polynomials, which was necessary to obtain the invariant values. Our Reeb space visualization reveals the topological information necessary for evaluating the invariants, and, in doing so, the topological $\mathcal{B}_+$-equivalence itself. Previously, the visualization was found to serve as an introductory learning tool for studying examples of singular fibers. The present article goes further to demonstrate professional use cases.

1 Introduction

As with the Morse theory, of particular interest in differential topology is the topological shapes of *fibers* $g^{-1}(v)$ for some smooth map g with a possibly vectored value v, together with the topological transition of fibers when varying v (see Fig. 1, left) [26]. Though many theories assume manifolds without boundaries, those with boundaries inevitably arise in nature, and, hence, studies have been going on. So far,

D. Sakurai (✉)
Kyushu University, 6-1 Kasugakoen, Kasuga, Fukuoka, Japan
e-mail: d.sakurai@computer.org

T. Yamamoto
Tokyo Gakugei University, 4-1-1 Nukuikita-machi, Koganei-shi, Tokyo, Japan
e-mail: yamush@u-gakugei.ac.jp

I. Hotz et al. (eds.), *Topological Methods in Data Analysis and Visualization VI*, Mathematics and Visualization, https://doi.org/10.1007/978-3-030-83500-2_10

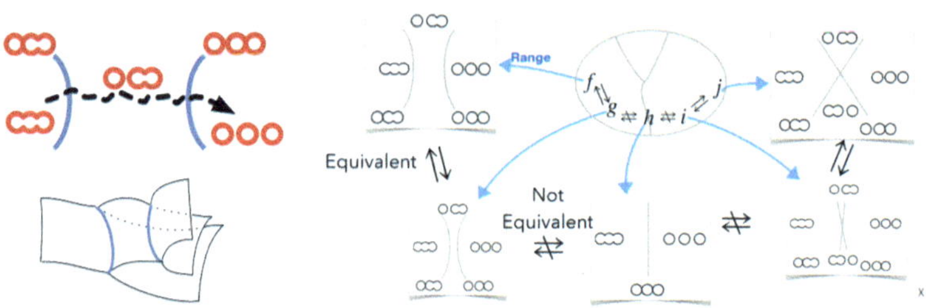

Fig. 1 Left: The range of some map g (on top) and the Reeb space (bottom) The range can be viewed as the projection of Reeb space of g with respect to the value $g(c)$ of the connected component c of each fiber. The range contains singular values (in blue), by crossing which the fiber g^{-1} (red) undergoes a topological change. Right: A schematic illustration of the space of maps $M \to R^2$ of a bounded 3-manifold domain. The $\mathcal{B}_+$-equivalence partitions the space of maps g into the quotient space

we have identified topological transitions of fibers [29] and have been developing a prototype visualization system for constructing examples of them [28, 31] with additional interest on potential impact to visualization for generalizing the topological analysis of isocontours to maps with a multi-dimensional range [6].

Throughout this article, maps are assumed to be smooth and the domain manifold to be compact. We also restrict our interest to bivariate fields $g\colon M \to R^2$, a simple case mapping a bounded 3-manifold M into the plane.

We wish to understand when, and when not, two maps are identical in terms of the topology of fibers. One goal of our series of studies is a topological classification of bounded bivariate fields. The classification directly results from when we regard two maps to be equivalent.

In the literature, we can find an equivalence relation for maps with boundary, called the *$\mathcal{B}$-equivalence* [20]. However, this equivalence, thus also the resulting classification, considers whether t wo maps can be morphed into each other *smoothly*. That is, the fibers must be morphed from one field into the other via a map that is differentiable for infinite times. Topological equivalence, however, is a looser distinction that ignores the differentiability in the morphing.

This chapter thus introduces a new topological equivalence, namely the *topological $\mathcal{B}_+$-equivalence*, or simply the *$\mathcal{B}_+$-equivalence*. (Find details in Sect. 5.) We compare the $\mathcal{B}_+$-equivalence to $\mathcal{B}$-equivalence in Sect. 6.

It is handy to view the classification as a space partition problem. For, e.g., polynomial maps $f = (\Sigma_k a_k x^k, \Sigma_l b_l x^l)$, the space can be understood as the vector space $\mathbb{R}^\infty$, where each map is located at the point $(a_0, b_0, a_1, b_1, ...)$. The $\mathcal{B}_+$-equivalence partitions the space of maps into cells, each inhabited by equivalent maps (Fig. 1, right). The partitioned map space is called the *quotient space*. In this view, questions about the classification of maps turns into those about the structure of the partition.

As with the $\mathcal{B}$-equivalence, two maps are equivalent when one map can be morphed into the other back and forth using some auxiliary maps Φ and ψ. Though Φ and ψ can be *proven* to exist, *finding* them for a particular pair of maps is too difficult to practice. Hence, we approximate the equivalence with *invariants* that are easier to obtain.

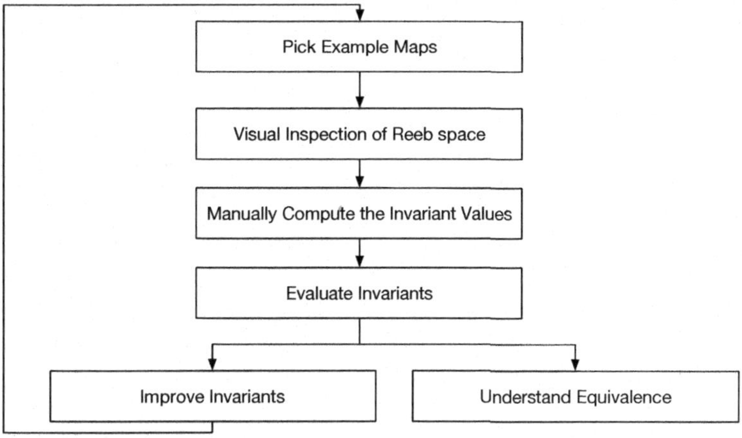

Fig. 2 Our workflow

For two equivalent maps, their values of invariants become identical. Therefore, if the invariant values of two fields differ, we learn that these fields have different topological characteristics. This also means that an invariant must be evaluated in terms of how well it reflects the actual, stricter, equivalence relationship.

We have obtained a few invariants (see Sect. 5) for the $\mathcal{B}_+$-equivalence in a separate yet ongoing mathematical study. Similar studies have been conducted for invariants concerned with unbounded manifolds, and, as they did not employ machine-aided visualization, they discovered convenient formulae that convert the polynomial expression of example maps into their invariant values, taking advantage of perturbation [13, 14, 21, 27, 35] (see Sect. 3.2).

As our study is in its infancy, we have yet to find such formulae. In fact, the quest for such a find has the risk of wasting time. This is especially because (i) such a task may be unfeasible and (ii) an invariant may turn out to be uninformative.

This is the major reason this work employs visualization to gather hints on the characteristics of the invariants. Using visualization we can evaluate our invariants and, consequentially, the $\mathcal{B}_+$-equivalence itself (see Fig. 2). The evaluations are conducted by visualizing example maps and comparing the indicated (in)equivalence due to the $\mathcal{B}_+$-equivalence to those due to the $\mathcal{B}$-equivalence. For the visualization we use the *jcnet* [31], which previously helped novices investigate fibers in $g\colon [-1, 1]^3 \to R^2$.

As our interest lies specifically in the topological singularity of fibers at the boundary, we localize our example maps in the H^3 domain, where $H^3 = \{(x, y, z) \in R^3 | z \geq 0\}$. A localized map is called a *map germ*. Indeed, an arbitrary bivariate field around the boundary can be approximated with spatially distributed map germs. The localization also lets us use jcnet without adaptation to general manifold domains.

The system visualizes the *Reeb space* [11], a space obtained by contracting each *fiber component* to a point. A fiber component is a connected component of a fiber. An example of a Reeb space is given in the left half of Fig. 1. The Reeb space is also known as the Stein factorization [16].

Even though our numerical computations are not a mathematical proof by itself, the hints obtained are valuable. In most cases, it is infeasible to draw this kind of visualization manually, as it involves solving polynomial equations. The results helped our communication with other mathematicians during nation-wide research gatherings. In this sense, the application we introduce is the first example of using the Reeb space for leading mathematical studies. The outcome in this article indicates that the visual approach using the Reeb space can be used for evaluating related invariants that utilize topological singularities.

2 Related Work in Visualization

The concept of Reeb space originates in the Reeb graph [25]. It is a special case of the Reeb space, for which the range is 1 dimensional. If the domain is free of voids, the Reeb graph is known to become free of cycles. For such a domain, the Reeb graph is specifically called the contour tree.

Bajaj et al. [3] first introduced to the visualization community the contour tree. They navigated contours in scalar field data, which were e.g. computational tomography results. Carr et al. [8] proposed the standard serial algorithm, while a recent parallel algorithm [15] is rapidly gaining popularity with its easily accessible implementation in the Topology ToolKit framework [34]. The Reeb graph has been used in visualization and computer graphics for shape analysis [4, 24].

Topological invariants have been used to extract a contour configuration from scalar fields. Pascucci et al. [23] used Betti numbers to obtain homological information. Takahashi et al. [32] extended Morse indices in order to describe the topological transitions of isocontours to describe the inclusions of one in another.

Later, Edelsbrunner et al. [11] extended the Reeb graph into the Reeb space. Their algorithm subdivides the range with the image of the edges of the simplicial complex embedded in the domain. For bivariate fields $g : \mathbb{R}^3 \to \mathbb{R}^2$ with boundary, Tierny and Carr [33] proposed a more efficient algorithm. These Reeb spaces can also be examined on the fly [30], avoiding the cost of full and explicit acquisition of the Reeb space.

The Joint Contour Net (JCN) [6] is a quantized approximation of the Reeb space. JCNs were used for Reeb space visualization [31] for studying the fiber topology of polynomial maps. Other uses include the analysis of nuclear scissions [9]. While the work targeted learners, we use the visualization to aid obtaining new mathematical results. Though a demonstration had been done for resolving the degeneracy of cusps [28], the present article shows the first application of aiding an active and professional study in mathematics.

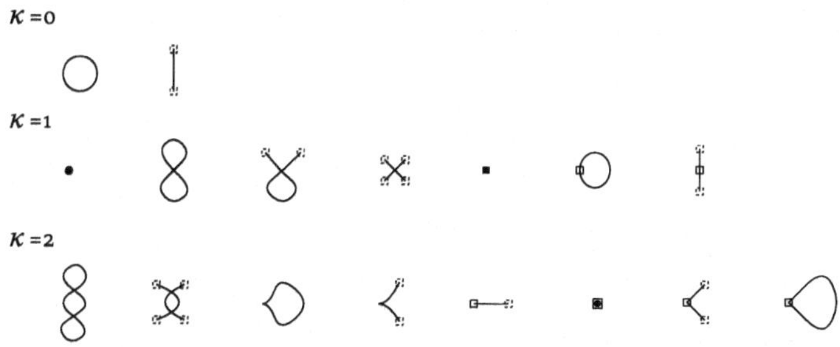

Fig. 3 Examples of the topological shape of a fiber component from $\kappa = 0, 1, 2$. Squared dots indicate the singular point. The black ones indicate a point inside the domain, while the white ones those at the boundary. The topology in $\kappa = 0$ is regular while that in $\kappa > 0$ is singular. The full list can be found in the literature [29]

In this article, we compute the JCN instead of the Reeb space since the JCN implementation had been readily tuned to capture the topology of our maps of interest [31]. In particular, by controlling the granularity of quantization, we get rid of small artificial topological singularities that might arise due to tessellation.

Reeb spaces can be combined with a scatterplot to navigate the *fiber surface* analysis, which is a brushing-and-linking technique using surfaces composed with a set of fibers [7, 33].

Given a time-varying scalar field, considering time as the second axis of the range gives a special case of Reeb space for the thus constructed bivariate field [5, 10]. Lehmann and Theisel [19] revealed that the singular curves appear in the continuous scatterplot [2]. The *Safari* interface [18] for time-varying field analysis is similar in design to the interface we used in this study. An alternative approach for extending the topological analysis to multi-fields is through the Pareto optimality [17].

3 Set-Up

In this section, we clarify the requirements for our visualization, shedding lights on the mathematical preliminaries. As stated, we want to understand the equivalence between smooth maps of bounded 3-manifolds into the plane, and we can localize these maps into map germs of the upper-half Euclidean domain $H^3 = \{(x, y, z) \in \mathbb{R}^3 \mid z \geq 0\}$. Thanks to the Taylor expansion, we can imagine our germs to be in polynomial expression $\sum a_i x^i (a_i, x \in \mathbb{R}, i \in \mathbb{N})$, which suggests the map space is identical to the coefficient space $\mathbb{R}^\infty$ and perturbation means a movement therein.

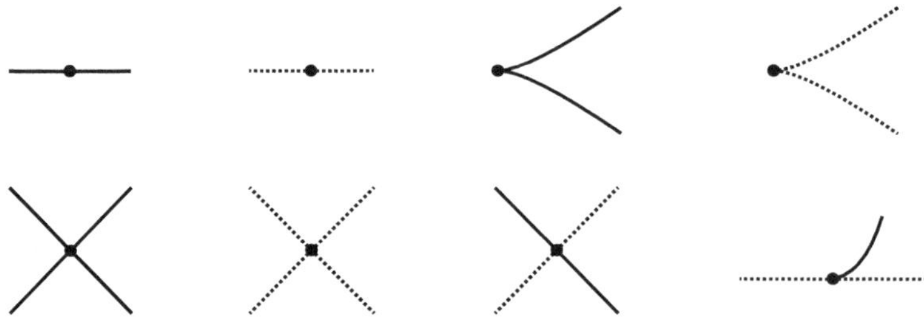

Fig. 4 List of local forms of $f(S(f) \cup S(f|_\partial))$ for a stable map f of a 3-manifold with boundary into a surface without boundary. Solid lines and dotted lines denote $f(S(f))$ and $f(S(f|_\partial))$, respectively

3.1 The $\mathcal{B}_+$-Equivalence and Fiber Topology

Maps $g, g' \colon M^3 \to \mathbb{R}^2$ are *$\mathcal{B}_+$-equivalent* if they can be transformed to each other preserving the topology, together with those of the boundary – or, formally, if there exists a homeomorphism germ $\Phi \colon M^3 \to M^3$ preserving the boundary of M^3 and the interior of M^3 and a homeomorphism $\psi \colon \mathbb{R}^2 \to \mathbb{R}^2$ preserving orientation such that $g \circ \Phi = \psi \circ g'$. If g and g' are map germs, the domain manifold is set to H^3 without loss of generality.

The topology of fibers for *stable maps*, i.e. maps whose fibers stay topologically equivalent after some sufficiently small perturbation, has been classified for certain configurations of $M^n \to \mathbb{R}^{n-1}$. Specifically, Saeki [26] established those for $n = 4$ with orientable manifolds and $n = 3$. We did so for $n = 5$ and non-orientable case of $n = 4$ [36]. The classification was later extended to bounded manifolds for $n = 3$ [29] (see Fig. 3). Locally around a point in the domain, the polynomial expression of a map may be transformed into the canonical form associated with a topological category. If not, the map is unstable, and such topology of a fiber (component) is said to be *degenerate*. Any unstable map can be approximated with a stable map that is close enough in the map space $\mathbb{R}^\infty$.

The classifications for $\kappa = 0$ and 1 in Fig. 3 match those for contour topology. This is because the 1-D fiber $g^{-1} = (g_1^{-1}(v_1), g_2^{-1}(v_2))$ can be embedded in the 2-D surface $g_1^{-1}(v_1)$. As we vary v_1, these 2-D and 1-D manifolds both morph continuously. Consequently, we can see components in $\kappa = 1$ mapped in the range as a curve called a *fold* (Fig. 4). A fold is *definite* iff the component vanishes by some perturbation; otherwise *indefinite*. If a fold is induced by the boundary we call it a *boundary fold*.

Folds may intersect with each other in the range, and this may not be resolved by perturbation (as explained by e.g. Edelsbrunner et al. [10]). For the intersection points, we find the corresponding topology of fiber components, which are in the category of $\kappa = 2$. The points connecting two folds, as found in Fig. 5, are called *cusps*, be they boundary folds or not. The hitting point of a fold and boundary fold is called a B_2 point.

3.2 Invariants

We saw that the topology is captured by the $\mathcal{B}_+$-equivalence. While a stable map is equivalent to all their close-enough neighbors in the map space $\mathbb{R}^\infty$, an unstable map is not so by definition. The map is thus more complex, requiring efforts to understand its characteristics. In the quotient space (Fig. 1), regions of stable maps are split by borders, whose points are unstable ones. The structure of these partitions is thus in the interest of studies on the $\mathcal{B}_+$-equivalence.

In the schematic picture (Fig. 1), the quotient space is shown as the stable maps occupying the rooms while unstable maps the walls. The cells' and walls' dimensionality is measured by their *codimension*. Roughly speaking, the codimension of a map is the dimensionality lacked for its containing wall to become a cell (although the cells and walls are both generally infinite; consult a book s.a. [1] for details). Cells having lower codimension are split by higher ones in-between, much like two 2-D cells are split by a 1-D wall in (Fig. 1). Research questions surround the cells and their codimension, which together contain information of the structure of the quotient space.

In order to assign an invariant value to unstable maps, which are actually located at the walls of the quotient space, the maps are perturbed into stable ones. For this purpose, we have guaranteed that the invariant values are independent of the choice of the perturbation. The benefit of the perturbation is that we can take advantage of the well-studied characteristics of the fiber topology – as with Morse functions, a proper perturbation resolves degeneracy.

Historically, the approach to utilize perturbation for finding invariants is old and new. It is well known that for a holomorphic map germ $g\colon (\mathbb{C}^n, 0) \to (\mathbb{C}, 0)$ defining an isolated singularity at 0, the Milnor number $\mu(g)$ of g which is the number of critical points of a Morse function near f is a topological invariant of f. Consult Milnor [21] for details. Later, Wall [35], Fukuda and Ishikawa [13], Gaffney and Mond [14] and Ohsumi [22] followed the strategy for $g\colon (\mathbb{R}^n, 0) \to (\mathbb{R}, 0)$ and $g\colon (\mathbb{R}^2, 0) \to (\mathbb{R}^2, 0), g\colon (\mathbb{C}^2, 0) \to (\mathbb{C}^2, 0), g\colon (\mathbb{R}^n, 0) \to (\mathbb{R}^{2n-1}, 0)$ respectively. The study closest to ours is that on $g : \mathbb{R}^n \to \mathbb{R}^2$ [27]. The boundary was not considered in any of these studies. On the other hand, the characterization of stable maps on manifolds with boundary is obtained recently. It is natural to investigate map germs defined on manifolds with boundary by utilizing perturbation.

To verify the stability of the perturbed map is a time-consuming task as it involves manual analysis of polynomials. Perturbation is thus another beneficiary of visualization.

Requirement 1. Perturb unstable maps into stable ones.

As demonstrated in Fig. 5, our invariants α, β and γ can be computed from the structure of the Reeb space and the type of the singularities of fibers. Hence, this is a requirement for evaluating the invariant:

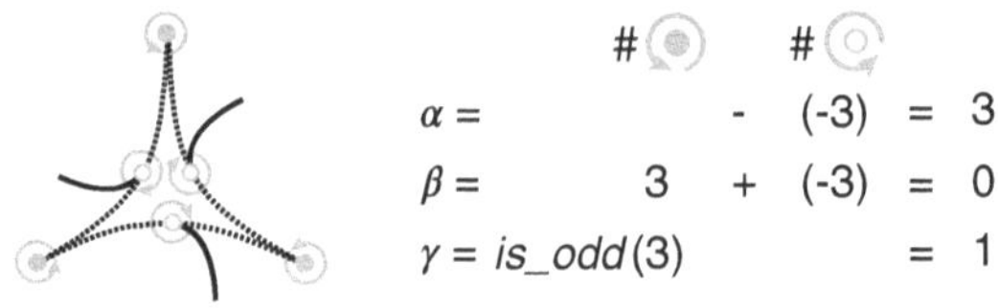

Fig. 5 Given a map germ, our invariants can be computed from its Reeb space after perturbation. In this illustration of a part of a Reeb space, solid curves consist of folds induced by the singular points inside of the domain and the dotted ones by the boundary. The calculation is done by counting boundary cusps (white circles) and B_2 points (gray circles) together with their orientation, which is indicated with arrows

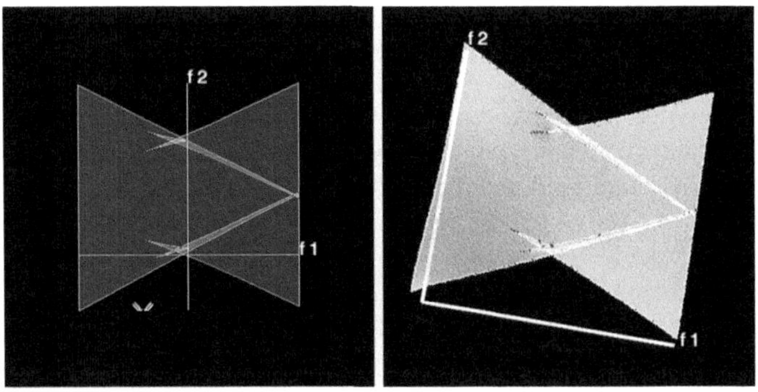

Fig. 6 The configuration of a Reeb space (Fig. 5) visualized in jcnet. Left: The visualized range shows the folds with color-coded identification of singularity types. Right: The configuration of the Reeb space in a 3-D arrangement

Requirement 2. Determine the singularity of fibers.

The two requirements above can be satisfied by our visualization system. While the numerical computation is an approximation, the hints and insights obtained are useful for speculating potential improvements to the invariants.

4 Reeb Space Visualization and Computation

We visualize the Reeb spaces of some example maps and compute the invariant values. As stated, this allows us to evaluate the invariants by exploring maps judged to be similar. Maps distinguished from each other are by the invariant values are, in fact, $\mathcal{B}_+$-inequivalent. In other words, they lie in a different partition in the quotient space.

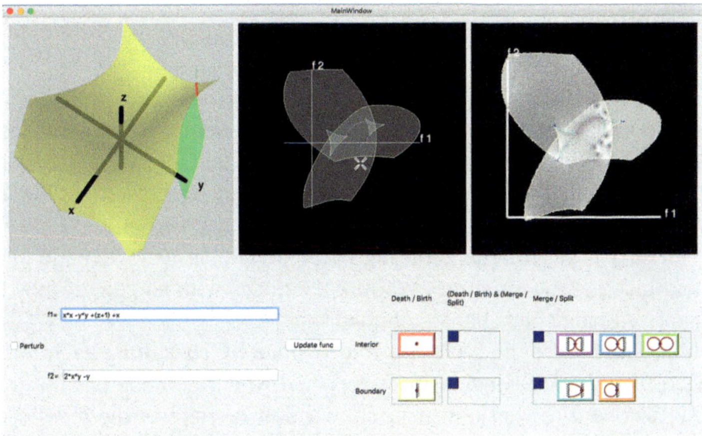

Fig. 7 The user interface of the jcnet program. In addition to the views for the range and Reeb space, the domain view is available. The inverse image of two component functions g_1 and g_2 are rendered as yellow and green surfaces. The fiber is drawn in red

4.1 Visualization

To determine the singularity of fibers (Requirement 1), the folds' configuration like in Fig. 5 is desirable. The system visualizes the configuration in 2-D and 3-D as in Fig. 6. The 2-D visualization is rendered the *range view* while the 3-D one in the *Reeb space view*. The full interface is equipped with an additional view for studying the domain (Fig. 7) and some controls for auxiliary interactions. In case folds in the range overlap, the view for the Reeb space helps seeing the continuity of the singularity of the fibers.

If the structure of the Reeb space becomes too complicated to be investigated in the Reeb space view, we can look at the domain view and manually track the continuity of fiber components. Though this last option may take some time to complete, we have not encountered a map that is infeasible to acquire the invariant values.

For perturbing the map (Requirement 2), the user interface lets us define and change a map g through polynomial expressions, and shows the domain, range and Reeb space together with the types of folds. Perturbation can be performed either by updating the polynomials or by editing the inverse images of component functions. To allow the latter, implicit surface modeling is employed [31].

We must also check whether the degeneracy of the fibers is resolved. For folds, this can be done primarily in the range view since folds with degenerate fiber topology is detected by the system and given special colors. For points where different folds intersect as in Fig. 4, the meeting points shall be manually compared with the classification and decided if it includes degeneracy. Though this observation task is manual, it is a trivial one that finishes instantly.

4.2 Computation

The computation of the JCN is done through the rasterization algorithm [6]. The map is discretized into a tetrahedral grid and interpolated linearly. The tetrahedra is then projected in the range, which is quantized into equidistant intervals $\{[v_i, v_i + \delta_1)\}_i \times \{[w_j, w_j + \delta_2)\}_j$ $(\forall i, j \in \mathbb{N}, v_k, w_k, \delta_k \in \mathbb{R})$. The algorithm rasterizes the tetrahedra in this pixel space. To contract the fiber component mapped in to the pixels in this way, we identify tetrahedra neighboring in the domain as a single node in the JCN, which is in fact a graph structure. We link the nodes with an edge if they neighbor in the quantized range and share a tetrahedron.

The identification of the type of a fold is done by consulting each node in this graph. Here, topological singularity is reported when the node has no or multiple edges [31] to one direction out of up, down, left or right in the pixel space. The JCN is computed on the boundary, too, for the mesh restricted into a triangular one. This is for identifying singularities there, and the two JCNs are integrated to finally determine the type of singularity as found in Fig. 3.

5 Invariants for Bounded Map Germs

From now on, the word "fold" refers exclusively to non-boundary folds unless stated otherwise. The proof of our results in this section, which by itself is out of scope of this article, will be published in a forthcoming paper. See Fig. 8. A map germ $g\colon (H^3, 0) \to (\mathbb{R}^2, 0)$ is called *cone-like* if there exist sufficiently small $\varepsilon > 0$ and $\delta > 0$ such that

1. $(D_\delta^3 \cap \{y \geq 0\}) \cap g^{-1}(S_\varepsilon^1)$ is a manifold with boundary,
2. $g_\partial = g|_{(D_\delta^3 \cap \{y \geq 0\}) \cap g^{-1}(S_\varepsilon^1)}\colon (D_\delta^3 \cap \{z \geq 0\}) \cap g^{-1}(S_\varepsilon^1) \to S_\varepsilon^1$ is a stable map,
3. $g|_{(S_\delta^2 \cap \{z > 0\}) \cap g^{-1}(D_\varepsilon^2)}\colon (S_\delta^2 \cap \{z > 0\}) \cap g^{-1}(D_\varepsilon^2) \to D_\varepsilon^2$ is an immersion, and
4. $g|_{(D_\delta^3 \cap \{z \geq 0\}) \cap g^{-1}(D_\varepsilon^2 \setminus \{0\})}\colon (D_\delta^3 \cap \{z \geq 0\}) \cap g^{-1}(D_\varepsilon^2 \setminus \{0\}) \to D_\varepsilon^2 \setminus \{0\}$ is a proper C^∞ stable map, and it is equivalent to $g_\partial \times \mathrm{id}_{(0,\varepsilon]}\colon (D_\delta^3 \cap \{z \geq 0\}) \cap g^{-1}(S_\varepsilon^1) \times (0, \varepsilon] \to S_\varepsilon^1 \times (0, \varepsilon]$.

For a generic map germ $g'\colon (\mathbb{R}^3, 0) \to (\mathbb{R}^2, 0)$ and a generic upper half space H^3, the map germ $g = g'|_{(H^3,0)}$ is cone-like. Note that the set of nongeneric map germs has infinite codimension [12].

Remark 5.1

A C^∞ map $f\colon M \to Q$ on a 3-manifold with boundary into a surface without boundary is *stable* if and only if it has only folds, cusps, swallowtails, boundary folds, boundary cusps or B_2 singularities. $f|_{S(f) \cup S(f|_\partial)}\colon S(f) \cup S(f|_\partial) \to Q$ may only have multi-germs shown in Fig. 4. Here, fold and cusp singularity (or boundary fold, boundary cusp and B_2 singularity) are a singular point of f (resp. $f|_\partial$) around which f is locally of the form $(x, y, z) \mapsto (x, y^2 \pm z^2)$ and $(x, y, z) \mapsto (x, y^3 + xy + z^2)$

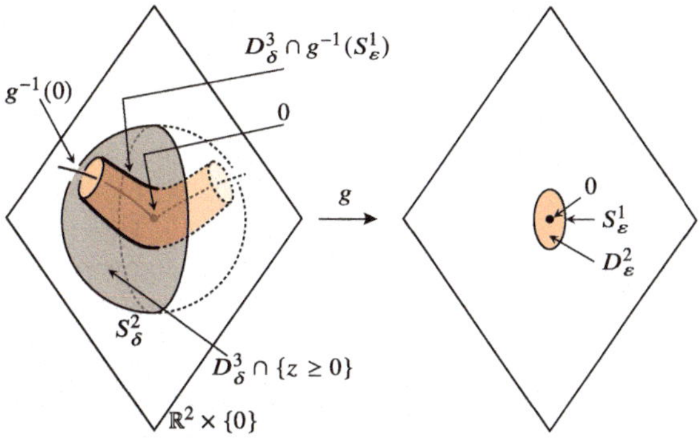

Fig. 8 A cone-like map germ $g : (H^3, 0) \to (\mathbb{R}^2, 0)$. The red cylinder is $g^{-1}(S^1_\varepsilon)$

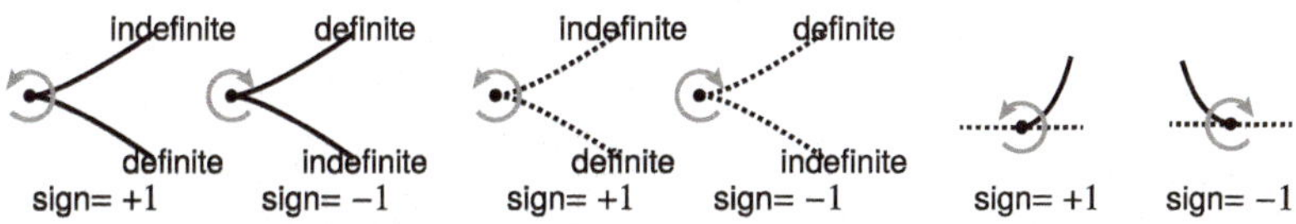

Fig. 9 The signs of cusps, boundary cusps, and B_2 points of f are defined through the orientation of their Reeb space configuration

(resp. $(x, y, z) \mapsto (x, y^2 \pm z), (x, y^3 + xy + z), (x, y^2 \pm z^2 + xz)$). In particular, if the sign is positive (or negative), the singularities are definite. In this case, $S(f)$ and $S(f|_\partial)$ are a closed non-singular 1-dimensional submanifold of M [20, 29]. □

Let p be a cusp, boundary cusp, or B_2 point of a stable map $f : M \to \mathbb{R}^2$ of a 3-manifold with boundary into the plane. We define the sign of p in the following manner. First, we orient $\mathbb{R}^2$ in the counter-clockwise manner. At the image of a cusp point, boundary cusp point or B_2 point p, the coincident curves may be configured only in the ways shown in Fig. 9. If the image around $f(p)$ are in the left hand side of Fig. 9, then the sign of p is defined to be $+1$; otherwise, the sign of p is equal to -1. Then, the algebraic number of cusps (or boundary cusps, B_2) of f is the total of signs of cusps (resp. boundary cusps, B_2) of f.

Finally, we introduce the invariants we have obtained.

Theorem 5.1. Let $g : (H^3, 0) \to (\mathbb{R}^2, 0)$ be a cone-like map germ. Then,

1. the difference of the algebraic number of cusps and the algebraic number of definite B_2 points, denote it α,
2. the total of the algebraic number of boundary cusps and the algebraic number of B_2 points, β,
3. the parity of the number of B_2 points, $\gamma \in \{0, 1\}$

of a stable perturbation $f : U \to \mathbb{R}^2$ of a representative $U \to \mathbb{R}^2$ of g are invariants of the topological $\mathcal{B}_+$-equivalence class of g. □

6 Outcome

We have evaluated the $\mathcal{B}_+$-equivalence with the jcnet. In the system, the domain has the bounding box $\{(x, y, z) \in \mathbb{R}^3| \ |x| = 1 \text{ or } |y| = 1 \text{ or } |z = 1|; 0 \leq x, y, z \leq 1\}$. To study the domain H^3, we choose the surface $z = -1$, and curves in the range introduced by $|x| = 1$, $|y| = 1$ and $z = 1$ were ignored. From the folds rendered in the range, we could determine those from $z = -1$ by looking at its pre-image in the domain view.

6.1 Comparing Forms of Equivalence Through Visual Investigation

We studied the relationship between $\mathcal{B}_+$-equivalence and the $\mathcal{B}$-equivalence. The two are similar concepts, but the $\mathcal{B}$-equivalence (1) additionally requires Φ and ψ to be diffeomorphisms, while (2) does not require ψ to preserve the orientation of the gradient vectors. Martins and Nabarro [20] classified map germs $g\colon (H^3, 0) \to (\mathbb{R}^2, 0)$ of the form $g(x, y, z) = (x, h(x, y, z))$ under $\mathcal{B}$-equivalence. Note that the corank of a map germ $g(x, y, z) = (x, h(x, y, z))$ is 0 or 1.

Table 1 shows that a selection of the classified map germs g_1 and g_2, g_3 (or g_1 and g_2, g_4) are distinguished by the combination of our invariants (α, β, γ), but g_3 and g_4 are not distinguished by the combination of our invariants (α, β, γ). The invariant values were computed from the visualization results in Fig. 10, 11, 12 and 13. To summarize, from the visualization we have learned that the $\mathcal{B}_+$-equivalence distinguishes a part of these categories (with a potential error due to numerical imprecision).

The selection corresponds to the map germs of codimension 1 found in the study by Martins and Nabarro. The codimension 1, among different codimensions, possesses the largest amount of the unstable maps.

On the other hand, for a map germ $g\colon (H^3, 0) \to (\mathbb{R}^2, 0)$ obtained by them, a versal unfolding of g induces a satisfactory perturbation $f\colon U \to \mathbb{R}^2$ of a representative $U \to \mathbb{R}^2$ of g.

Then, we can check the (α, β, γ)-value of g calculated from f coincides with those calculated with help of the jcnet for map germs in Table 1. Indeed, it shows that map germs g_1 and g_2, g_3, g_4 are not all $\mathcal{B}_+$ equivalent, suggesting a similarity between $\mathcal{B}$ and $\mathcal{B}_+$ with examples.

6.2 Germs of Corank 2

Figure 14 shows another example $g_5(x, y, z) = (x^2 - y^2 + z, 2xy)$. The germ was chosen since we were able to obtain the singularities manually under the restriction

Table 1 The combination of the invariants α, β and γ distinguish three map germs out of four of those with codimension 1 in the classification for the $\mathcal{B}$-equivalence.

$g_i(x, y, z)$	Perturbation ($t \neq 0$)	α	β	γ
$g_1(x, y, z) = (x, y^2 + z^3 + xz)$	$(x, y^2 + z^3 + xz + tz^2)$	-1	1	1
$g_2(x, y, z) = (x, y^2 - z^3 + xz)$	$(x, y^2 + z^3 + xz + tz^2)$	0	1	1
$g_3(x, y, z) = (x, yz + xy + y^3)$	$(x, yz + zy + y^3 + tz)$	0	0	1
$g_4(x, y, z) = (x, yz + xy - y^3)$	$(x, yz + zy + y^3 + tz)$	0	0	1

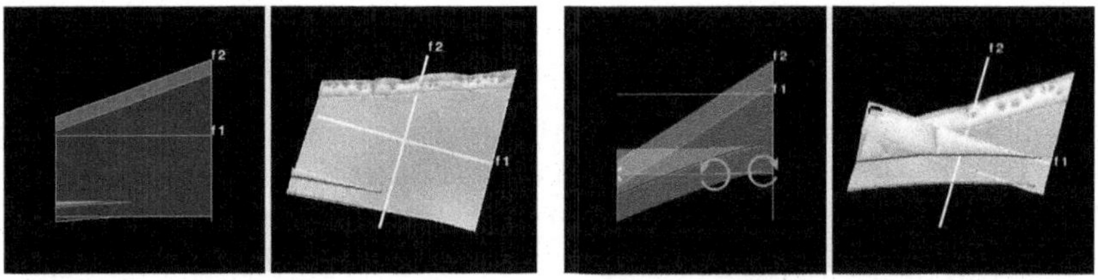

Fig. 10 Left: Germ $g_1(x, y, z) = (x, y^2 + z^3 + xz)$. Right: Perturbation of g_1

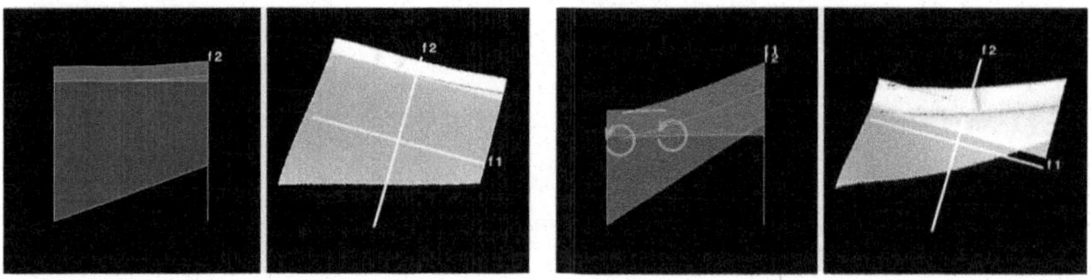

Fig. 11 Left: Germ $g_2(x, y, z) = (x, y^2 - z^3 + xz)$. Right: Perturbation of g_2

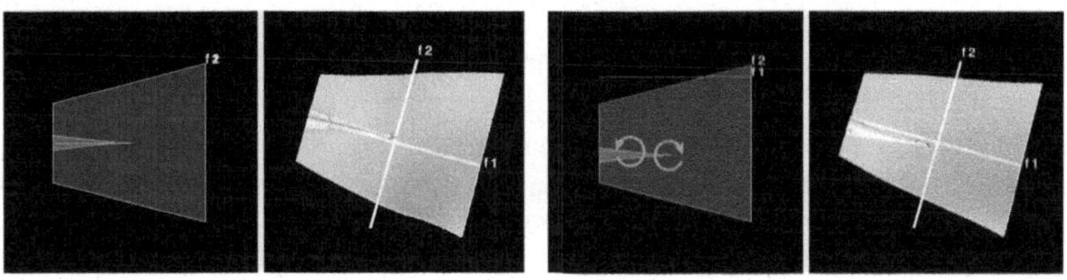

Fig. 12 Left: Germ $g_3(x, y, z) = (x, yz + zy + y^3)$. Right: Perturbation of g_3

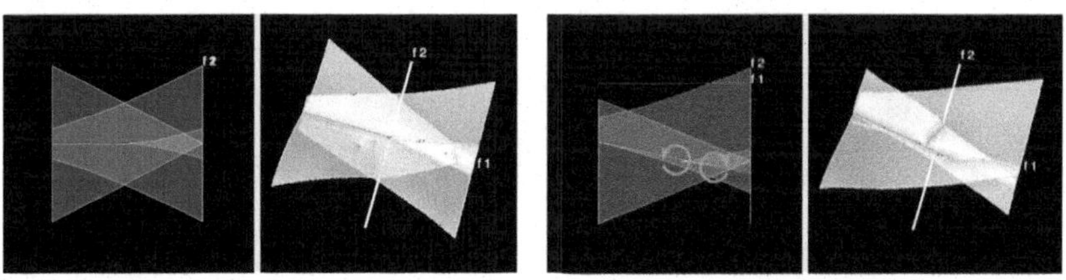

Fig. 13 Left: Germ $g_4(x, y, z) = (x, yz + zy - y^3)$. Right: Perturbation of g_4

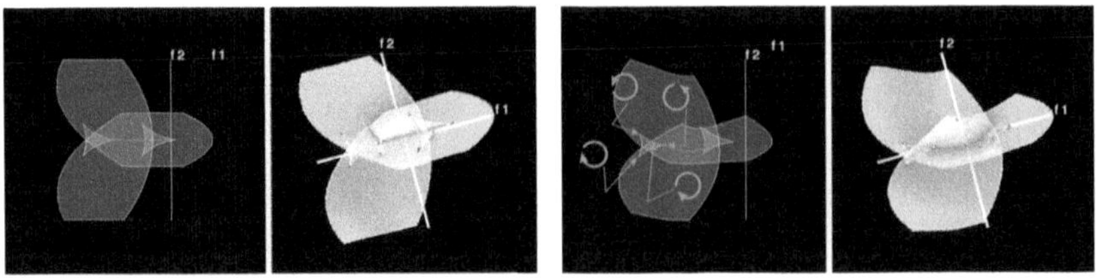

Fig. 14 Germ f obtained by perturbing $g_5(x, y, z) = (x^2 - y^2 + z, 2xy)$. Left: g_5 perturbed into $(x^2 - y^2 + z + x, 2xy - y)$ through equational transforms. Right: perturbed further with mouse interaction

of $z = 0$. It is especially difficult to evaluate (α, β, γ) for this germ manually since the corank of the map germ is 2, which indicates that the first component of the 2-D range cannot be simplified into x. From the visualization, we learned that the resolved germ has no cusp point, but three boundary cusp points and one B_2 point. This gave us $(\alpha, \beta, \gamma) = (0, 0, 1)$.

7 Discussion

We combined two perturbation schemes: equation input and mouse input (as introduced in [31]). Equational perturbation was used for rough perturbation via explicit alteration of perturbation parameters. We tended to find singularities that are not classified in Fig. 4, i.e. the perturbation leaves unresolved degeneracy. For detailed perturbation with immediate visual feedback, we used mouse interaction (Fig. 7) to resolve the degeneracies entirely.

Potentially, errors may arise when quantizing the field during the rasterization stage in the JCN computation. Thanks to the experiments, we are comfortable about the numerical precision of the results regarding our fields. Due to the simplicity of the polynomial expressions, we believe that the examples do not contain singularities so small that they are not captured during the quantization stage and hence invisible in the visualization.

Even though the numerical imprecision rules out our visualization as a mathematical proof on the correctness of the invariant values, the hints we get are valuable given the rapidity of acquisition and the amount of thus obtainable examples.

8 Conclusion

In this article, we investigated the $B+$-equivalence by visualizing Reeb spaces. We approximated the equivalence with three invariants, α, β and γ. We acquired a few canonical forms that are inequivalent to each other both in the B and B_+-equivalence.

We were also able to treat map germs of corank 2, whose treatment would be significantly more challenging than those of corank 0 and 1 without the Reeb space visualization.

Acknowledgment This work was supported by JSPS KAKENHI (Grant Number JP20K19809) and IMI Joint Use Research Program Workshop (II) 20200011 "Fiber Topology Meets Applications."

References

1. Arnold, V.I., Goryunov, V.V., Lyashko, O.V., Vasil'ev, V.A.: Singularities Theory I. Springer, Heidelberg (1998). https://doi.org/10.1007/978-3-642-58009-3
2. Bachthaler, S., Weiskopf, D.: Continuous scatterplots. IEEE Trans. Visual Comput. Graphics **14**(6), 1428–1435 (2008)
3. Bajaj, C.L., Pascucci, V., Schikore, D.R.: The contour spectrum. In: Proceedings of IEEE Visualization 1997, pp. 167–173. IEEE (1997)
4. Biasotti, S., Giorgi, D., Spagnuolo, M., Falcidieno, B.: Reeb graphs for shape analysis and applications. Theoret. Comput. Sci. **392**(1–3), 5–22 (2008)
5. Bremer, P.T., et al.: Topological feature extraction and tracking. J. Phys. Conf. Ser. **78**, 012007 (2007)
6. Carr, H., Duke, D.: Joint contour nets. IEEE Trans. Visual Comput. Graphics **20**(8), 1100–1113 (2014)
7. Carr, H., Geng, Z., Tierny, J., Chattopadhyay, A., Knoll, A.: Fiber surfaces: generalizing isosurfaces to bivariate data. Comput. Graph. Forum **34**(3), 241–250 (2015)
8. Carr, H., Snoeyink, J., Axen, U.: Computing contour trees in all dimensions. Comput. Geom. **24**(2), 75–94 (2003)
9. Duke, D., Carr, H., Knoll, A., Schunck, N., Nam, H.A., Staszczak, A.: Visualizing nuclear scission through a multifield extension of topological analysis. IEEE Trans. Visual Comput. Graphics **18**(12), 2033–2040 (2012)
10. Edelsbrunner, H., Harer, J., Mascarenhas, A., Pascucci, V., Snoeyink, J.: Time-varying Reeb graphs for continuous space-time data. Comput. Geom. **41**(3), 149–166 (2008)
11. Edelsbrunner, H., Harer, J., Patel, A.K.: Reeb spaces of piecewise linear mappings. In: Proceedings of the 24th Annual Symposium on Computational Geometry, pp. 242–250 (2008)
12. Fukuda, T.: Local topological properties of differentiable mappings II. Tokyo J. Math. **8**(2), 501–520 (1985)
13. Fukuda, T., Ishikawa, G.: On the number of cusps of stable perturbations of a plane-to-plane singularity. Tokyo J. Math. **10**(2), 375–384 (1987)
14. Gaffney, T., Mond, D.: Cusps and double folds of germs of analytic maps $\mathbb{C}^2 \to \mathbb{C}^2$. J. Lond. Math. Soc. **43**(2), 185–192 (1991)
15. Gueunet, C., Fortin, P., Jomier, J., Tierny, J.: Task-based augmented contour trees with Fibonacci heaps. IEEE Trans. Parallel Distrib. Syst. **30**(8), 1889–1905 (2019)
16. Hiratuka, J.T., Saeki, O.: Triangulating Stein factorizations of generic maps and Euler characteristic formulas. RIMS Kokyuroku Bessatsu **B38**, 61–89 (2013)
17. Huettenberger, L., Heine, C., Carr, H., Scheuermann, G., Garth, C.: Towards multifield scalar topology based on Pareto optimality. Comput. Graph. Forum **32**(3pt3), 341–350 (2013)
18. Kettner, L., Rossignac, J., Snoeyink, J.: The Safari interface for visualizing time-dependent volume data using Iso-surfaces and contour spectra. Comput. Geom. **25**(1), 97–116 (2003)
19. Lehmann, D.J., Theisel, H.: Discontinuities in continuous scatter plots. IEEE Trans. Visual Comput. Graphics **16**(6), 1291–1300 (2010)
20. Martins, L.F., Nabarro, A.C.: Projections of hypersurfaces in $\mathbb{R}^4$ with boundary to planes. Glasgow Math. J. **56**(1), 149–167 (2014)

21. Milnor, J.W.: Singular Points of Complex Hypersurfaces. Princeton University Press, Princeton (1968)
22. Ohsumi, M.: Whitney's umbrellas in stable perturbations of a map germ $(\mathbb{R}^n, 0) \to (\mathbb{R}^{2n-1}, 0)$. Tokyo J. Math **29**(2), 475–493 (2006)
23. Pascucci, V., Cole-McLaughlin, K.: Efficient computation of the topology of level sets. In: Proceedings of IEEE Visualization 2002, pp. 187–194. IEEE Computer Society Press (2002)
24. Pascucci, V., Scorzelli, G., Bremer, P.T., Mascarenhas, A.: Robust on-line computation of Reeb graphs: simplicity and speed. ACM Trans. Graph. **26**(3), 58 (2007)
25. Reeb, G.: Sur les points singuliers d'une forme de pfaff complètement inté grable ou d'une fonction numérique. Comptes Rendus de l'Acadèmie des Sciences de Paris **222**, 847–849 (1946)
26. Saeki, O.: Topology of Singular Fibers of Differentiable Maps. Lecture Notes in Mathematics, vol. 1854. Springer, Heidelberg (2004). https://doi.org/10.1007/b100393
27. Saeki, O.: Cobordism of Morse functions on surfaces, the universal complex of singular fibers and their application to map germs. Algebraic Geom. Topol. **6**(2), 539–572 (2006)
28. Saeki, O., et al.: Visualizing multivariate data using singularity theory. In: Wakayama, M., et al. (eds.) The Impact of Applications on Mathematics, Proceedings of the Forum of Mathematics for Industry 2013, pp. 51–65. Springer, Tokyo (2014). https://doi.org/10.1007/978-4-431-54907-9_4
29. Saeki, O., Yamamoto, T.: Singular fibers of stable maps of 3-manifolds with boundary into surfaces and their applications. Algebr. Geom. Topol. **16**(1379–1402) (2016)
30. Sakurai, D., Ono, K., Carr, H., Nonaka, J., Kawanabe, T.: Flexible Fiber Surface: A Reeb-Free Approach, pp. 187–201. Springer, Cham (2019). https://doi.org/10.1007/978-3-030-43036-8_12
31. Sakurai, D., Saeki, O., Carr, H., Wu, H.Y., Yamamoto, T., Duke, D., Takahashi, S.: Interactive visualization for singular fibers of functions $f : R^3 \to R^2$. IEEE Trans. Visual Comput. Graphics **22**(1), 945–954 (2016)
32. Takahashi, S., Takeshima, Y., Fujishiro, I.: Topological volume skeletonization and its application to transfer function design. Graph. Models **66**(1), 24–49 (2004)
33. Tierny, J., Carr, H.: Jacobi fiber surfaces for bivariate Reeb space computation. IEEE Trans. Visual Comput. Graphics **23**(1), 960–969 (2017)
34. Tierny, J., Favelier, G., Levine, J.A., Gueunet, C., Michaux, M.: The topology ToolKit. IEEE Trans. Visual Comput. Graphics **24**(1), 832–842 (2017)
35. Wall, C.T.C.: Topological invariance of the Milnor number mod 2. Topology **22**, 345–350 (1983)
36. Yamamoto, T.: Euler number formulas in terms of singular fibers of stable maps. In: Real and Complex Singularities Proceeding of Australian-Japanese Workshop, pp. 345–350. World Scientific Publishing (2007)

Topological Feature Search in Time-Varying Multifield Data

Tripti Agarwal, Amit Chattopadhyay, and Vijay Natarajan

Abstract A wide range of data that appear in scientific experiments and simulations are multivariate or multifield in nature, consisting of multiple scalar fields. Topological feature search of such data aims to reveal important properties useful to the domain scientists. It has been shown in recent works that a single scalar field is insufficient to capture many important topological features in the data, instead one needs to consider topological relationships between multiple scalar fields. In the current paper, we propose a novel method of finding similarity between two multifield data by comparing their respective fiber component distributions. Given a time-varying multifield data, the method computes a metric plot for each pair of histograms at consecutive time stamps to understand the topological changes in the data over time. We validate the method using real and synthetic data. The effectiveness of the proposed method is shown by its ability to capture important topological features that are not always possible to detect using the individual component scalar fields.

Keywords Multifield topology · Features · Fiber-component · Distribution · Comparison measure · Time-varying

T. Agarwal · A. Chattopadhyay (✉)
International Institute of Information Technology, Bangalore, India
e-mail: a.chattopadhyay@iiitb.ac.in

T. Agarwal
e-mail: tripti.agarwal@iiitb.org

V. Natarajan
Indian Institute of Science, Bangalore, India
e-mail: vijayn@iisc.ac.in

I. Hotz et al. (eds.), *Topological Methods in Data Analysis and Visualization VI*,
Mathematics and Visualization, https://doi.org/10.1007/978-3-030-83500-2_11

1 Introduction

Scientists understand different physical phenomena by studying the interrelationships between features in different fields. It has been observed and shown that such multifield or multivariate data can reveal many important phenomena about an experiment that are impossible to study using a single scalar field data [10, 13]. Development of tools and techniques for extracting and visualizing features in multifield data is an important topic of research interest [18]. Topology-based methods have been shown to be extremely effective in this context. During the previous two decades, topological analysis of shapes and data was mostly driven by scalar topology, using contour tree, Reeb graph, Morse-Smale complex and their variants [6]. Such techniques have also been extended for time-varying scalar field data by defining different topology-aware similarity measures between two scalar fields [4, 29, 33].

Generalization of the techniques to time-varying multifield data is challenging and requires further development in both theory and computational methods. More recently, new tools have been proposed for understanding and visualizing multifield data—Reeb space [16], Jacobi set [7, 14, 15], Joint Contour Net [9, 13] and Pareto analysis [21]. Extending these methods to time-varying multifield data requires the development of techniques for comparative analysis and visualization. For example, developing a comparative measure between two Reeb spaces is a challenging open problem. In this paper, we consider a simpler feature descriptor of a multifield, namely its fiber-component distribution or histogram. Using this, we take the first step towards a topology-aware distance measure between two multifields in terms of the distance between their fiber-component distributions. Our contribution in the current paper is as follows:

- We introduce simple topology-aware distance measures between two multifields based on their fiber-component distributions or histograms in the range space. We prove the metric properties of the proposed distance measures.
- We show that the proposed measures capture significant or interesting events in time-varying phenomena, not possible using a study of individual fields. We validate the method by experimenting on a time-varying synthetic data where topological features are known in advance.
- We show effectiveness of our method by experimenting on previously studied nuclear-scission data [13] and re-explain how scission events are captured. We also apply our method in capturing important feature in the orbital data of Pt-CO interaction.

Section 2 discusses related works on scalar and multifield data analysis. Section 3 describes different data structures or representations used for understanding and visualizing multifield data. Section 4 introduces our proposed topology-aware distance measures and describes important properties of the measure. Section 5 discusses the implementation details and Sect. 6 and Sect. 7 describe various results of experiments on synthetic and real data. The experiments are conducted on nuclear scission, fission, and molecular orbital density data of Pt-CO interaction. Finally, Sect. 8 presents conclusions and lists some limitations of the method.

2 Related Work

Feature extraction in time-varying data is a well studied topic and several approaches have been proposed. We describe a few relevant approaches here.

Various similarity measures between scalar fields have been studied to analyze repeating patterns and similar arrangements in the data. Hilaga et al. studied topological shape matching using a multiresolution Reeb Graph (MRG) [20]. Saikia et al. propose a method for finding repeating topological structure in a scalar data using a data structure called the extended branch decomposition graph (eBDG) [33]. In a following paper [34] the authors describe a histogram feature descriptor to compare subtrees of merge trees against each other. Narayanan et al. define a distance measure between extremum graphs to compare two scalar fields [29].

Many other comparison measures have been proposed in the literature for finding the distance between graphs or topological data structures. Bauer et al. have proposed a functional distortion metric on Reeb Graph and show its stability properties [4]. A survey on graph edit distance by Gao et al. [17] discusses different inexact graph matching algorithms for the application in pattern analysis. Sridharamurthy et al. propose an edit distance between merge trees for feature visualization in time-varying scalar data [37]. Thomas et al. propose a multiscale symmetry detection technique in scalar fields using contour clustering and studying the similarity between them [38]. In related works, different distance metrics between the merge trees have been proposed to provide a similarity between the corresponding scalar fields [5, 28].

Other techniques that are not based on topological analysis have also been proposed in the literature for tracking and visualizing time-varying features. Ji et al. [22] proposed a global optimization algorithm for time-varying data and resolved the problems of volume overlapping and aggregate-attribute criteria by using the earth mover's distance. A branch-and-bound approach was used for the global cost evaluation. The resultant approach and the metric was able to track features accurately and efficiently. Lee et al. [26] proposed a time activity curve (TAC) to visualize time-varying features.

However, topological feature search in time-varying multifield data is a comparatively new area of research and only few works can be found in the literature. Duke et al. [13] propose a joint contour net (JCN) based visualization technique for detecting nuclear scission feature in the time-varying multifield density data. It has been observed that direct visualization of the topological features using JCNs does not scale to large data sizes because the JCN structure can be extremely complicated. In this paper, our method replaces this JCN visualization technique by a histogram comparison method.

3 Background

In this section, we discuss a few tools and techniques from the literature that are required to describe our proposed distance measure.

3.1 *Histogram and Isosurface Statistics, Continuous Scatter Plot*

A histogram visualizes the distribution of the samples of a scalar field using a bar graph that is constructed by binning the samples in the field range. Histograms provide a measure of importance of isovalues based on the statistics of sample points. Carr et al. [8] show that histograms represent the spatial distribution of scalar fields with a nearest neighbourhood interpolation. Moreover, they show that isosurface statistics, such as the area of isosurfaces [3], betters represent the distribution of a scalar field.

Bivariate histograms represent two fields together. These histograms consist of bins of possibly different shapes such as square, triangle or hexagonal [35]. Square shaped bins of the histogram consist of the count for each pair of values defined on the axes. This count can be used to calculate the variance and bias from the integrated mean square error by using appropriate formulae. The square bins can be stretched to a rectangular shape based on the scale defined on the axes.

The density function corresponding to a collection of continuous input fields is well represented by a continuous scatter plot. Unlike histograms, continuous scatter plots do not depend on the bin sizes. Bachthaler et al. [2] describe a mathematical model for generic continuous scatter plots of maps from n-D spatial domain to m-D data domain. Lehmann et al. [27] describe algorithm for detecting discontinuities in the continuous scatter plots that reveal important topological features in the data.

3.2 *Multifield Topology and Jacobi Set*

A multifield on a d-manifold $\mathbb{M}\ (\subseteq \mathbb{R}^d)$ with r component scalar fields $f_i : \mathbb{M} \rightarrow \mathbb{R}$ $(i = 1, \ldots, r)$ is a *map* $\mathbf{f} = (f_1, f_2, \ldots, f_r) : \mathbb{M} \rightarrow \mathbb{R}^r$. In differential topology, $\mathbf{f}$ is considered to be a *smooth map* when all its partial derivatives of any order are continuous. A point $\mathbf{x} \in \mathbb{M}$ is called a *singular point* (or *critical point*) of $\mathbf{f}$ if the rank of its differential map $\mathbf{df_x}$ is strictly less than $\min\{d, r\}$ where $\mathbf{df_x}$ is the $r \times d$ Jacobian matrix whose rows are the gradients of f_1 to f_r at $\mathbf{x}$. And the corresponding value $\mathbf{f}(\mathbf{x}) = \mathbf{c} = (c_1, c_2, \ldots, c_r)$ in $\mathbb{R}^r$ is a *singular value* . Otherwise if the rank of the differential map $\mathbf{df_x}$ is $\min\{d, r\}$ then $\mathbf{x}$ is called a *regular point* and a point $\mathbf{y} \in \mathbb{R}^r$ is a *regular value* if $\mathbf{f}^{-1}(\mathbf{y})$ does not contain a singular point.

The inverse image of the map $\mathbf{f}$ corresponding to a value $\mathbf{c} \in \mathbb{R}^r$, $\mathbf{f}^{-1}(\mathbf{c})$ is called a *fiber* and each connected component of the fiber is called a *fiber-component*

[31, 32]. In particular, for a scalar field these are known as the *level set* and the *contour*, respectively. The inverse image of a singular value is called a *singular fiber* and the inverse image of a regular value is called a *regular fiber*. If a fiber-component passes through a singular point, it is called a *singular fiber-component*. Otherwise, it is known as a *regular fiber-component*. Note that a singular fiber may contain a regular fiber-component. Topology of a multifield data is usually studied based on its fiber-topology [12].

Jacobi set is used to study topological relationship between two or multiple scalar fields. Jacobi set $\mathbb{J}_{\mathbf{f}}$ of a multifield $\mathbf{f}$ is the closure of the set of all its singular points, i.e. $\mathbb{J}_{\mathbf{f}} = cl\,\{\mathbf{x} \in \mathbb{M} : \text{rank } \mathbf{df_x} < \min\{d, r\}\}$. Alternatively, the Jacobi set is the set of critical points of one component field (say f_i) of $\mathbf{f}$ restricted to the intersection of the level sets of the remaining component fields [14]. Intuitively, Jacobi set of two generic Morse functions $f_1, f_2 : \mathbb{M} \to \mathbb{R}$ is the set of points where gradients of the individual fields are parallel, i.e. $\mathbb{J} = \{\mathbf{x} \in \mathbb{M} : \nabla f_1(\mathbf{x}) \times \nabla f_2(\mathbf{x}) = \mathbf{0}\}$. Jacobi set plays a central role in the design of a comparison measure between two or multiple scalar fields [15].

3.3 Reeb Space and Joint Contour Net

Similar to the Reeb graph of a scalar field, the Reeb space parameterizes the fiber-components of a multifield and its topology is described by the standard quotient space topology [16]. A Jacobi structure has been defined as a projection of the Jacobi set on the Reeb space, by the quotient map [12]. Figure 1c illustrates a Reeb space with Jacobi structure (in red) corresponding to a bivariate field.

Joint Contour Net (JCN) [9] gives a practical algorithm for approximating a Reeb space. A JCN is built in four stages. The first step of the JCN algorithm constructs all the *contour fragments* in each cell of the entire mesh corresponding to a quantization of each component field. In the second step, the *joint contour fragments* are computed by computing the intersections of these contour fragments for the component fields in a cell. The third step is to construct an adjacency graph (dual graph) of these joint contour fragments where a node in the graph corresponds to a joint contour fragment and there is an edge between two nodes if the corresponding joint contour fragments are adjacent. Finally, the JCN is obtained by collapsing the neighbouring redundant nodes with identical isovalues. Thus, each node in the JCN corresponds to a *joint contour slab* or quantized fiber-component and an edge represents the adjacency between two quantized fiber-components. We use the JCN implementation for computing the quantized fiber-components and its histogram, see Fig. 1d.

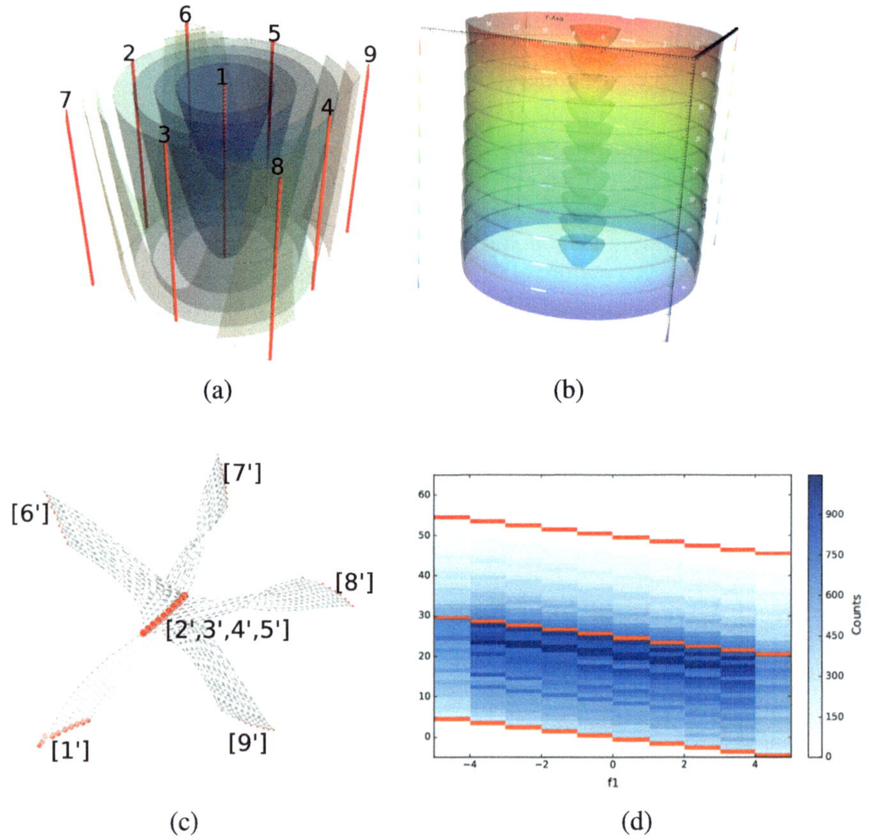

Fig. 1 Figure shows a bivariate synthetic data and corresponding structures to understand its topology. **a** Paraboloid and height field with Jacobi set (red), total 9 connected components of the Jacobi set are numbered as 1 to 9 **b** Singular fiber-components that pass through the Jacobi set points, **c** Reeb space (JCN) with Jacobi structure (in red). Jacobi structure components that are the projection of the Jacobi set components on the Reeb space are shown by the corresponding dashed numbers. **d** Histogram with singular values (bins)

3.4 Histogram Distance Measures

Different measures have been proposed in the literature to study the distance between two histograms [30]. The measures may be classified into two types based on how they are computed—bin-to-bin measures or cross-bin measures. In the former type, bins with the same indices are compared. We list below, a few examples of measures for finding distance between two histograms H and K with bin count h_i and k_i respectively.

Minkowski-form Distance:

$$d_{L_r}(H, K) = \left(\sum_i |h_i - k_i|^r\right)^{1/r} \tag{1}$$

Commonly used Minkowski-form distances are d_{L_1}, d_{L_2} and d_{L_∞}. These are often used to compute dissimilarity between two color images.

Histogram Intersection:

$$d_{\cap}(H, K) = 1 - \frac{\sum_i \min(h_i, k_i)}{\sum_i k_i} \tag{2}$$

This distance can capture the partial matches when the areas of the two histograms are not equal.

Kullback-Leibler (KL) Divergence:

$$d_{KL}(H, K) = \sum_i h_i \log \frac{h_i}{k_i} \tag{3}$$

This is designed from an information-theoretic viewpoint. The measure is non-symmetric and sensitive to histogram binning.

One example of a cross-bin dissimilarity measure is the

Quadratic-form Distance:

$$d_A(H, K) = \sqrt{(\mathbf{h} - \mathbf{k})^T \mathbf{A}(\mathbf{h} - \mathbf{k})}, \tag{4}$$

where $\mathbf{h}$ and $\mathbf{k}$ are vector representations of H and K, respectively. The matrix $\mathbf{A} = [a_{ij}]$ is the similarity matrix where a_{ij} denote the similarity between the i-th bin of H with the j-th bin of K [30].

4 Our Method

Let us consider two continuous multifields $\mathbf{f} = (X_1, X_2, \ldots, X_r)$ and $\mathbf{g} = (Y_1, Y_2, \ldots, Y_r)$ over a d-dimensional compact domain $\mathbb{D} \subseteq \mathbb{R}^d$ where each of X_i and Y_i, $(i = 1, 2, \ldots, r)$ are real-valued scalar fields in the domain We consider comparing multifields $\mathbf{f}$ and $\mathbf{g}$ at two consecutive time steps of a time-varying multifield data where topological features vary continuously over time. A fiber of the multifield $\mathbf{f}$ corresponding to a parametric point $\mathbf{c} = (c_1, c_2, \ldots, c_r)$ is the preimage $\mathbf{f}^{-1}(\mathbf{c}) = X_1^{-1}(c_1) \cap X_2^{-1}(c_2) \cap \ldots \cap X_r^{-1}(c_r)$. A connected component of the fiber is called a fiber-component. Fiber-component topology is used to study multifield topology, similar to the use of contour topology for scalar field studies. The Reeb space is a generalization of the Reeb graph. It captures the fiber-component topology corresponding to a multifield. However, Reeb space structure is rather complicated and computing an effective distance measure between two Reeb spaces for comparing corresponding multifields is an open problem.

In the current work, we consider the change in fiber-component distribution over parametric space to capture the change in topology in two multifields. We observe that a change in the number of fiber-components that correspond to a point in the parametric space implies a change (birth or death) in number of sheets of the Reeb space, as described in Sect. 4.2. Therefore, to study the topological changes from $\mathbf{f}$ to $\mathbf{g}$ we first consider the fiber-component distributions as the feature-descriptors of the respective multifields. Next, we propose few simple distance measures between the fiber-component distributions to capture the difference in terms of topological features.

4.1 Fiber-Component Distribution over the Range Space

Let $\mathbf{f} = (X_1, X_2, \ldots, X_r)$ be a continuous multifield from a d-dimensional compact domain $\mathbb{D} \subseteq \mathbb{R}^d$ to the r-dimensional range space $R_\mathbf{f} = [a_1, b_1] \times [a_2, b_2] \times \ldots \times [a_r, b_r]$, $a_i, b_i \in \mathbb{R}$. Define the function $N : R_\mathbf{f} \to \mathbb{N}$ as $N(\mathbf{x}) = |\mathbf{f}^{-1}(\mathbf{x})|$ for $\mathbf{x} \in R_\mathbf{f}$, where $|\mathbf{f}^{-1}(\mathbf{x})|$ represents the number of connected components in the fiber $\mathbf{f}^{-1}(\mathbf{x})$. In other words, $N(\mathbf{x})$ maps each point $\mathbf{x}$ of $R_\mathbf{f}$ to the corresponding number of fiber-components of $\mathbf{f}$. We assume that N is a bounded function for multifields $\mathbf{f}$ defined over a compact domain $\mathbb{D}$. To compute the total number of fiber-components, we partition the range $R_\mathbf{f}$ into a union of m^r sub-boxes by introducing the partitions of the intervals: $a_i = x_0^{(i)} < x_1^{(i)} < \ldots < x_m^{(i)} = b_i$ for $i = 1, 2, \ldots, r$. Let $\mathbf{x}_{i_1 i_2 \ldots i_r}$ be a point in the sub-box $B_{i_1 i_2 \ldots i_r} = [x_{i_1-1}^{(1)}, x_{i_1}^{(1)}] \times [x_{i_2-1}^{(2)}, x_{i_2}^{(2)}] \times \ldots \times [x_{i_r-1}^{(r)}, x_{i_r}^{(r)}]$ for $i_1, i_2, \ldots, i_r = 1, 2, \ldots, m$ with volume $\Delta V_{i_1 i_2 \ldots i_r}$. Then, $\mathbf{N}$, defined as the sum of number of fiber-components over all points in $R_\mathbf{f}$ is equal to

$$\mathbf{N} = \lim_{\text{all } \Delta V_{i_1 i_2 \ldots i_r} \to 0} \sum_{i_1, i_2, \ldots, i_r = 1}^{m} N(\mathbf{x}_{i_1 i_2 \ldots i_r}) \Delta V_{i_1 i_2 \ldots i_r} = \int_{R_\mathbf{f}} N(\mathbf{x}) d\mathbf{x}. \tag{5}$$

The function N is bounded and hence integrable. Next, we define a density function of the fiber-component distribution as:

$$\mathfrak{p}_\mathbf{f}(\mathbf{x}) = \frac{N(\mathbf{x})}{\mathbf{N}} \text{ for } \mathbf{x} \in R_\mathbf{f}, \tag{6}$$

where

$$\int_{R_\mathbf{f}} \mathfrak{p}_\mathbf{f}(\mathbf{x}) d\mathbf{x} = 1.$$

In practice, to compute the fiber-component distribution over the range space, we first discretize the continuous multifield $\mathbf{f} = (X_1, X_2, \ldots, X_r)$ in the r-dimensional range space. Let field X_i be discretized (quantized) uniformly at the values $x_1^{(i)} < x_2^{(i)} < \ldots < x_{m_i}^{(i)}$ for $i = 1, 2, \ldots, r$. We denote this discrete range space as spec $(R_{\mathbf{f}}) = I_1 \times I_2 \times \ldots \times I_r$, the Cartesian product of $I_i = \{x_1^{(i)}, x_2^{(i)}, \ldots, x_{m_i}^{(i)}\}$ $(i = 1, 2, \ldots, r)$. Then we compute the frequency distribution of the corresponding fiber-components over this discrete range space (spectrum). The probability mass function of the corresponding discrete probability distribution is given by

$$p_{\mathbf{f}}(\mathbf{x}) = \frac{\tilde{N}_{\mathbf{x}}}{\tilde{\mathbf{N}}}, \text{ where } \mathbf{x} \in \text{spec}(R_{\mathbf{f}}). \tag{7}$$

Here, $\tilde{N}_{\mathbf{x}}$ counts the number of fiber-components at the parametric point $\mathbf{x} = (x_{i_1}^{(1)}, x_{i_2}^{(2)}, \ldots, x_{i_r}^{(r)})$ in spec$(R_{\mathbf{f}})$ (for $i_1 = 1, 2, \ldots, m_1$; $i_2 = 1, 2, \ldots, m_2$; ...; $i_r = 1, 2, \ldots, m_r$) and $\tilde{\mathbf{N}}$ is the sum of number of fiber-components of $\mathbf{f}$ over all points in the discrete range space spec$(R_{\mathbf{f}})$. Note that $p_{\mathbf{f}}$ defines a probability mass function (p.m.f.) since $p_{\mathbf{f}}(\mathbf{x}) \geq 0$ and

$$\sum_{\mathbf{x} \in \text{spec}(R_{\mathbf{f}})} p_{\mathbf{f}}(\mathbf{x}) = 1.$$

We note, for a piecewise-linear multifield on a triangulated domain, when the quantization level goes to infinity then the corresponding sequence of JCNs converges to the actual Reeb space [12]. Therefore, the discrete distribution in (7) converges to the continuous distribution in (6). Alternatively, one can define p.m.f. using $A_{\mathbf{x}}$ by measuring the size of the quantized fiber-components at the parametric point $\mathbf{x} \in$ spec$(R_{\mathbf{f}})$ and A is the total measure of all the fiber-components over spec$(R_{\mathbf{f}})$. Thus we have

$$p_{\mathbf{f}}(\mathbf{x}) = \frac{A_{\mathbf{x}}}{A}, \text{ where } \mathbf{x} \in \text{spec}(R_{\mathbf{f}}). \tag{8}$$

In the proposed distance measure that we will describe next, we consider the definitions in (6) and (7) because they capture the topological changes in the fibers of the multifield.

4.2 Distance Between Two Fiber-Component Distributions

Let us consider two multifields $\mathbf{f}_1 = (X_1, X_2, \ldots, X_r)$ and $\mathbf{f}_2 = (Y_1, Y_2, \ldots, Y_r)$ over the domain $\mathbb{D} \subseteq \mathbb{R}^d$. Let $R_{\mathbf{f}_1}$ and $R_{\mathbf{f}_2}$ be the range spaces of $\mathbf{f}_1$ and $\mathbf{f}_2$, respectively. We note that the range spaces $R_{\mathbf{f}_1}$ and $R_{\mathbf{f}_2}$ may be different but restrict our attention to the case when they are almost equal. To define our distance measures between the fiber-component distributions of $\mathbf{f}_1$ and $\mathbf{f}_2$, first we extend the range spaces $R_{\mathbf{f}_1}$ and $R_{\mathbf{f}_2}$ to an equal range R. We define R as: $R = R_1 \times R_2 \times \ldots \times R_r$ where $R_i =$

range $X_i \cup$ range Y_i for $i = 1, 2, \ldots, r$. This extended range R is considered as the common domain of fiber-component distributions of both $\mathbf{f}_1$ and $\mathbf{f}_2$. The fiber-component distributions of $\mathbf{f}_1$ on the part $R \setminus R_{\mathbf{f}_1}$, corresponding to which $\mathbf{f}_1$ has no data, is filled with zeros. Similarly fiber-component distributions of $\mathbf{f}_2$ on $R \setminus R_{\mathbf{f}_2}$ is filled with zeros.

For the continuous case: let $\mathfrak{p}_{\mathbf{f}_1}$ and $\mathfrak{p}_{\mathbf{f}_2}$ be the density functions of the fiber-component distributions of $\mathbf{f}_1$ and $\mathbf{f}_2$, respectively, over the extended range R. Let $\mathbf{P}_1$ and $\mathbf{P}_2$ be the corresponding distribution functions. Then we define a point-wise distance measure between $\mathbf{P}_1$ and $\mathbf{P}_2$ as:

$$d_q(\mathbf{P}_1, \mathbf{P}_2) = \left(\int_R |\mathfrak{p}_{\mathbf{f}_1}(\mathbf{x}) - \mathfrak{p}_{\mathbf{f}_2}(\mathbf{x})|^q d\mathbf{x} \right)^{1/q} \tag{9}$$

for any real number $q \geq 1$. In particular for $q = 1, q = 2$ or $q = \infty$ we get similar distance measures of practical importance.

For the discrete case, let the range space R be discretized (quantized) as $\text{spec}(R) = I_1 \times I_2 \times \ldots \times I_r$ where $I_i = \{x_1^{(i)}, x_2^{(i)}, \ldots, x_{m_i}^{(i)}\}$. Let $\mathbf{P}_1 = \{p_{\mathbf{x}}^{(1)} : \mathbf{x} \in \text{spec}(R)\}$ and $\mathbf{P}_2 = \{p_{\mathbf{x}}^{(2)} : \mathbf{x} \in \text{spec}(R)\}$ be the fiber-component distributions of $\mathbf{f}_1$ and $\mathbf{f}_2$, respectively, over the discrete range space $\text{spec}(R)$. Then we define the point-wise distance measure between the distributions $\mathbf{P}_1$ and $\mathbf{P}_2$ as:

$$d_q(\mathbf{P}_1, \mathbf{P}_2) = \left(\sum_{\mathbf{x} \in \text{spec}(R)} |p_{\mathbf{x}}^{(1)} - p_{\mathbf{x}}^{(2)}|^q \right)^{1/q}. \tag{10}$$

for any real number $q \geq 1$. In particular, for $q = 1, q = 2$ and $q = \infty$ we have

$$d_1(\mathbf{P}_1, \mathbf{P}_2) = \sum_{\mathbf{x} \in \text{spec}(R)} |p_{\mathbf{x}}^{(1)} - p_{\mathbf{x}}^{(2)}| \tag{11}$$

$$d_2(\mathbf{P}_1, \mathbf{P}_2) = \left(\sum_{\mathbf{x} \in \text{spec}(R)} |p_{\mathbf{x}}^{(1)} - p_{\mathbf{x}}^{(2)}|^2 \right)^{1/2} \tag{12}$$

and

$$d_\infty(\mathbf{P}_1, \mathbf{P}_2) = \sup_{\mathbf{x} \in \text{spec}(R)} |p_{\mathbf{x}}^{(1)} - p_{\mathbf{x}}^{(2)}|. \tag{13}$$

These distance measures are motivated from the observation that the point-wise difference $|\tilde{N}_{\mathbf{x}}^{(1)} - \tilde{N}_{\mathbf{x}}^{(2)}|$ captures the number of changes in fiber-components between two multifields at consecutive time steps for $\mathbf{x} \in \text{spec}(R)$. Note that each fiber-component of a multifield corresponds to exactly one sheet of its Reeb space. So, the difference in number of fiber-components captures the number of possible changes in Reeb space sheets containing the parameter value $\mathbf{x}$. Thus, $|\tilde{N}_{\mathbf{x}}^{(1)} - \tilde{N}_{\mathbf{x}}^{(2)}|$ captures the number of births or deaths of sheets containing the parameter value $\mathbf{x}$ of the corresponding Reeb spaces.

4.3 Weighted Distance for the Singular Values

Singular fibers capture the topological changes in the evolution of fibers in a multifield. The image of a singular fiber in the parametric space is called a singular value. Because of importance of the singular values compare to regular values, we propose a variant to the distance measure that weights the singular values differently,

$$d_q^{\mathbb{S}}(\mathbf{P}_1, \mathbf{P}_2; \omega) = \left[\omega \sum_{\mathbf{x} \in \mathbb{S}} |p_{\mathbf{x}}^{(1)} - p_{\mathbf{x}}^{(2)}|^q + \sum_{\mathbf{x} \notin \mathbb{S}} |p_{\mathbf{x}}^{(1)} - p_{\mathbf{x}}^{(2)}|^q \right]^{1/q}. \tag{14}$$

Here, $\mathbb{S}$ is the set of singular values in the discrete range space spec(R) and $q \geq 1$. Moreover, $\omega > 1$ is the weight parameter to impose more importance to the singular values than the regular values. We observe from our experiments on different datasets that increasing the weight ω increases the prominence of the events that correspond to topological changes when we plot weighted distances over time. Figure 1d shows a fiber-component histogram with the singular values (in red) corresponding to the bivariate field in Fig. 1a.

4.4 Metric Space Properties of the Distance Measures

It is important to show that the proposed distance measures between two distributions satisfy the metric space properties for the space $\mathscr{P}_R$ of all possible fiber-component distributions corresponding to different multifields with range R. Let us first show that $(\mathscr{P}_R, d_q)$ is a metric space.

1. **Non-negativity.** Note d_q is real-valued, finite and non-negative.
2. **Identity.** We note that for two distributions $\mathbf{P}_1, \mathbf{P}_2 \in \mathscr{P}_R$, $d_q(\mathbf{P}_1, \mathbf{P}_2) = 0$ if and only if $\mathbf{P}_1 = \mathbf{P}_2$, since $\sum_{\mathbf{x} \in \text{spec}(R)} |p_{\mathbf{x}}^{(1)} - p_{\mathbf{x}}^{(2)}|^q = 0$ implies $p_{\mathbf{x}}^{(1)} = p_{\mathbf{x}}^{(2)}$ for all $\mathbf{x} \in$ spec(R).
3. **Symmetry.** It is straight-forward to show that $d_q(\mathbf{P}_1, \mathbf{P}_2) = d_q(\mathbf{P}_2, \mathbf{P}_1)$. This implies the symmetry property of d_q.
4. **Triangle inequality.** To show the triangle inequality of d_q we consider three fiber-component distributions $\mathbf{P}_1$, $\mathbf{P}_2$ and $\mathbf{P}_3$. Note, for $q = 1$, $|p_{\mathbf{x}}^{(1)} - p_{\mathbf{x}}^{(3)}| \leq |p_{\mathbf{x}}^{(1)} - p_{\mathbf{x}}^{(2)}| + |p_{\mathbf{x}}^{(2)} - p_{\mathbf{x}}^{(3)}|$. For $q \geq 1$, using Minkowski inequality [19] we can show that $d_q(\mathbf{P}_1, \mathbf{P}_3) \leq d_q(\mathbf{P}_1, \mathbf{P}_2) + d_q(\mathbf{P}_2, \mathbf{P}_3)$.

Similar properties can be proved for the other distance measures $d_q^{\mathbb{S}}$, d_1, d_2 and d_∞. However, note the above metric properties hold in the space of fiber-component distributions, not necessarily in the space of actual multifields.

5 Implementation

We implement the distance measures described in the previous section using Visualization Toolkit (VTK) [24] under the Joint Contour Net [9] implementation framework. The implementation works for a generic pair for multifields but is particularly designed for time-varying multifields. We note that the range spaces of two multifields at two consecutive time steps are not necessarily the same and may vary slightly. We expand the range of both multifields by considering their component wise union and use zero-padding to compute the histogram as described in Sect. 4.2. Next, we describe the four main steps of our implementation.

I. **Computing Fiber-Components:** First, we discretize or quantize the common range of the multifields into finite numbers of bins. Then corresponding to each bin-value, we compute the quantized fiber-components as described in the JCN algorithm [9]. In other words, compute the contour slabs in each cell for each of the scalar fields and then find intersection of the slabs to get the fragments. Finally an adjacency graph is computed from the fragments to obtain quantized fiber-components. Each quantized fiber-component corresponds to a node of the JCN.

II. **Computing Fiber-Component Histograms:** Next, we compute the r-dimensional fiber-component histogram corresponding to each multifield on the range space. We use the same binning as used for the quantized fiber-component computation. Each bin in the range is populated with the corresponding fiber-components. We compute the number of fiber-components in each bin for the fiber-component histogram computation. A color map specifying the number of all the nodes is shown in Fig. 1d. The color map is chosen over a range of blue values. Light blue shows fewer number of nodes (fiber-components), and as the color darkens the number of nodes (fiber-components) also increases.

III. **Computing Singular Values of Multifields:** To compute singular values first one needs to compute the singular points or the Jacobi set in the domain of the multifield and then the corresponding range values of those points are actually the singular values. In the current implementation we first compute the Jacobi structure using a multi-dimensional Reeb graph (MDRG) as described in [11, 12] and then project them in the histogram-bins and call those bins as singular bins. We note that a singular bin of the histogram may contain both singular and regular fiber-components (nodes). In the histogram plot Fig. 1d, the red colored bins indicate the singular bins and blue are the regular bins. For the singular bins of the histogram the singular, regular and total nodes (singular and regular together) are stored separately for further computation.

IV. **Computing Distance Metrics between Histograms:** The above three steps are performed for multifields at all the time stamps or sites, and the corresponding histograms are stored in different files. A python script is then implemented to compute the corresponding probability density from the histogram. Then the distance metrics between two probability densities at the consecutive time steps are computed as in Sects. 4.2 and 4.3. The distance metric $d_q^{\mathbb{S}}$ (as in Eq. 14) is computed for different values of q and ω. This metric is computed using the singular and regular nodes. Note that if $q = 1$ and $\omega = 1$ the metric $d_q^{\mathbb{S}}$ is same as d_1. To validate the experiment d_1 is calculated using all the nodes (regular and singular nodes together). Along with the measures that we have proposed we even calculated the distance measures for the already defined metrics for histogram comparison as defined in Sect. 3.4. The values for these distance metrics are stored and then used to create a comparison line plot. The values were also used to check the metric properties defined in Sect. 4.4. We also calculated the simple root mean square distance for bivariate data for experimental comparison.

6 Applications

We now describe applications of the proposed comparison driven feature search method to four different datasets, namely (i) a synthetic data consisting of two polynomial functions, (ii) the scission data of plutonium atom, (iii) fission data of Fermium atom and (iv) the DFT data of carbon monoxide and platinum (CO-Pt) molecular bond.

6.1 Synthetic Data

We generate a synthetic bivariate field whose components are the height field $f_1(x, y, z) = z$ and the paraboloid field $f_2(x, y, z) = x^2 + y^2 - z$. Both fields are defined on an axis-aligned box $[-5.5, 4.5] \times [-5.5, 4.5] \times [-5.5, 4.5]$ and sampled on a grid of size $20 \times 20 \times 20$. Next, we generate a sequence of multifield data by incrementally translating the domain-box along each of the three axes with small magnitude 0.05, i.e. if (C_x, C_y, C_z) and (c_x, c_y, c_z) are respectively the coordinates of a point on the box before and after the translation, then $C_x = c_x + 0.05,\ C_x = c_y + 0.05,\ C_z = c_z + 0.05$. In total, we create 21 bivariate datasets. To create the consecutive datasets, we begin with the domain $[-5.5, 4.5] \times [-5.5, 4.5] \times [-5.5, 4.5]$ and then apply the above described sequence of translations 21 times until we obtain the domain of the final dataset, namely $[-4.5, 5.5] \times [-4.5, 5.5] \times [-4.5, 5.5]$. The major topological feature is expected in the dataset corresponding to the domain $[-5, 5] \times [-5, 5] \times [-5, 5]$ (which is symmetric about origin) because of degenerate intersections of the fiber-components with the boundary of the box.

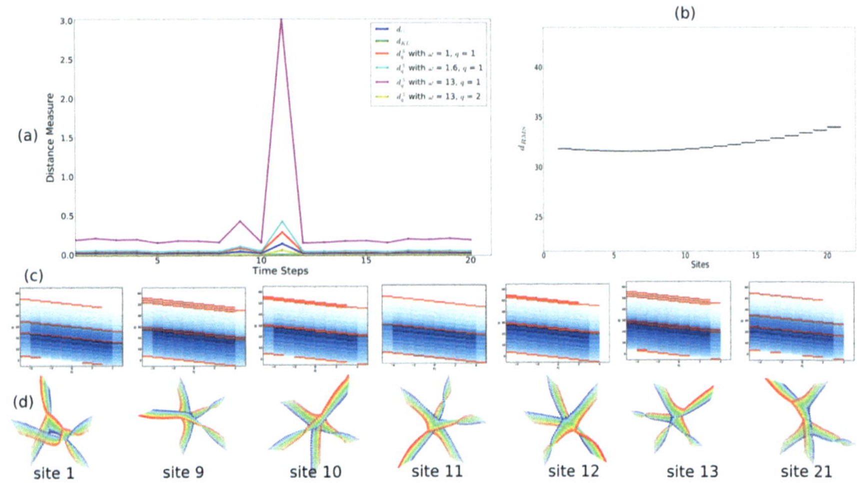

Fig. 2 Plots of distance measures between consecutive sites in a series of bivariate (height, paraboloid) fields. **a** Various distance measures show a peak at site 11, indicating a topological change. The proposed metric $d_q^{\mathbb{S}}$ also exhibits a peak, more significant than other distance measures. **b** Root-mean-square plot is not able to capture the topological change. This indicates the need for a topological data structures for multifield data that captures topological changes. **c** Fiber-component distributions for selected sites. Singular values are highlighted in red. Blue nodes indicate regular nodes and the shades of blue indicate the number of nodes in a particular bin (light indicates low). **d** Corresponding Reeb spaces. The height field is mapped to color (blue is low and red is high)

Observations and Results

We compute the fiber-component histograms for each dataset in the series and plot the distance between two consecutive datasets, see Fig. 2. The distance peaks at site 11 as expected. The red color in the histograms indicates singular nodes and blue color indicates regular nodes. The number of regular nodes in a particular bin is mapped to different shades of blue. Colors in the Reeb space indicate the height field value. Although various distance measures are able to capture the topological change, the peak was not sharp enough. The peak is most prominent using the $d_q^{\mathbb{S}}$ metric and increased weight for singular nodes. Note that all the subsequent experiments are done with $\omega = 13$ in order to keep the consistency in our experiments for all the datasets. If the value of ω is increased better peaks can be obtained and the value is not dependent on the chosen dataset.

Comparison with the Root Mean Squares Metric

To show the usefulness of the proposed metrics, we compute the distance between two multifields by directly extending the root mean square metric. The root mean square distance between two multifields $\mathbf{f} = (f_1, \ldots, f_r)$ and $\mathbf{g} = (g_1, \ldots, g_r)$ can be generalized as the square root of the mean of the sum of the difference between consecutive component fields:

$$d_{RMS} = \sqrt{\frac{1}{m}\sum_{i=1}^{m}\{(f_1(x_i) - g_1(x_i))^2 + \cdots + (f_r(x_i) - g_r(x_i))^2\}}.$$

Here m is the number of data points in the domain. Figure 2(b) shows the root mean square distance metric plot. We observe that the rms metric is not capable of capturing the topological change. This further motivates the study of measures such as the one proposed in this paper for comparing multifield data.

6.2 Plutonium Atom Dataset

Nuclear Density Functional Theory (DFT) is an approach to understand the nuclear fission occurring in a nucleon-nucleon interaction in atomic nuclei. Nuclear fission is a process by which an atom's nucleus splits into two or more fragments. The splitting of the nucleus can be identified as stretching the core, hence it involves some deformation. This deformation can be a crucial indicator of the topology of the atom's nucleus. An important problem in nuclear fission study is the accurate identification of points in a continuous high dimensional manifold where the core is split. The time when the atom breaks into multiple fragments is known as nuclear scission. At this time the topology of the atom changes in terms of the number of components. Physicists typically identify this phenomenon via tedious manual process. Previous works have described a visual approach to identification of scission [13]. However, these methods require the inspection of the geometry of the Reeb space for all time steps. Further, the Reeb space is a complex structure that is difficult to examine. We aim to detect the key time steps that correspond to topological changes by plotting a graph of the distance measure over time.

The dataset consists of nuclear densities of plutonium atom which represents the internal structure of a heavy nucleus. The dataset is a multifield data consisting of spatial density of proton, the spatial density of neutrons and spatial density of nucleons (protons + neutrons) in the nucleus. These densities, represented as **p**, **n** and **t** are sampled on a $40 \times 40 \times 66$ grid. The dataset available to us is a negative log transformed sample at 14 different time steps, namely [665, 670, 675, 680, 686, 687, 688, 689, 690, 692, 693, 694, 695, 699]. The time step where the nuclear scission occurs is reported in earlier work [13] and confirmed by physicists. We use sufficiently small slab width to capture the topological change. We use the following parameters in our experiments: **p** (slab width 8) and **n** (slab width 2), **p** (slab width 8) and **t** (slab width 2), **n** (slab width 2) and **t** (slab width 2).

Observations and Results

We experiment with all combination of proton, neutrons and nucleon density considering two fields at a time. The plots in Fig. 3 show the distance measure for the first combination, **p** (slab width 8) and **n** (slab width 2). We observe a sudden change between time steps 690 and 692. The d_1 distance was typically in the range of 0.0 to

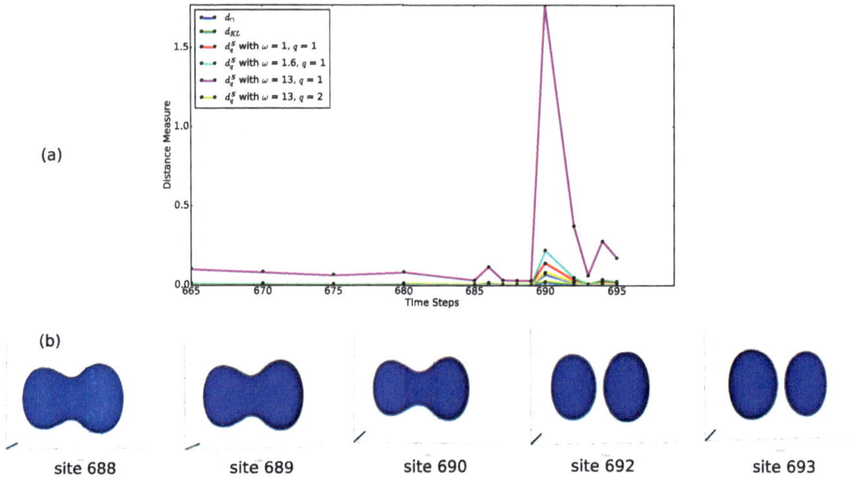

Fig. 3 Plots of the distance measures for the scission data for the plutonium atom. **a** Distance measure between fields at consecutive time steps vs. the time step in the range [665–699]. The proposed distance measure $d_q^{\mathbb{S}}$ exhibits a prominent peak between time step 690–692, which indicates a significant change. **b** Geometry of the plutonium atom at various time steps. The point of scission is between site 690–692 and can be seen in the geometry

0.02, but at nuclear scission, the measure increases to 0.1. This is due to the change in the number of quantized fiber-components in the range space. After scission, the distance measure dropped down to small values because the number of fiber-components does not change after the split. Figure 3(a) shows a comparison with other bin-to-bin measures that are also able to capture the topology change but the peak is not as prominent. We plot the measure $d_q^{\mathbb{S}}$ for different values of q and weights. As the weight for singular values is increased, the peak becomes more prominent and as q is increased the plot becomes smoother. Figure 3 shows the highest peak in the plot using weight $\omega = 13$ (for singular bins) and $q = 1$.

6.3 *Fermium Atom Dataset*

We experiment with another scission dataset, namely that of the Fermium-258 atom. In this dataset, our goal is again to find the point where nuclear scission occurs. As described in the literature [13], this dataset consists of three different types of data viz. aEF: asymmetric elongated fission, sCF: symmetric compact fission and sEF: symmetric elongated fission. The dataset that was made available is the sCF data and was sufficient to detect the topological change where the fermium nucleus scission happens symmetrically. The sCF dataset consists of three fields i.e. proton density (**p**), neutron density (**n**) and total density (**t**) defined on a $19 \times 19 \times 19$ sized grid. The field is available at 56 regularly spaced time steps. Time steps 20–40 were chosen

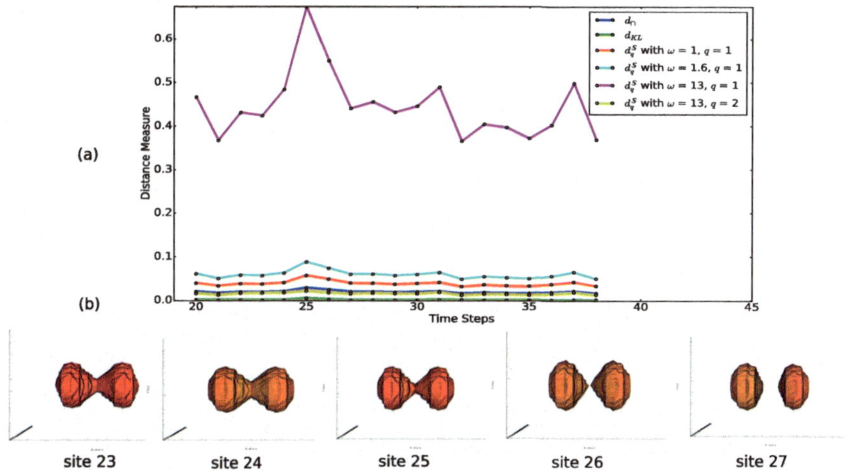

Fig. 4 Plots of the distance measures for the scission data for the fermium-258 atom. **a** Distance measure between fields at consecutive time steps vs. the time step in the range [20, 39]. The proposed distance measure $d_q^{\mathbb{S}}$ exhibits a prominent peak at time step 26, which indicates a significant change. **b** Geometry of the fermium-258 atom at various time steps. The point of scission is at site 26 and can be seen in the geometry

for analysis. Choosing the slab width was still an issue, and we end up working with the same slab width as that for Plutonium atom data, namely **p** (slab width 8) and **n** (slab width 2), **p** (slab width 8) and **t** (slab width 2), **n** (slab width 2) and **t** (slab width 2).

Observations and Results
The same set of experiments were done using the fermium-258 atom dataset. Figure 4 shows the plots with proton and neutron density data from time step 20 to 39. We observe a topological change at time step 26. Other bin-to-bin histogram metrics, e.g. the KL divergence and the histogram intersection, exhibit a much smaller peak as compared to the proposed $d_q^{\mathbb{S}}$ distance.

6.4 *Chemistry Data: Pt-CO Bond*

Adsorption of gas molecules on metal surfaces has various applications including heterogeneous catalysis, electrochemistry, corrosion, and molecular electronics [23, 36]. Particularly, the adsorption of the CO molecule on platinum surfaces has attracted attention of a wide scientific community, due to its role in the areas of automobile emission, fuel cells and other catalytic processes [1, 25]. Therefore, an atomic-level understanding of the CO molecule interacting with the Pt surface is of utmost importance. In this study, we have considered seven Pt atoms representing a platinum surface which interacts with a CO molecule. As the CO molecule approaches towards

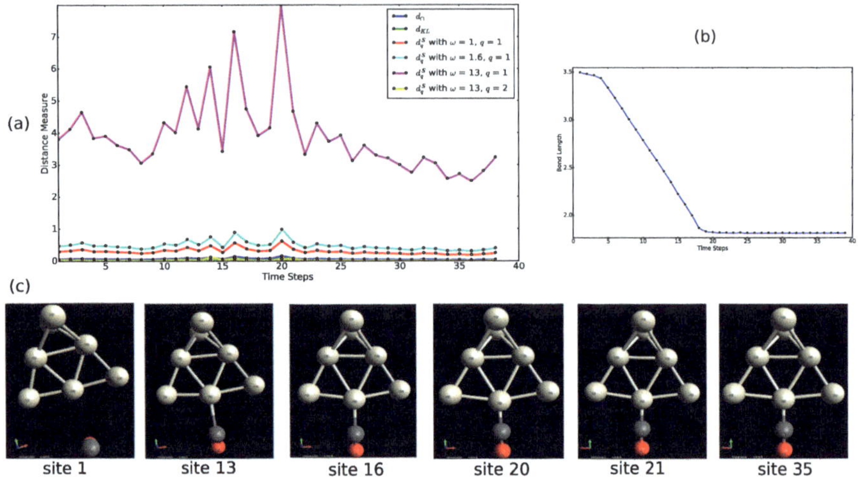

Fig. 5 Plots of the distance measures for the orbital density data of Pt-CO bond at different time steps. **a** Distance measure between fields at consecutive time steps vs. the time step in the range [0, 39]. The plots are for two field values, HOMO and LUMO and the highest peak is obtained at time stamp 21. The proposed distance measure $d_q^{\mathbb{S}}$ exhibits a prominent peak, which indicates a significant change. **b** Pt-CO Bond length vs time. Bond length stabilizes at time step 21. **c** Geometry of the Pt-CO bond creation at various time steps, visualized using the tool Avogadro. Although the bond is visible at time step 13, the bond length is not stable at this site

one of the Pt atoms, the CO bond starts weakening, and Pt-CO bond formation takes place. Quantum mechanical computations were used to generate the electron density distribution corresponding to the highest occupied molecule orbital (HOMO), lowest occupied molecular orbital (LUMO) and HOMO−1. The electron density distribution was computed for varying distance between the carbon atom of the CO molecule and the Pt atom. The Pt-CO bond forms when the distance between the Pt atom and the CO molecule becomes $\sim 1.83A$. This Pt-CO dataset consists of orbital density for orbital numbers 69, 70 and 71. Orbital number 70 corresponds to HOMO, orbital number 71 to LUMO and orbital number 69 to HOMO−1.

Observations and Results

Figure 5 shows different plots for the Pt-CO dataset. At site 21, we get the most stable bond length between Pt and CO molecule. We observe that although the bond is formed at site 13 (as validated by the geometry), the bond-length is not stable. The bond length stabilizes at site 21 and does not change much later. We observe a sharp peak in the plot of the proposed $d_q^{\mathbb{S}}$ distance. This peak corresponds to the formation of the stable bond.

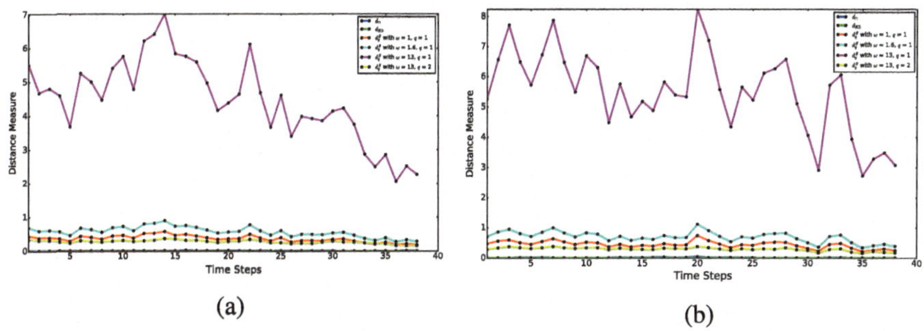

Fig. 6 Distance plot for scalar data for Pt-CO bond detection dataset. **a** Plot for orbital density 69 (HOMO−1). The highest peak is at site 16. **b** Plot for orbital density 70 (HOMO). Significant peak is at site 21

7 Single Scalar Field vs. Multifield

We now describe an experiment to demonstrate the importance of studying tools for multifield data over single scalar field analysis tools. Consider the Pt-CO molecular dataset. Using only orbital 69 (HOMO-1) data the highest peak in the distance measure plot is obtained at site 16 (Fig. 6. Distance plots for orbital 70 (HOMO) exhibit the highest peak at site 21. On the other hand, using two fields together, i.e. orbital data 69 and 70, or orbital data 70 and 71, or orbital data 69 and 71, we observe the highest peak is always at site 21. Some topological changes may not be captured using a bivariate data and we may need to consider more than two fields to detect the changes.

8 Conclusions and Future Work

We propose the use of fiber-component distribution as a topological feature-descriptor for multifield data. We describe a novel method for extracting topological features from time-varying multifield data based on a distance measure defined between fiber-component distributions. This method is simple and a first step towards the development of a more accurate topological comparison measure between two Reeb spaces. We show effectiveness of our method by applying it on several datasets, both synthetic and real data. While the method captures important changes, it flags a few unimportant ones also. For example, in the plot for the Pt-Co data, we observe additional peaks. Such false positives are a key drawback of the current method. To overcome such issues in future we want to explore distance measures between two Reeb spaces. Overall, the proposed distance measures can be used to quickly identify interesting time-steps and intervals. The Reeb space could be studied in a subsequent step for detailed analysis. The distance measure can also be computed for sub-domains thereby allowing for finer grained analysis.

Acknowledgements We would like to thank all authors of the paper [13] for sharing the nuclear scission data. We also thank Prof. Brijesh Kumar Mishra, IIIT-Bangalore for his expert help in understanding the Pt-CO interaction data. This work is partially supported by the Science and Engineering Research Board, India (SERB/CRG/2018/000702), IIIT-Bangalore, a Swarnajayanti Fellowship from the Department of Science and Technology, India (DST/SJF/ETA-02/2015-16), a Mindtree Chair research grant, and the Robert Bosch Centre for Cyber Physical Systems, Indian Institute of Science, Bangalore.

References

1. Gasteiger, A.H., Markovic, N., Ross, P.: H2 and CO electrooxidation on well-characterized Pt, Ru, and Pt-Ru. 2. Rotating disk electrode studies of CO/H2 mixtures at 62.degree.C. J. Phys. Chem. **99** (1995). https://doi.org/10.1021/j100045a042
2. Bachthaler, S., Weiskopf, D.: Continuous scatterplots. IEEE Trans. Vis. Comput. Graph. **14**(6), 1428–1435 (2008)
3. Bajaj, C.L., Pascucci, V., Schikore, D.R.: The contour spectrum. In: Proceedings of the 8th Conference on Visualization 1997, VIS 1997, pp. 167-ff. IEEE Computer Society Press, Los Alamitos (1997). http://dl.acm.org/citation.cfm?id=266989.267051
4. Bauer, U., Ge, X., Wang, Y.: Measuring distance between reeb graphs. In: Proceedings of the Thirtieth Annual Symposium on Computational Geometry, p. 464. ACM (2014)
5. Beketayev, K., Yeliussizov, D., Morozov, D., Weber, G.H., Hamann, B.: Measuring the distance between merge trees. In: Topological Methods in Data Analysis and Visualization III, pp. 151–165. Springer (2014)
6. Biasotti, S., et al.: Describing shapes by geometrical-topological properties of real functions. ACM Comput. Surv. **40**, 12:1–12:87 (2008)
7. Bremer, P.T., Bringa, E.M., Duchaineau, M., Laney, D., Mascarenhas, A., Pascucci, V.: Topological feature extraction and tracking. J. Phys. Conf. Ser. **78**, 012007 (2007)
8. Carr, H., Duffy, B., Denby, B.: On histograms and isosurface statistics. IEEE Trans. Vis. Comput. Graph. **12**(5), 1259–1266 (2006). https://doi.org/10.1109/TVCG.2006.168
9. Carr, H., Duke, D.: Joint contour nets. IEEE Trans. Vis. Comput. Graph. **20**(8), 1100–1113 (2014). https://doi.org/10.1109/TVCG.2013.269
10. Carr, H., Geng, Z., Tierny, J., Chattopadhyay, A., Knoll, A.: Fiber surfaces: generalizing isosurfaces to bivariate data. Comput. Graph. Forum **34**(3), 241–250 (2015)
11. Chattopadhyay, A., Carr, H., Duke, D., Geng, Z.: Extracting Jacobi structures in reeb spaces. In: Elmqvist, N., Hlawitschka, M., Kennedy, J. (eds.) EuroVis - Short Papers, pp. 1–4. The Eurographics Association (2014). https://doi.org/10.2312/eurovisshort.20141156
12. Chattopadhyay, A., Carr, H., Duke, D., Geng, Z., Saeki, O.: Multivariate topology simplification. Comput. Geom. Theory Appl. **58**, 1–24 (2016)
13. Duke, D., Carr, H., Schunck, N., Nam, H.A., Staszczak, A.: Visualizing nuclear scission through a multifield extension of topological analysis. IEEE Trans. Vis. Comput. Graph. **18**(12), 2033–2040 (2012)
14. Edelsbrunner, H., Harer, J.: Jacobi sets of multiple morse functions. In: Foundations of Computational Matematics, pp. 37–57. Cambridge University Press, Minneapolis (2004)
15. Edelsbrunner, H., Harer, J., Natarajan, V., Pascucci, V.: Local and global comparison of continuous functions. In: Proceedings of the conference on Visualization, pp. 275–280 (2004)
16. Edelsbrunner, H., Harer, J., Patel, A.K.: Reeb spaces of piecewise linear mappings. In: SoCG, pp. 242–250 (2008)
17. Gao, X., Xiao, B., Tao, D., Li, X.: A survey of graph edit distance. Pattern Anal. Appl. **13**(1), 113–129 (2010)
18. Hansen, C., Chen, M., Johnson, C., Kaufman, A., Hagen, H. (eds.): Scientific Visualization. Mathematics and Visualization. Springer, London (2014)

19. Hardy, G.H., Littlewood, J.E., Pólya, G.: Inequalities, 2nd edn. Cambridge University Press, Cambridge (1952). ISBN 0-521-35880-9
20. Hilaga, M., Shinagawa, Y., Kohmura, T., Kunii, T.L.: Topology matching for fully automatic similarity estimation of 3d shapes. In: Proceedings of the 28th Annual Conference on Computer Graphics and Interactive Techniques, pp. 203–212. ACM (2001)
21. Huettenberger, L., Heine, C., Carr, H., Scheuermann, G., Garth, C.: Towards multifield scalar topology based on pareto optimality. Comput. Graph. Forum **32**(3.3), 341–350 (2013)
22. Ji, G., Wei Shen, H.: Feature tracking using earth movers distance and global optimization, pacific graphics (2006)
23. Kendrick, I., Kumari, D., Yakaboski, A., Dimakis, N., Smotkin, E.: Elucidating the ionomer-electrified metal interface. J. Am. Chem. Soc. **132** (2010). https://doi.org/10.1021/ja1081487
24. Kitware Inc.: The Visualization Toolkit User's Guide (2003)
25. Kresse, G., Gil, A., Sautet, P.: Significance of single-electron energies for the description of CO on Pt(111). Phys. Rev. B **68** (2003). https://doi.org/10.1103/PhysRevB.68.073401
26. Lee, T.Y., Shen, H.W.: Visualizing time-varying features with TAC-based distance fields. In: 2009 IEEE Pacific Visualization Symposium (2009)
27. Lehmann, D.J., Theisel, H.: Discontinuities in continuous scatter plots. IEEE Trans. Vis. Comput. Graph. **16**(6), 1291–1300 (2010). https://doi.org/10.1109/TVCG.2010.146
28. Morozov, D., Beketayev, K., Weber, G.: Interleaving distance between merge trees. Discrete Comput. Geom. **49**(22–45), 52 (2013)
29. Narayanan, V., Thomas, D.M., Natarajan, V.: Distance between extremum graphs. In: 2015 IEEE Pacific Visualization Symposium (PacificVis), pp. 263–270. IEEE (2015)
30. Rubner, Y., Tomasi, C., Guibas, L.J.: The earth mover's distance as a metric for image retrieval. Int. J. Comput. Vis. **40**(2), 99–121 (2000)
31. Saeki, O.: Topology of Singular Fibers of Differentiable Maps. Springer, Heidelberg (2004)
32. Saeki, O., et al.: Visualizing Multivariate Data Using Singularity Theory, Mathematics for Industry, vol. 1, chap. The Impact of Applications on Mathematics, pp. 51–65. Springer, Tokyo (2014)
33. Saikia, H., Seidel, H.P., Weinkauf, T.: Extended branch decomposition graphs: structural comparison of scalar data. Comput. Graph. Forum **33**(3), 41–50 (2014)
34. Saikia, H., Seidel, H.P., Weinkauf, T.: Fast similarity search in scalar fields using merging histograms. In: Topological Methods in Data Analysis and Visualization, pp. 121–134. Springer (2015)
35. Scott, D.W.: A note on choice of bivariate histogram bin shape. J. Off. Stat. **4**(1), 47–51 (1988)
36. Somorjai, G.A., Li, Y.: Introduction to Surface Chemistry and Catalysis. Wiley, Hoboken (2010)
37. Sridharamurthy, R., Bin Masood, T., Kamakshidasan, A., Natarajan, V.: Edit distance between merge trees. IEEE Trans. Vis. Comput. Graph. **1** (2018). https://doi.org/10.1109/TVCG.2018.2873612
38. Thomas, D.M., Natarajan, V.: Multiscale symmetry detection in scalar fields by clustering contours. IEEE Trans. Vis. Comput. Graph. **20**(12), 2427–2436 (2014)

Tensor Fields for Data Extraction from Chart Images: Bar Charts and Scatter Plots

Jaya Sreevalsan-Nair, Komal Dadhich, and Siri Chandana Daggubati

Abstract Charts are an essential part of both graphicacy (graphical literacy), and statistical literacy. As chart understanding has become increasingly relevant in data science, automating chart analysis by processing raster images of the charts has become a significant problem. Automated chart reading involves data extraction and contextual understanding of the data from chart images. In this paper, we perform the first step of determining the computational model of chart images for data extraction for selected chart types, namely, bar charts, and scatter plots. We demonstrate the use of positive semidefinite second-order tensor fields as an effective model. We identify an appropriate tensor field as the model and propose a methodology for the use of its degenerate point extraction for data extraction from chart images. Our results show that tensor voting is effective for data extraction from bar charts and scatter plots, and histograms, as a special case of bar charts.

Keywords Chart images · Spatial locality · Local features · Chart data extraction · Positive semidefinite second-order tensor fields · Structure tensor · Tensor voting · Saliency maps · Topological analysis

1 Introduction

Charts fall in the intersection of graphicacy (graphical literacy), and statistical literacy. The recent popularity of statistical analysis in data science and ubiquitousness of learning algorithms have made chart graphicacy relevant. Chart comprehension is an outcome of chart graphicacy. Students often face difficulties in comprehending

---

J. Sreevalsan-Nair (✉) · K. Dadhich · S. C. Daggubati
Graphics-Visualization-Computing Lab, International Institute of Information Technology Bangalore, Karnataka, India
e-mail: jnair@iiitb.ac.in

K. Dadhich
e-mail: komal.dadhich@iiitb.ac.in

S. C. Daggubati
e-mail: daggubati.sirichandana@iiitb.ac.in

I. Hotz et al. (eds.), *Topological Methods in Data Analysis and Visualization VI*, Mathematics and Visualization, https://doi.org/10.1007/978-3-030-83500-2_12

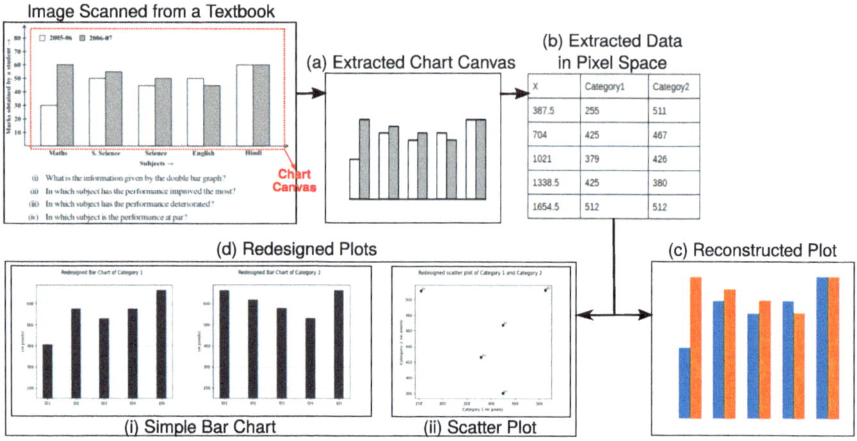

Fig. 1 Data extraction in pixel space from a chart image using our method, for chart reconstruction and chart redesign. The input image is from a Grade-8 mathematics textbook (The textbook is available at NCERT http://ncert.nic.in/textbook/textbook.htm)

new chart types when learning new concepts in the visual representation of data, e.g., grouped bar charts [6]. Chart images embedded in documents, articles and books often become strenuous to analyze further due to unavailability of source data. Each *entity* in a grouped bar corresponds to an ordered proximal placement of a *group* of category-wise bars. Such charts give better communicative signals of (inter-category, intra-entity) trends or relations than those of (intra-category, inter-entity) [6]. In an example of different academic years (category) grouped together for different subjects (entity) (Fig. 1), the trends in a subject are (inter-category, intra-entity), and those in an academic year are (intra-category, inter-entity). Combinations of intra- and inter-category/entity analysis, e.g., "what is the difference between the lowest scores in the two academic years?" require complex interpretation. The inter-category, intra-category, and combination questions are best communicated by grouped bar charts, a set of simple bar charts, and a scatter plot, respectively (Fig. 1). Such chart redesign is possible only with the data extracted from the original charts, that are usually available in raster image format scanned from subject textbooks, e.g., mathematics, science, economics, etc. Thus, we are interested in automatic data extraction from the chart images.

In 1999, Kimura, a Japanese educator, had proposed a six-level scheme for statistical ability using graphs (charts) [2]. Even though traditionally "graph" and "chart" are interchangeably used, we use the latter to disambiguate from the graph data structures used in computer science. The scheme progresses from a basic Level-A with four sub-levels (A1-A4) to the advanced Level-F, where the knowledge from a chart combined with other relevant information is used for new inferences. The sub-levels A1 refers to the basic reading of chart image by reading title, unit and values which further is enhanced with sub-level A2 by enabling the students to read key features, e.g., minimum and maximum values and value differences etc. The sub-level A3 is

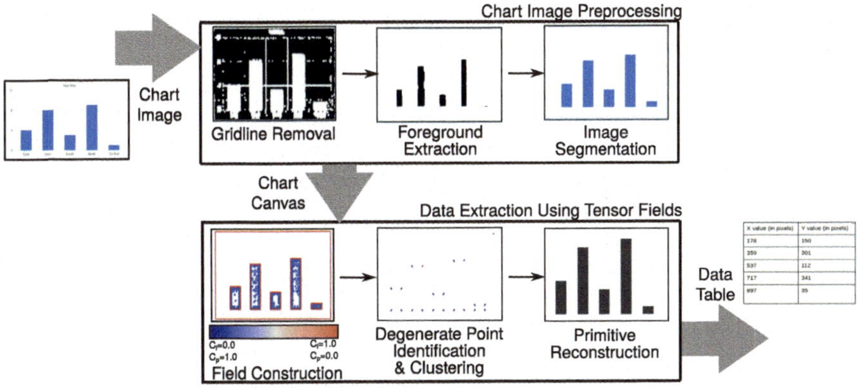

Fig. 2 Our proposed workflow of data extraction from a chart image using positive semidefinite second-order tensor field of local geometric descriptors (The input image has been downloaded from https://excelnotes.com/how-to-add-gridlines-to-a-chart/)

for comparing information from two different charts, and A4 is in reading trends in charts. Levels B-F involves knowledge outside of the chart itself to understand the chart, e.g. data sources, complex statistical computations, current affairs, etc. The computational model encompassing the six levels is needed for completely automated chart interpretation that can address a pertinent need of assistive solutions for chart graphicacy for the differently-abled, including the visually impaired [6, 8].

Appropriate computational models complement the machine interpretation of charts [32] and learning algorithms for chart interpretation. The focus of the models discussed here, are on understanding charts using WYSIWYG (what you see is what you get) features from its images. In general, the challenges for developing such models for chart images lie in the vastness of the charts' design space, attributed to the different types, formatting, applications, and usability patterns. Despite the recent advances in the use of deep learning for chart interpretation [8, 9], these solutions do not still cover the vast design space of charts [25]. The use of machine learning in chart interpretation has been limited in its applications owing to the inadequacy of a single model. The learning models generated using procedurally generated datasets do not account for artefacts present in readily available images (e.g., those available on the internet) [9]. Reuse of object detection models, e.g. YOLO that have been developed for all image types leads to detection errors in chart images, accounting for about 21.6% cases of data extraction failure [8]. A generalized computational model for chart image processing that enables feature extraction can ideally complement existing learning models.

In the cognitive science of chart interpretation, the global precedence principle states that local properties (color, geometry) of a visual object are processed only after its global properties [37]. However, computationally building global properties from local ones is more tractable [12] than vice-versa. Structure tensor and tensor voting are both effective for pixel-wise local structure estimation in images [30]. Tensor voting has also been used to generate a positive semidefinite second-order tensor field in 3D point clouds for local geometric descriptors, that used for semantic classification [36]. Similarly, we propose to use these tensor fields for local structure estimation in charts for data extraction (Fig. 2), using its degenerate points. We focus on two different chart types, namely, simple bar charts and scatter plots, that have a bijective mapping between data items and geometric objects (*i.e.,* bars and scatter points, respectively). But they differ in their source data distributions, namely, univariate and bivariate distributions, respectively. Similar work has so far been dealt with either a single chart types [9, 35] or multiple types of a univariate distribution [8, 32]. Thus, our contributions are in:

- using a positive semidefinite second-order tensor field as a computational model for processing chart images, with a focus on bar charts and scatter plots,
- identifying an appropriate tensor field representing the geometric information from the images for data extraction,
- using degenerate points of the tensor field for data extraction.

Overall, our contributions result in a computational model to automate levels A1 and A2 in Kimura's scheme, from chart images, as they are the only levels pertaining to the geometric information.

2 Related Work

We discuss the relevant state-in-the-art in characterizing chart interpretation and automated systems for chart analysis.

Chart Interpretation and Analysis: While there have been several studies done in chart interpretation in parts, or whole for few chart types, there is still a gap in standardized and generalized methods for chart interpretation, which will work for more than one type of chart [25]. Liu *et al.* concluded that prior human knowledge is an important element, which makes a case for machine learning algorithms. Our work will be useful in generating effective feature vectors for supervised learning models, which perform chart classification, chart segmentation, and chart segment classification, etc., [8]. Chart interpretation has been researched extensively in cognitive science [15] and document engineering [16]. The findings in the latter continue to be extensively used in separating text and graphics in charts using Optical Character Recognition (OCR) techniques [8, 32].

Our proposed method has been developed based on the information gathered from the vast literature in cognitive science in the context of chart analysis. The use of proximity of displayed variables validates our use of spatial locality-based approaches for an integral display, specifically the scatter plot [19]. Wickens and colleagues have extensively studied separable and integral displays [19] and considered bar charts as separable displays. Separable displays are known generally to be not effective for information integration. However, bar chart analysis was found to be useful for information integration [39], e.g., when using the proximity of center points of top-line of bars to visualize trends. Bar charts and other charts have been found to demonstrate properties of integral, configural, and separable/object displays depending on the mapping of the variable [5]. Unlike the use of spatial proximity in chart analysis, we use spatial proximity differently in bar charts to identify "corners" of the top line of bars and scatter plots to locate the scatter points. The spatial locations are crucial for data extraction, thus substantiating the use of spatial locality.

Systems for Chart Analysis: There have been several systems recently developed for various aspects of chart interpretation: text localization and recognition [32]. A more comprehensive system for the visually impaired takes in chart images in a webpage through a browser extension and outputs a data table and accessible interactive charts [8]. Choi *et al.* perform data extraction differently for bar charts, pie charts, and line charts. For bar charts, a combination of a Darknet neural network outcomes for object detection and an OCR system for text extraction is used. For pie charts, a combination of color to item matching using the OCR system and finding the proportion of pixels corresponding to a particular color within a bounding box of the circle is used. For line charts, pixels with the same color and across spans between consecutive tick values are used to identify y-values at each x-tick mark. However, since the geometric information captured in these different methods is extracted from image data, e.g., color, we hypothesize that a geometry-aware method can be generically used for identifying key pixels in the chart images. FigureSeer [35] is another end-to-end framework for summarization of line charts. It is implemented using a convolutional neural network that compares image patches. For data extraction, an optimal solution for the path-finding problem is used. Beagle [4] is a system for classifying charts found as visualizations on the internet. ReVision [34], Scatteract [9], and a bar chart-based method [1] are examples of automated systems for data extraction from selected chart type(s).

Feature Extraction: There has been an even lesser focus on generating feature vectors for chart analysis. In ReVision [34], a feature vector is created using a clustering method for determining codebook patches in an image and finding similar patches to determine a histogram of activated patches. The feature vector has been further used for chart classification. The data extraction has been done using pixel-based methods, driven by the knowledge of chart layouts for bar charts and pie charts. Our proposed tensor field-based method is generically used across two different chart types to identify critical points to localize "objects" for value extraction. These critical points can serve as features.

Overall, chart analysis is a problem solved in pieces, and our work is novel in geometric analysis of chart canvas for data extraction. Our work would be the closest to the use of Hough transform in image space to recognize bar charts [42], in the context of deriving features in image space and finding generic patterns.

3 Background on Local Geometric Descriptors

Local geometric descriptors, which encode the local geometry of each entity in a dataset, can be represented in the form of positive semidefinite second-order tensor fields. Descriptors such as structure tensor [22] have been used for corner detection, shape analysis, and feature tracking, and tensor voting [27] has been used in images widely for segmentation [18].

Structure Tensor: Structure tensor T_s encodes the directionality of the gradient of the local neighborhood. T_s is computed from the gradient tensor, $T_g = G^T G$, using the gradient vector, $G = \left[\frac{\partial I}{\partial x}\ \frac{\partial I}{\partial y}\right]$, at a pixel with intensity I.

Difference kernels, such as Sobel operator, are used for determining discretized differences required for computing G in images. Applying Gaussian convolution to T_g gives T_s. For a Gaussian function with zero mean and standard deviation ρ, $T_s = G_\rho * T_g$, where $*$ is a 2D convolution operator, in the case of 2D images.

Tensor Voting: Tensor voting is a technique of determining global perceptual information organization by garnering votes at each entity based on the normal tensors of its local neighbors [27]. It is especially effective for applications requiring a global context, such as image segmentation. Here, the votes are aggregated component-wise, namely, stick-, plate-, and ball-tensors, in 3D data, and stick- and ball-tensors in 2D. Thus, tensor voting gives a positive semidefinite aggregated tensor T_v of propagated votes, that are positive semidefinite normal tensors.

Gradient information can be used to compute tensor voting by using T_g to initialize the stick-tensor component specifically [30]. For grey-scale images, the T_g is computed for any one of the RGB channels; and for color images, the T_g is computed separately for the three channels, and the tensor votes across all neighbors and all channels are aggregated by summation. Additionally, the tensor voting in natural images has been approximated to stick-tensor votes alone, since the percentage of pixels with a ratio of eigenvalues $\left(\frac{\lambda_0}{\lambda_1} > 0.1\right)$, where eigenvalues of T_v at each pixel are, such that, $\lambda_0 \geq \lambda_1$, has been found to be considerably low ($\sim$10%) [30]. Recently, the closed-form analytical solution of tensor voting has been determined [41].[1] The tensor vote cast at x_i by x_j using a second-order tensor K_j in d-dimensional space is:

$$S_{ij} = c_{ij} R_{ij} K_j R'_{ij}, \text{ where } R_{ij} = (I_d - 2r_{ij}r_{ij}^T), \text{ and } R'_{ij} = (I_d - \frac{1}{2}r_{ij}r_{ij}^T)R_{ij}.$$

I_d is the d-dimensional identity matrix; direction vector $r_{ij} = \hat{d_{ij}}$, where the distance vector $d_{ij} = x_j - x_i$; and $c_{ij} = \exp\big(-\big(\sigma_d^{-1}.\|d_{ij}\|_2^2\big)\big)$, where σ_d is the scale parameter. The gradient tensor T_g can be used as K_j [30]. In the closed form, if a generic tensor field is used as K_j, then a voting field is not required [41]. Thus, tensor voting is no longer separately computed as the different components of a stick-, plate-, and ball-voting fields, and aggregated. Once S_{ij} is computed for a point with each of its

[1] This is the version of the paper with clarification of comments given on a previous publication [40].

neighbors in a (von Neumann or 4-) neighborhood $\mathbb{N}_4$. Then the votes are aggregated across all the neighbors using summation. Using $\sigma_d = 4$, based on neighborhood size, we get the aggregated positive semidefinite second-order tensor as:

$$T_v = \sum_{k=0}^{(d-1)} \sum_{j \in \mathbb{N}_4} S_{ij}(d).$$

Anisotropic Diffusion: Since T_v propagates the normal tensor votes, the aggregated tensor is in the normal space. However, we need tensors that encode geometry of objects (such as bars and points) in the image, implying the tensors must be tangential to the object boundaries. Thus, the tensor T_v must be transformed into tangent space, which can be done using anisotropic diffusion [36, 38]. Adapting to the 2D case, using eigenvalues of T_v, $\lambda_0 \geq \lambda_1$, and corresponding eigenvectors v_0 and v_1, respectively, the tensor after anisotropic diffusion using diffusion parameter δ, is:

$$T_{\text{v-ad}} = \sum_{k=0}^{1} \lambda'_k . v_k v_k^T, \text{ where } \lambda'_k = \exp\Big(-\frac{\lambda_k}{\delta}\Big).$$

Diffusion parameter ($\delta = 0.16$) is a widely used parameter setting [36, 38].

Saliency Map Computation: The eigenvalue decomposition of the local geometric descriptor of a 3D point determines if the point is part of either a line-, surface, or point-type feature [20, 24]. Similarly, in 2D image space, the eigenvalue decomposition of the local structure estimator, such as $T_{\text{v-ad}}$, determines the probabilistic geometric classification of pixels to the line- and point-type features. Hereafter, we use "local structure estimator" and "local geometric descriptor" interchangeably. The saliency maps [27] summarize the likelihood of the point or pixel belonging to these classes at the end of the tensor voting process.

Thus the likelihood of a pixel being a point-type feature, C_p, can be derived from the junction-map and that of a line-type feature, C_l, from the curve-map. It is a probabilistic classification, and hence, $C_l + C_p = 1.0$.

Adapting the computation of anisotropy in superquadric tensor glyphs [21] has been found suitable for saliency maps of 3D points [36]. Similarly, in 2D images, we get the saliency maps at each pixel, namely curve-map C_l and junction-map C_p, as:

$$C_l = \frac{\lambda_0 - \lambda_1}{\lambda_0 + \lambda_1} \text{ and } C_p = \frac{2\lambda_1}{\lambda_0 + \lambda_1}.$$

where we have eigenvalues of $T_{\text{v-ad}}$ of the pixel with $\lambda_0 \geq \lambda_1$. Thus, bringing together tensor field topology and probabilistic geometric classification, a pixel with $C_p \approx 1.0$ is a degenerate point, attributed to the anisotropic local neighborhood.

4 Our Proposed Method

The aim of our work is twofold—firstly, we propose the use of local structure estimation to generate a computational model for chart interpretation, and secondly, we show the effectiveness of this model for data extraction from simple bar charts and scatter plots. We run experiments on a variety of charts to study the effectiveness of our model. For this work, we focus exclusively on data extraction purely using geometry, without extracting contextual information from text, axes, and legend. Hence, we work with only with the relevant region in the chart image, which warrants defining chart image components. The local geometric descriptors of the pixels in this region are the tensor field, and it is used as a computational model for chart images. We explain the observed behavior of the proposed tensor model for geometric entities, namely bars and scatter points, and its degenerate points. The understanding of the field is important for deciding the data extraction workflow (Fig. 2) and performing the error analysis of the model. This section has three main parts on chart image components, our proposed computational model, and the data extraction workflow.

A. Chart Image Components

Here, we connect the characteristics of chart images with tensor field topology. Hence, we first define the terminology pertaining to the charts and descriptions of relevant parts of the charts. As mentioned in Sect. 1, the digitized image format of a chart is called the *chart image*.

Definition 1. The *chart data* is the data used for plotting in a chart. The chart data is a uni- and bi-variate distribution in simple bar charts and scatter plots, respectively.

Definition 2. A two-dimensional geometric primitive or mark (in information visualization parlance) in the chart that encodes chart data is defined as a *chart object*. Bars and scatter points are chart objects in bar charts and scatter plots, respectively.

Definition 3. Two pixels p and q are said to be *connected* if $q \in N_4(p)$ or $a \in N_8(p)$, in 4-neighborhood (N_4) or 8-neighborhood (N_8) of p. A set of connected pixels in an image is defined as a *connected component*. A connected component in a chart image is called a *chart-image-component*.

Definition 4. The *chart canvas* is the rectangular region of the chart image corresponding to the plot, which is the graphical representation of chart data.

Definition 5. A chart-image-component that corresponds to a chart object and is a subset of chart canvas is defined as a *chart-object-component*, respectively. Connected components of more than one chart object components are called *chart-object-clusters*. For sake of simplicity, we refer to isolated chart-object-components also as chart-object-clusters.

Definition 6. The *component boundary* of a chart-object-cluster is a chart-image-component whose pixels belong to the chart-object-cluster, and one or more their N_8

neighbors do not belong to the chart-object-cluster. The *borders* applied to individual chart objects using the plotting tool are subsets of the component boundary.

Definition 7. The *component interior* of a chart-object-cluster is the chart-image-component whose pixels and their entire N_8 neighborhoods belong to the chart-object-cluster. The union set of component boundary and component interior of all chart-object-cluster in chart canvas becomes the *foreground* of the chart canvas.

Definition 8. The *component exterior* of a chart-object-cluster is the difference set of the chart canvas and the union of the component boundary and the component interior of the chart-object-cluster. The intersection set of component exteriors of all chart-object-clusters in chart canvas is the *background* of the chart canvas.

B. Computational Model

The size (height) and position of the chart objects encode the information from chart data in bar charts and scatter plots, respectively. Methods for detection and localization of chart-image-components rely on object detection methods used in image processing [8]. For objects found in any raster image, its boundary and position can be extracted using the local descriptors of pixels using edges and junctions [23]. In our case, we are interested in identifying and localizing chart-object-clusters in bar charts and scatter plots for data extraction. We propose identifying component boundaries for the chosen chart types. In the chart canvas, the gradient change along the component boundaries is captured effectively using the gradient tensor T_g. Thus, we propose to use local geometric descriptors derived from T_g as options for tensor fields to model the chart canvas. The descriptors are the structure tensor, T_s, and those from the tensor voting field after anisotropic diffusion, $T_{\text{v-ad}}$. The use of positive semidefinite second-order tensor fields formalizes a generalized model across different chart types, with the potential scope of global-local feature extraction.

The component boundary of chart-object-clusters manifests as line-type features, *i.e.*, with high C_l of the local geometric descriptors. However, the pixelated boundaries also lead to occurrences of point-type features, where C_p is high. The point-type features are synonymous with the degenerate points in tensor fields, where the eigenvalues of the tensor are equal to each other [10]. In local geometric descriptors of 3D point clouds, these have been loosely referred to as point-type features [36], or as critical points [20]. The junction points in images are known to have high ball saliency [28] or high junction saliency [13]. The junctions are also synonymous with degenerate points in tensor fields, as they are the intersection of multiple edges and, thus, have anisotropy in its local neighborhood. Overall, the degenerate points are relevant here for extracting the component boundary in both the two chart types.

- In bar charts, the corners of the bars are captured using junction maps or degenerate points. With the knowledge of the orientation of the bars (vertical/horizontal), it is straightforward to determine the distance between degenerate points in the corners along the appropriate axis, which gives the size of the bar in pixel space.

- In scatter plots, the centroid of the degenerate points in the component boundary of the chart-object-clusters gives the location of the chart object, *i.e.,* the scatter point, in pixel space. The centroid of degenerate points in the boundary can be treated equivalent to the centroid of the component interior of the scatter points.

Thus, the degenerate points of the proposed tensor field encode the information in the chart data of our chosen chart types. We label the pixels with $C_p > \tau_{cp}$ for a threshold τ_{cp} as degenerate points. For our experiments, we use $\tau_{cp} = 0.6$.

C. Data Extraction Workflow

We first look at data extraction from the tensor fields, and then we look at the requirement for chart image preprocessing for improving the tensor fields themselves, as shown in the blocks in Fig. 2.

Data Extraction Using Tensor Fields: In the tensor field of the locality of the corners of bars or the component boundary of scatter points, the degenerate points tend to form local clusters, some of them tend to be weak. Since our requirement for extracting specific locations of geometric primitives, we perform two-step postprocessing of degenerate points. Firstly, We discard weak degenerate points by thresholding based on a tensor invariant, namely the trace. Here, we consider the degenerate points with unity-based normalized trace $T < \tau_{wd}$, for a threshold τ_{wd}, as weak. In our experiments, we have used $\tau_{wd} = 0.005$ for bar charts, and $\tau_{wd} = 0.01$ for scatter plots. Secondly, we use density-based clustering to consolidate the degenerate points in the corners of the bar as well as to separate individual scatter points. As most clustering algorithms depend on the hyperparameters based on cluster output, e.g., the number and shape of clusters, we require a method that can work without the prior knowledge of such variables. Hence, we choose DBSCAN (Density-Based Spatial Clustering of Applications with Noise) [11] for clustering degenerate points. The DBSCAN algorithm is based on this intuitive notion of "clusters" and "noise," where each point in a cluster must have at least a minimum number of local neighbors. DBSCAN output is influenced by the distance between clusters (ϵ) and the minimum number of points in a cluster (n). In our method, we have found that these parameters of DBSCAN need to be evaluated for each image individually owing to the variability present in the chart images. We use our visualization of degenerate points to decide these values for each image.

For visualizing the tensor field of the chart canvas, we use ellipsoid glyphs colored based on saliency maps. We also use a dot plot to visualize the saliency maps at all pixels in the chart canvas image and superimpose degenerate points on the original chart image to identify DBSCAN parameters. We use the colorblind safe divergent color palette, namely the coolwarm palette, for C_l values from 0 to 1. This color map also reveals C_p values, as $C_l + C_p = 1.0$, thus, identifying the blue pixels as degenerate points. We also use the visualization to evaluate the performance of T_s and $T_{\text{v-ad}}$ tensor fields as an appropriate computational model.

Our workflow (Algorithm 1) for data extraction from tensor fields includes tensor field computation in a chart canvas and chart-type-dependent data extraction. Since we do not have the contextual information of the plots, the data is extracted in pixel space. We use the set of centroids of degenerate point clusters D_{cq} for data extraction. For bar charts, we sort the centroids of first by x, and then by y. We use a scanline algorithm in increasing value of x to determine missing points based on the pattern of occurrence of corners by repeating x- and y-values. We thus use a *rule-based occurrence pattern* to find missing points, e.g., simple bar charts, and histograms have distinct patterns. Finding the range of y-intervals at each unique x value gives the univariate distribution of the chart data. On the other hand, for scatter plots, the centroids of points in D_{cq} provide the bivariate distribution of the chart data. Since we do not have a mechanism for filling missing values in scatter plots in our current work, we encounter omission errors (type-2 errors or false negatives), unlike in the case of bar charts.

Chart Image Preprocessing: We need to preprocess the chart images before computing the tensor field for two reasons. Firstly, since our proposed method works on a raster image, it is imperative that the image must be of high resolution. While this is guaranteed by chart images procedurally/programmatically generated, the images readily available from various sources or scanned from the textbook could be of lower quality. Conventional image processing includes compression, formatting, smoothening, etc. One of the undesirable outcomes of these processing methods is aliasing, which affects our raster-based methods. However, we can preprocess the chart image by performing antialiasing on the component boundary. Secondly, we require only the chart canvas from the chart image for data extraction. Thus, we remove all other chart components, namely, axes, gridlines, legend, and text for data extraction. Overall, we preprocess for antialiasing and chart canvas extraction.

We perform tasks such as grid removal and adding a border to bars to preprocess the chart images, implemented using image binarization followed by morphological operations (Algorithm 2). Morphological operations are a set of image processing operations that exploit the property of shape. They are used routinely for extracting connected components from images that are useful in representation and description of region shape. We first check if the chart objects are filled or not, by binarizing and checking the percentage of filled pixels. If the chart objects are not filled, we perform object-fill, as our proposed tensor fields are effective for data extraction only in charts with filled chart objects. The morphological operations for foreground extraction involve erosion for white noise removal, dilation for expanding shrunk objects, and appropriate distance transformations. We perform the morphological opening of the binarized image and dilation for background extraction. The foreground is used for extracting connected components, which is followed by the watershed algorithm used for segmentation. The segmented image is the subset of chart canvas, *i.e.*, the chart-object-clusters.

These morphological operations remove formatted borders, if present, of all chart-object-clusters. However, the differences in color between component interior, component exterior, and component boundary manifest as strong degenerate points in corners or junctions. Such degenerate points help in improving the accuracy of the data extraction. Hence, in our image preprocessing procedure, we further perform

Algorithm 1: Data extraction using tensor fields from chart images

```
Input: Chart image C_i, chart-type C_t
Output: Data table D
Initialize D ← ∅
Initialize S_deg-pt ← ∅                       // Set of degenerate points

Initialize D_cq ← ∅            // Cluster centroids of degenerate points

C_c ← chart-canvas-extraction(C_i)                        // Algorithm 2

for pixel i in C_i do
    N ← find-N_8-local-neighborhood(i)
    T_geom ← compute-tensor(descriptor-type, N)       // Local geometric
     descriptor

    C_l, C_p ← compute-saliency-map(T_geom)
    /* Check if the pixel is a strong degenerate point               */

    if C_p > τ_cp and trace(T_geom) > τ_wd then
        S_deg-pt ← set-union(S_deg-pt, i)

C_deg-pt ← DBScan(S_deg-pt)                  // Cluster degenerate points

for cluster q in C_deg-pt do
    C_q ← compute-centroid(q)
    D_cq ← set-union(D_cq, C_q)

if C_t is bar-chart then
    D_cq ← set-union(D_cq, missing-points)      // Rule-based occurrence
     patterns

    D_cq ← sort-first-by-x-and-sort-second-by-y(D_cq)
    for unique x-value in (x,y) in D_cq do
        δ_y ← find-y-intervals(x, D_cq)
        D ← set-union(D, (x, δ_y))                   // Add to data table

else if C_t is a scatter-plot then
    for (x,y) in D_cq do
        D ← set-union(D, (x,y))                      // Add to data table

return D
```

contouring to identify component boundary and add a border of predetermined pixel-width. This ensures borders for all chart-object-clusters, uniformly across input chart images. We have found 2-pixel-wide borders to be effective.

Algorithm 2: Chart canvas extraction

Input: Chart image C_i
Output: Chart canvas C_c
$C_b \leftarrow$ binarize(C_i)
$n_w \leftarrow$ count-white-pixels(C_b)
$n_b \leftarrow$ count-black-pixels(C_b)
if $n_w < 0.2 * (n_w + n_b)$ **then**
    $C_i \leftarrow$ object-fill(C_i)
    $C_b \leftarrow$ binarize(C_i)
`/* Morphological operations for foreground */`
$C_{fg} \leftarrow$ morphology-foreground-extraction(C_b)
$C_b \leftarrow$ morphology-opening(C_b) `// Morphological opening operation`
$C_{bg} \leftarrow$ morphology-dilation(C_b) `// Morphological dilation for background`
components, component-labels $\leftarrow$ find-connected-components(C_{fg})
for *pixel i in* C_{bg} **do**
    component-labels(i) := 0 `// Setting component label for background`
$C_{\text{segmented}} \leftarrow$ morphology-watershed(C_i, component-labels) `// Object extraction`
object-edges $\leftarrow$ Canny-edge-detection($C_{\text{segmented}}$) `// Edge detection`
contours := contouring(object-edges, pixel-width=2) `// Contour extraction`
for *pixel i in contours* **do**
    color($C_{\text{segmented}}$(i)) := black `// Adding 2-pixel width border to objects`
return $C_{\text{segmented}}$

Error Analysis of Extracted Data: We qualitatively compare the visualizations of the chart of the same type plotted using the extracted data and the original chart for each experiment. For quantitative evaluation, we must first determine the performance of chart reading, which is tied to the accuracy with which quantitative information can be "decoded" from the chart specifiers, such as geometry [7]. In our case, we observe that our proposed model is effective only if the error in data extraction is minimal. Hence, we determine the error between our extracted data and the ground truth. However, for quantitative comparisons, we run into the issue of our data being extracted in image space, owing to the loss of context of the chart provided by the

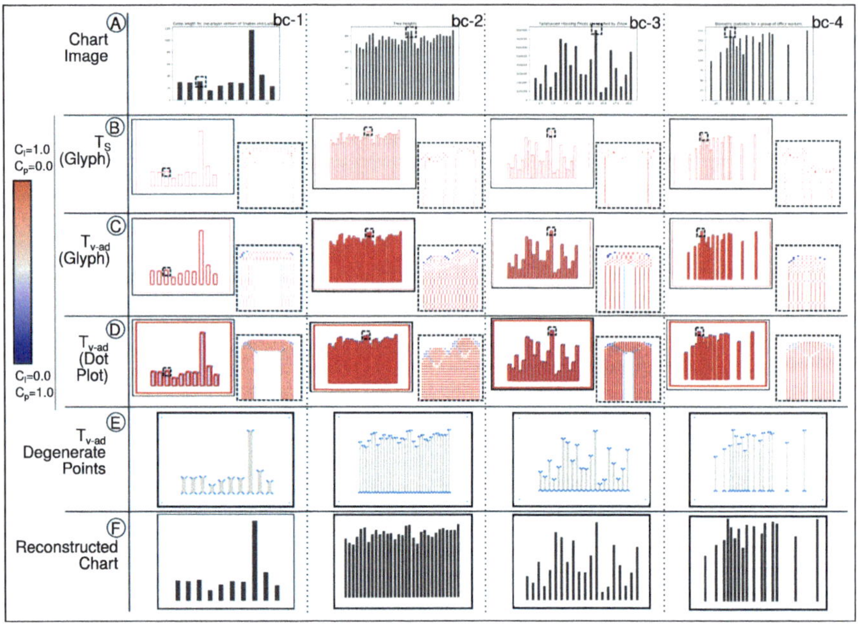

Fig. 3 Our proposed data extraction method from programmatically generated raster images of bar charts from readily available table data ($\mathbf{DS_{Tbl}}$), using saliency map of structure tensor T_s and tensor voting $T_{\text{v-ad}}$, to determine degenerate points

text, axes, and legend. Overall, we compare distributions of extracted data with the ground truth by using appropriate distance measures.

Earth mover's distance, d_{EMD}, is a measure of cross-bin distances between histograms of distributions, which uses ground distance measures [33]. We compute d_{EMD} between the univariate distributions for the extracted data and original data in the case of bar charts, $d_{\text{EMD-BC}}$. In the case of scatter plots, we compute the d_{EMD} between the 2D point clouds of the extracted data and original data, $d_{\text{EMD-SP}}$.

5 Experiments and Results

For experiments, we have performed data extraction using tensor fields on bar charts and scatter plots, as well as on histograms, as a special case of bar charts. We have used three sets of data for our experiments. The dataset descriptions are available at the project GitHub webpage.[2] The dataset $\mathbf{DS_{Tbl}}$ contains multivariate table datasets that are publicly available, and $\mathbf{DST_{Img}}$ contains chart images that are publicly available. We have programmatically generated charts for $\mathbf{DS_{Tbl}}$ using Python

[2] https://github.com/GVCL/Tensor-field-framework-for-chart-analysis.

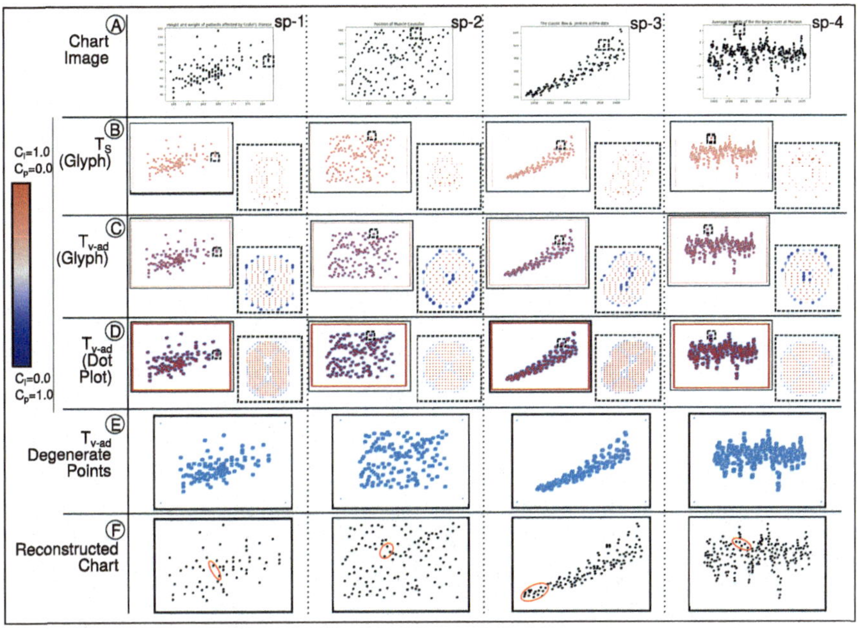

Fig. 4 Our proposed data extraction method from programmatically generated raster images of scatter plots from readily available table data ($\mathbf{DS_{Tbl}}$), using saliency map of structure tensor T_s and tensor voting $T_{\text{v-ad}}$, to determine degenerate points. The reconstruction errors are marked in red in row F

Table 1 Reconstruction error using Earth Mover's Distance of distributions of normalized values in data tables from source and that reconstructed from chart images

Bar Chart ($\mathbf{DS_{Tbl}}$)		Scatter Plot ($\mathbf{DS_{Tbl}}$)		Histogram ($\mathbf{DS_{Tbl}}$)	
(Figure 3)	d_{EMD}	(Figure 4)	d_{EMD}	(Figure 6)	d_{EMD}
bc-1	3.5e-4	sp-1	5.4e-2	hg-1	1.9e-2
bc-2	4.4e-3	sp-2	1.1e-2	hg-2	6.3e-3
bc-3	2.6e-3	sp-3	5.5e-2	hg-3	3.9e-2
bc-4	3.0e-3	sp-4	2.7e-2		

library, `matplotlib.pyplot` [17], and stored them in .png image format. We have specifically used this library, as it generates high-resolution images, compared to a plotting tool, such as Microsoft®Excel®. We have reported the dataset sources in the Acknowledgements.

For all chart images of test datasets, we have constructed the tensor fields and reconstructed data. For bar charts, we have tested with datasets with a large number of bars (*i.e.,* thinner bars), non-uniformly placed bars, smaller set of bars, and bar charts with large variation in the bar heights. For scatter plots, we have tested for positive, negative, and zero correlation data, with a large number of scatter points, with overlapping scatter points. As a special case of simple bar charts, we have used a dataset for histograms. In the case of histograms, we have tested with datasets with variations in the number of bins, with close-to-zero frequencies in some of the

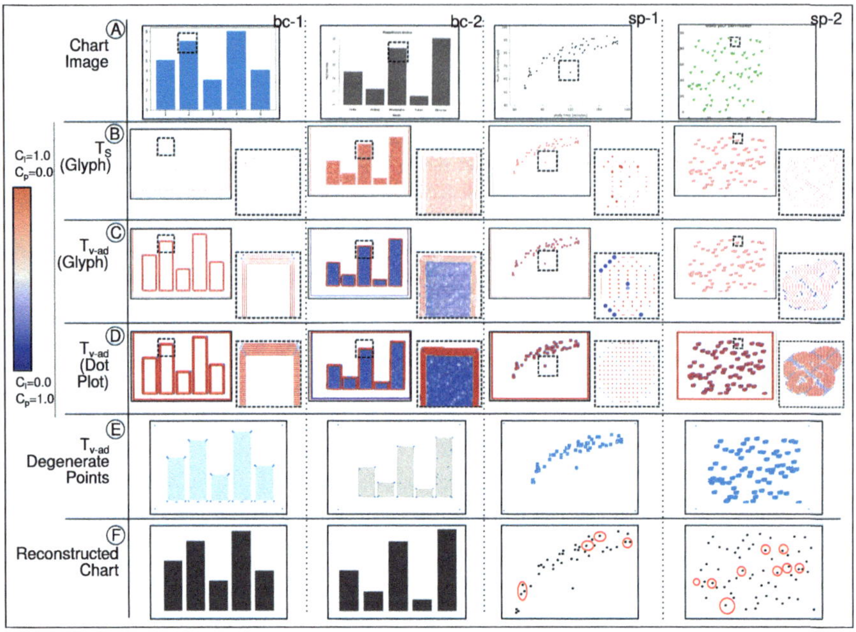

Fig. 5 Our proposed data extraction method from raster images of bar charts and scatter plots from readily available image data (**$\mathbf{DST_{Img}}$**), using saliency map of structure tensor T_s and tensor voting $T_{\text{v-ad}}$, to determine degenerate points. The reconstruction errors are marked in red in row F

histogram bins, with several large variations in frequencies (*i.e.*, several peaks and valleys in the histogram), and with the close-to-normal distribution. Since we have the ground truth for dataset **$\mathbf{DS_{Tbl}}$**, we have performed an error analysis of the same. For the error analysis of histograms, we have determined the Earth mover's distance between the extracted data and the frequency table of the original data, $d_{\text{EMD-HG}}$.

Our results are provided in Figs. 3, 4, 5, 6, and Table 1. We describe aspects of our results related to the tensor field model, data extraction, and error analysis here.

Tensor Field Model: Starting from the pixel-wise gradient tensors in the chart images, the tensor voting field after anisotropic diffusion gives stronger degenerate points (*i.e.*, higher C_p values) than the structure tensor. This is uniformly observable in all the experiments (Figs. 3, 4, 5 and 6). Hence, we choose to use $T_{\text{v-ad}}$ further for data extraction.

The glyph-based tensor field visualization of the subsampled grid (Figs. 3, 4, 5 and 6, row D) is effective for locating strong degenerate points. The dot plots (Figs. 3, 4, 5 and 6, rows B-C) of the tensor fields using color map based on the saliency map are effective for overall tensor field visualization.

The thickness of the bar influences tensor field modeling. In thicker bars (Fig. 3), we observe that in the component interior of the bar object has a region of zero-tensor near the centroid. We refer to this as the region of "homogeneity", where tensors have

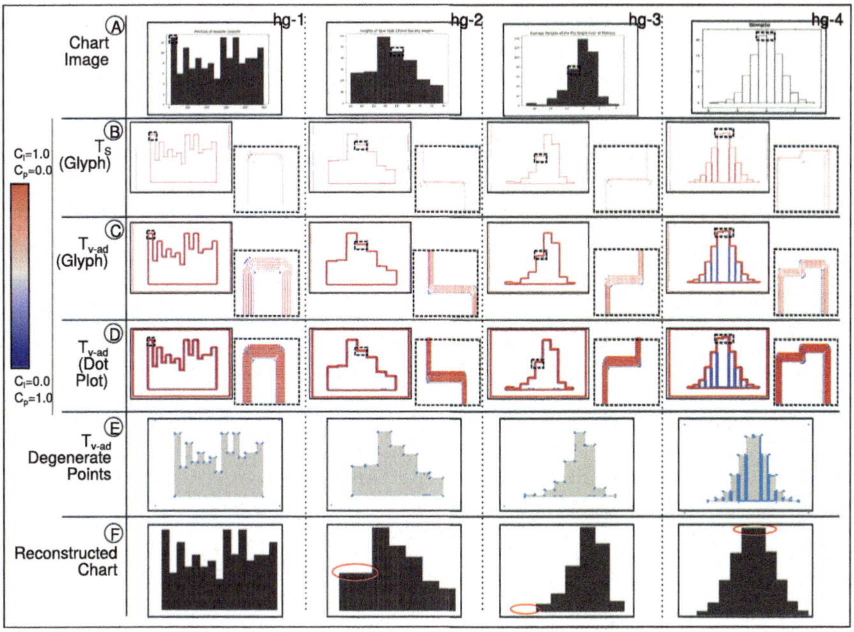

Fig. 6 Our proposed data extraction method from raster images of histograms, using saliency map of structure tensor T_s and tensor voting $T_{\text{v-ad}}$, to determine degenerate points, of (a) readily available table data ($\mathbf{DS_{Tbl}}$) {hg-1, hg-2, hg-3}, and (b) chart images ($\mathbf{DST_{Img}}$), {hg-4}. The reconstruction errors are marked in red in row F

zero-gradient value. The homogeneous region near the center of the bar grows with its thickness.

Data Extraction: Our results for bar charts (Figs. 2, 3 and, 5) show that our method can extract data for a variety of charts stored in both image format and those programmatically plotted from available data tables. Using morphological operations has been effective in removing aliasing effects in low-quality chart images (Fig. 2 and `bc-2` in Fig. 5). In addition to preprocessing, postprocessing by filtering out the weak degenerate points helps in denoising. We have used filtering threshold $\tau_{wd} = 0.003$ for histograms, different from bar charts.

Error Analysis: In the case of scatter plots, our proposed data extraction method suffers from omission (type-2) errors, *i.e.,* false negatives (Figs. 4 and 5, row F). This happens when scatter points are clustered together or when the mark of the scatter point intersects the axes (Fig. 5, `sp-2`). Our model requires clustering of degenerate points to localize a single scatter point, and the second level of clustering to separate clusters of scatter points. DBSCAN parameters have to be modified for each run depending on the pixel-distances between the scatter points.

We compare our result of data extraction in scatter plots with that of Scatteract [9] (Fig. 5, `sp-2`), which had several false positives (type-1 errors). The type-1 errors

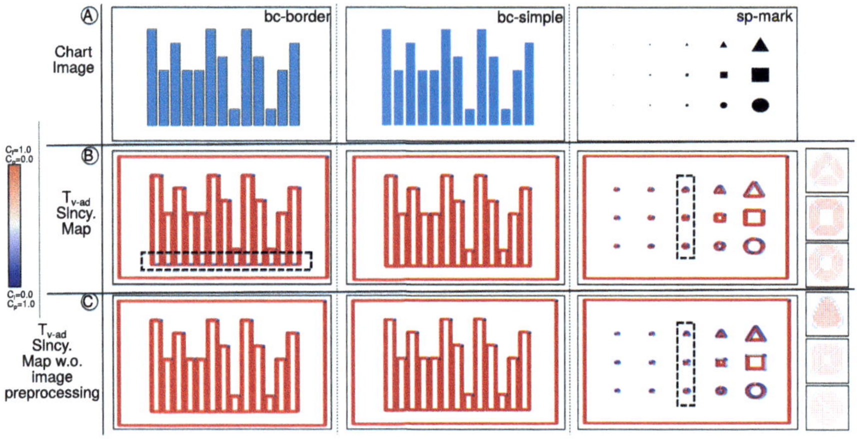

Fig. 7 The differences in the saliency (Slncy.) map of tensor voting with anisotropic diffusion $T_{\text{v-ad}}$ with and without image preprocessing on chart objects: (left, middle) bars in bar chart with and without formatted border, (right) marks in scatter plot

have been attributed to each "clover" mark being detected as three different scatter points. In our case, we have detected the "clover" marks as centroids of clusters. Thus, we avoided false positives but not omission errors.

As a special case of bar charts, histograms perform well with our proposed data extraction method (Fig. 6). We observe errors in the case of histograms with two adjacent bars with similar heights (Fig. 6, `hg-2` and, `hg-4`), as well as with bars that are close to the x-axis (Fig. 6, `hg-3`).

The error analysis using d_{EMD} shows higher errors in the case of scatter plots than the bar charts, or histograms, owing to the type-2 errors (Table 1). In the case of histograms, the histograms of tailed distributions containing ~0-frequency bins at the tails cause larger errors (Fig. 6, `hg-3`).

Discussion

We explain the different factors that influence our tensor field construction to evaluate the robustness of such a field. We also discuss the alternative school of thought, the Gestalt theory of perceptual organization for chart interpretation.

Geometry: Our results in bar charts (Fig. 3) show the effect of bar width in the tensor field computation. In a similar vein, we tested the influence of the size of the mark of scatter points on the tensor field (Fig. 7, `sp-mark`). We observe and confirm that the homogeneous regions grow larger in the center/centroid of chart-object-clusters with the size of the connected component(s). The position of the chart objects, and hence, that of the chart-object-clusters, is ideally the centroid of the homogeneous region. However, the homogeneous region itself does not carry any information, and the degenerate points occur in the non-homogeneous region surrounding the homogeneous region in the chart-object-cluster. Thus, the position of the chart-object-cluster is approximated as the centroid of the clusters of the degenerate points in the non-homogeneous region, as done in our proposed method (Algorithm 1). Overall, the chart object size does not affect our data extraction results.

For scatter plots, we observe that all shapes of scatter points have degenerate points in their pixelated format (Fig. 7, `sp-mark`). While shapes with inherent corners reinforce degenerate point clusters, we also observe that the distribution of these clusters is not uniform in such cases. Hence, this introduces minor errors in the position of the chart object. We have already discussed the type-2 errors in our method owing to overlapping or clustered scatter points. It may be noted that not all type-2 errors lead to a change in correlation, which is an of-studied statistical measure using scatter plots. We expect similar type-2 errors in bar charts where thin bars are closely placed. The degenerate-point-clustering algorithm needs to be evaluated and modified to handle the chart object localization and separation automatically.

Our method currently works for charts with separable geometric objects, such as bars, scatter plots, and with corners. In such charts, the tensor fields produce degenerate points whose location corresponds directly to the value of the data to be extracted. Further study needs to be undertaken on the role of degenerate points in the case of charts where there is non-linear mapping of the geometric object properties to data, e.g. sector area in the case of pie charts.

Color: Since the tensor field is computed on raster images which have a color attribute, the color model and the color difference functions used for generating the gradient tensor influence the local geometric descriptors. The CIELab color model has been recommended to be the model for computing tensor voting for color image denoising [29]. In our work, we have used either the grayscale or 1-color palette, which requires only one of the 3 channels in the RGB model for computing gradients. The influence of color and in a related way, texture are yet to be studied in depth in chart image processing for data extraction.

Image Preprocessing: We have observed that the morphological operations performed to address the influence of aliasing and other image formatting help in reducing noise in the data (Fig. 5, `bc-1` and, `bc-2`). Our image preprocessing procedure reintroduces border for all chart-object-clusters, owing to which we expect generalized behavior of tensor fields and its degenerate points in the bar and scatter point objects. However, we observe that in the case of bar charts with formatted borders, our new borders introduce degenerate points in the base of the bar objects (Fig. 7, `bc-border`, row B). This experiment was conducted on programmatically generated bar charts (Fig. 7, `bc-border`, `bc-simple`) to remove influences of other image processing or compression artefacts.

Our method is designed for raster images using gradient tensor field processing for edge detection. The data extracted from these images is interpreted using the shapes formed by these edges, and hence, the data is WYSIWYG. That also implies that our method works only for discernible images. Our method, thus, fails in the case of images with noisy content and/or of low resolution, which cause the images to be indiscernible.

Machine Learning Models: While image processing needed for computer vision applications is done using robust machine learning models, chart image processing is an understudied area. Most models work for specific chart type(s) [8, 9], and unified models for larger classes of chart types are still a gap. This area can be better served using appropriate machine modeling constructs. The characterization of the chart images and study of feature extraction from such data enables the construction of appropriate models that handle sparsity and geometric characterization in images appropriately. In terms of sparsity, the machine learning models must also account for the data-ink ratio in chart images. The images used for our experiments, the data-ink ratio has been less than 10%, which is considerably low.

Emergent Features: There are two different kinds of perceptual information, namely local and emergent features. Local features are those computed in local proximity in an image, and emergent features are non-decomposable and higher-order ones, which provide an overview/global understanding of the image. For instance, in a scatter plot, a single scatter point is a local feature, whereas the grouping of the scatter points implies the type of correlation, and hence is an emergent feature. These features have their respective theories governing how they contribute to the perceptual understanding of images. The Gestalt theory of perceptual organization states that emergent features enable in the perceptual understanding of the image, and the local theory emphasizes on the local features, instead. The Gestalt theory has been experimentally found to be a better model for perceptual understanding than the local theory for a pair of dots [14]. In cognitive science, this theory is rephrased as the Principle of Perceptual Organization [15], which says,

> "Ensure that groupings based on Gestalt principles, and emergent features more generally, are compatible with the tasks to be carried out with a display."

In our work, we have focused on the local theory. Our overarching goal is to model the Gestalt theory. We chose tensor voting for generating the tensor field for its extensibility to the Gestalt model, as tensor voting takes into consideration Gestalt principles of the perceptual organization [27].

6 Conclusions

Automating chart interpretation has been a long-standing challenge owing to the vast design space of charts and their raster images. We have demonstrated automating data extraction from chart images using a computational model that exploits the local structure in the raster images. The novelty of our work lies in identifying tensor voting after anisotropic diffusion, $T_{\text{v-ad}}$, as an effective local geometric descriptor for data extraction. We have used the positive semidefinite second-order tensor fields of the local geometric descriptors in the chart canvas and extracted the tensor topological features, namely degenerate points, for data extraction. We have shown how patterns of clustering of degenerate points correspond to the chart data, enabling data extraction. We focus on bar charts and scatter plots where the proximity of local features defines the charts' geometric objects. There are limitations to our work as the tensor field is dependent on the user-defined image characteristics. These include image resolution and styling features of the plots, e.g., glyph shapes and sizes, bar width, and borders.

Our current model achieves Levels-A1 and A2 in Kimura's six-level scheme of statistical ability [2] by performing data extraction with considerable accuracy. The future scope of our work is in automated clustering and cluster-classification of degenerate points and extending the usefulness of the tensor voting for Gestalt theory of perceptual organization [27] in charts. This extension will enable the computational model to achieve the Level-A in Kimura's six-level scheme of statistical ability [2].

Acknowledgements The authors acknowledge the financial support from the MINRO (Machine Intelligence and Robotics) grant from the Government of Karnataka, for this work. The authors are grateful to all members of GVCL and IIIT Bangalore for supporting this work. Several people have helped shape this work through feedback and discussions: anonymous reviewers, T. K. Srikanth, IIITB; Sindhu Mathai of Azim Premji University; Vidhya Y. and Supriya Dey of Vision Empower; Neha Trivedi, XRCVC; Vani, Pushpaja, Kalyani, and Anjana of Braille Resource Center, Matruchayya, that has shaped this work.

The datasets used in $\mathbf{DS_{Tbl}}$ are from CSV data repository https://people.sc.fsu.edu/~jburkardt/data/csv/csv.html for bar charts, and from R datasets archive [3] for histograms and scatter plots. The datasets used in $\mathbf{DST_{Img}}$ are from online documentation of package BoutrousLab.plotting.general [31], except for one histogram from a course project in "Computer Vision" course by Roberts [26]. The bar chart available at http://www.datasciencemadesimple.com/r-bar-chart/ has been printed and scanned to get the chart image bc-2 in $\mathbf{DST_{Img}}$.

References

1. Al-Zaidy, R.A., Giles, C.L.: Automatic extraction of data from bar charts. In: Proceedings of the 8th International Conference on Knowledge Capture, pp. 1–4 (2015)
2. Aoyama, K., Stephens, M.: Graph interpretation aspects of statistical literacy: a Japanese perspective. Math. Educ. Res. J. **15**(3), 207–225 (2003)
3. Arel-Bundock, V.: R Datasets (2012). https://vincentarelbundock.github.io/Rdatasets/datasets.html. Accessed 01 Oct 2019
4. Battle, L., Duan, P., Miranda, Z., Mukusheva, D., Chang, R., Stonebraker, M.: Beagle: automated extraction and interpretation of visualizations from the web. In: Proceedings of the 2018 CHI Conference on Human Factors in Computing Systems, p. 594. ACM (2018)
5. Bennett, K.B., Flach, J.M.: Graphical displays: implications for divided attention, focused attention, and problem solving. Hum. Factors **34**(5), 513–533 (1992)
6. Burns, R., Carberry, S., Elzer, S.: Modeling relative task effort for grouped bar charts. In: Proceedings of the Annual Meeting of the Cognitive Science Society, vol. 31 (2009)
7. Carswell, C.M., Wickens, C.D.: The perceptual interaction of graphical attributes: configurality, stimulus homogeneity, and object integration. Percept. Psychophysics **47**(2), 157–168 (1990)
8. Choi, J., Jung, S., Park, D.G., Choo, J., Elmqvist, N.: Visualizing for the non-visual: enabling the visually impaired to use visualization. In: Computer Graphics Forum, vol. 38, pp. 249–260. Wiley Online Library (2019)
9. Cliche, M., Rosenberg, D., Madeka, D., Yee, C.: Scatteract: Automated extraction of data from scatter plots. In: Ceci, M., Hollmén, J., Todorovski, L., Vens, C., Džeroski, S. (eds.)

Joint European Conference on Machine Learning and Knowledge Discovery in Databases, pp. 135–150. Springer, Cham (2017). https://doi.org/10.1007/978-3-319-71249-9_9

10. Delmarcelle, T., Hesselink, L.: The topology of symmetric, second-order tensor fields. In: Proceedings of the conference on Visualization 1994, pp. 140–147. IEEE Computer Society Press (1994)
11. Ester, M., Kriegel, H.P., Sander, J., Xu, X., et al.: A density-based algorithm for discovering clusters in large spatial databases with noise. KDD **96**, 226–231 (1996)
12. Guy, G., Medioni, G.: Inferring global perceptual contours from local features. In: Proceedings of IEEE Conference on Computer Vision and Pattern Recognition (CVPR), pp. 786–787. IEEE (1993)
13. Guy, G., Medioni, G.: Inference of surfaces, 3D curves, and junctions from sparse, noisy, 3D data. IEEE Trans. Pattern Anal. Mach. Intell. (TPAMI) **19**(11), 1265–1277 (1997)
14. Hawkins, R.X.D., Houpt, J.W., Eidels, A., Townsend, J.T.: Can two dots form a Gestalt? Measuring emergent features with the capacity coefficient. Vision Res. **126**, 19–33 (2016)
15. Hegarty, M.: The cognitive science of visual-spatial displays: implications for design. Top. Cogn. Sci. **3**(3), 446–474 (2011)
16. Huang, W., Tan, C.L.: A system for understanding imaged infographics and its applications. In: Proceedings of the 2007 ACM Symposium on Document Engineering, pp. 9–18. ACM (2007)
17. Hunter, J.D.: Matplotlib: a 2D graphics environment. Computi. Sci. Eng. **9**(3), 90–95 (2007). https://doi.org/10.1109/MCSE.2007.55
18. Jia, J., Tang, C.K.: Inference of segmented color and texture description by tensor voting. IEEE Trans. Pattern Anal. Mach. Intell. **26**(6), 771–786 (2004)
19. Jones, P.M., Wickens, C.D., Deutsch, S.J.: The display of multivariate information: an experimental study of an information integration task. Hum. Perform. **3**(1), 1–17 (1990)
20. Keller, P., et al.: Extracting and visualizing structural features in environmental point cloud LiDaR data sets. In: Pascucci, V., Tricoche, X., Hagen, H., Tierny, J. (eds.) Topological Methods in Data Analysis and Visualization, pp. 179–192. Springer, Heidelberg (2011). https://doi.org/10.1007/978-3-642-15014-2_15
21. Kindlmann, G.: Superquadric tensor glyphs. In: Proceedings of the Sixth Joint Eurographics-IEEE TCVG conference on Visualization, pp. 147–154. Eurographics Association (2004)
22. Knutsson, H.: Representing local structure using tensors. In: 6th Scandinavian Conference on Image Analysis, Oulu, Finland, pp. 244–251. Linköping University Electronic Press (1989)
23. Köthe, U.: Integrated edge and junction detection with the boundary tensor. IN: ICCV, vol. 3, pp 424–431 (2003)
24. Kumari, B., Sreevalsan-Nair, J.: an interactive visual analytic tool for semantic classification of 3D urban LiDAR point cloud. In: Proceedings of the 23rd SIGSPATIAL International Conference on Advances in Geographic Information Systems, p. 73. ACM (2015)
25. Liu, Y., Lu, X., Qin, Y., Tang, Z., Xu, J.: Review of chart recognition in document images. In: Visualization and Data Analysis 2013, vol. 8654, p. 865410. International Society for Optics and Photonics (2013)
26. Lozman, E., Rahemtulla, S., Schroeder, J.: Pre-processing and Edge Detection (1997). https://cs.stanford.edu/people/eroberts/courses/soco/projects/1997-98/computer-vision/edges.html. Accessed 01 Oct 2019. Online documentation of a project in the course by Eric Robers, Stanford
27. Medioni, G., Tang, C.K., Lee, M.S.: Tensor voting: theory and applications. In: Proceedings of RFIA, Paris, France 3 (2000)
28. Mordohai, P., Medioni, G.: Junction inference and classification for figure completion using tensor voting. In: 2004 Conference on Computer Vision and Pattern Recognition Workshop, pp. 56–56. IEEE (2004)
29. Moreno, R., Garcia, M.A., Puig, D., Julià, C.: Edge-preserving color image denoising through tensor voting. Comput. Vis. Image Underst. **115**(11), 1536–1551 (2011)
30. Moreno, R., Pizarro, L., Burgeth, B., Weickert, J., Garcia, M.A., Puig, D.: Adaptation of tensor voting to image structure estimation. In: New Developments in the Visualization and Processing of Tensor Fields, pp. 29–50. Springer (2012)

31. P'ng, C., Green, J.: A guide to data visualization using BoutrosLab.plotting.general (2019). https://labs.oicr.on.ca/files/6993/file/PlottingGuide_5.9.2.1.pdf. Accessed 01 Oct 2019
32. Poco, J., Heer, J.: Reverse-engineering visualizations: recovering visual encodings from chart images. In: Computer Graphics Forum, vol. 36, pp. 353–363. Wiley Online Library (2017)
33. Rubner, Y., Guibas, L.J., Tomasi, C.: The earth mover's distance, multi-dimensional scaling, and color-based image retrieval. In: Proceedings of the ARPA Image Understanding Workshop, vol. 661, p. 668 (1997)
34. Savva, M., Kong, N., Chhajta, A., Fei-Fei, L., Agrawala, M., Heer, J.: Revision: automated classification, analysis and redesign of chart images. In: Proceedings of the 24th Annual ACM Symposium on User Interface Software and Technology, pp. 393–402. ACM (2011)
35. Siegel, N., Horvitz, Z., Levin, R., Divvala, S., Farhadi, A.: FigureSeer: parsing result-figures in research papers. In: Leibe, B., Matas, J., Sebe, N., Welling, M. (eds.) Computer Vision. ECCV, pp. 664–680. Springer, Cham (2016). https://doi.org/10.1007/978-3-319-46478-7_41
36. Sreevalsan-Nair, J., Kumari, B.: Local geometric descriptors for multi-scale probabilistic point classification of airborne LiDAR point clouds. In: Schultz, T., Ozarslan E., Hotz, I. (eds.) Modeling, Analysis, and Visualization of Anisotropy. Mathematics and Visualization, pp. 175–200. Springer, Cham (2017). https://doi.org/10.1007/978-3-319-61358-1_8
37. Wagemans, J., Feldman, J., Gepshtein, S., Kimchi, R., Pomerantz, J.R., Van der Helm, P.A., Van Leeuwen, C.: A century of gestalt psychology in visual perception: II. Conceptual theoretical foundations. Psychol. Bull. **138**(6), 1218 (2012)
38. Wang, S., Hou, T., Li, S., Su, Z., Qin, H.: Anisotropic elliptic PDEs for feature classification. IEEE Trans. Vis. Comput. Graph. **19**(10), 1606–1618 (2013)
39. Wickens, C.D., Carswell, C.M.: The proximity compatibility principle: its psychological foundation and relevance to display design. Hum. Factors **37**(3), 473–494 (1995)
40. Wu, T.P., Yeung, S.K., Jia, J., Tang, C.K., Medioni, G.: A closed-form solution to tensor voting: theory and applications. IEEE Trans. Pattern Anal. Mach. Intell. **34**(8), 1482–1495 (2011)
41. Wu, T.P., Yeung, S.K., Jia, J., Tang, C.K., Medioni, G.: A closed-form solution to tensor voting: theory and applications. arXiv preprint arXiv:1601.04888 (2016)
42. Zhou, Y., Tan, C.L.: Hough-based model for recognizing bar charts in document Images. In: Document Recognition and Retrieval VIII, vol. 4307, pp. 333–340. International Society for Optics and Photonics (2000)

Part III: Topology for Geometric Data

A Fast Approximate Skeleton with Guarantees for Any Cloud of Points in a Euclidean Space

Yury Elkin, Di Liu, and Vitaliy Kurlin

Abstract The tree reconstruction problem is to find an embedded straight-line tree that approximates a given cloud of unorganized points in $\mathbb{R}^m$ up to a certain error. A practical solution to this problem will accelerate a discovery of new colloidal products with desired physical properties such as viscosity. We define the Approximate Skeleton of any finite point cloud C in a Euclidean space with theoretical guarantees. The Approximate Skeleton $\text{ASk}(C)$ always belongs to a given offset of C, i.e. the maximum distance from C to $\text{ASk}(C)$ can be a given maximum error. The number of vertices in the Approximate Skeleton is close to the minimum number in an optimal tree by factor 2. The new Approximate Skeleton of any unorganized point cloud C is computed in a near linear time in the number of points in C. Finally, the Approximate Skeleton outperforms past skeletonization algorithms on the size and accuracy of reconstruction for a large dataset of real micelles and random clouds.

1 Introduction: Reconstructions from Unorganized Clouds

Potential molecules for new colloidal products are tested by simulations that produce unorganized finite clouds of points (one point per molecule in Fig. 1). Molecules tend to form clusters (called *micelles*) whose shapes (degrees of branching, edge-lengths) affect physical properties of colloidal products, e.g. their viscosity.

These 3D micelles can have complicated branched shapes as in Fig. 7 and are visually analyzed by human experts who struggle to make reliable measurements

---

Electronic supplementary material The online version of this chapter (https://doi.org/10.1007/978-3-030-83500-2_13) contains supplementary material, which is available to authorized users.

Y. Elkin · D. Liu · V. Kurlin (✉)
Materials Innovation Factory and Computer Science Department, University of Liverpool, Liverpool L69 3BX, UK
Url: http://kurlin.org

I. Hotz et al. (eds.), *Topological Methods in Data Analysis and Visualization VI*, Mathematics and Visualization, https://doi.org/10.1007/978-3-030-83500-2_13

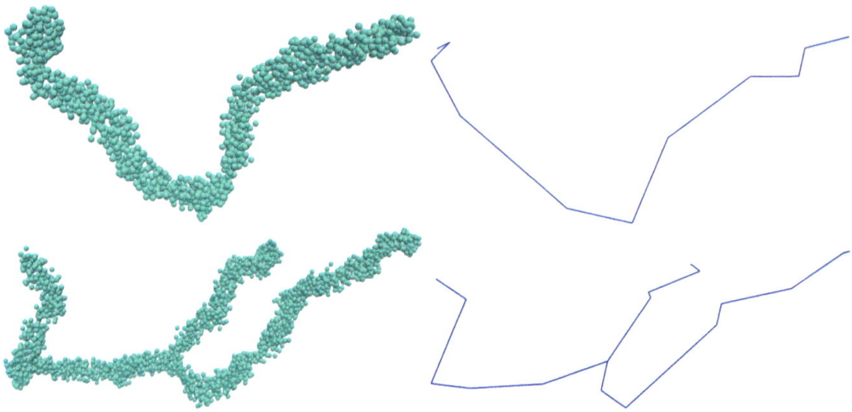

Fig. 1 **Left**: point clouds C from real micelles. **Right**: Approximate Skeletons $\mathrm{ASk}(C)$

quickly. To substantially speed-up the discovery of new molecules, we propose a new Approximate Skeleton $\mathrm{ASk}(C)$ to solve the following problem.

The Tree Reconstruction Problem. Given a point cloud $C \subset \mathbb{R}^m$ and an error ε, design a fast algorithm to build a straight-line tree $T \subset \mathbb{R}^m$ (see Definition 1) that has a minimum number of vertices and whose ε-offset (neighborhood) covers C.

The first (combinatorial) guarantee is for the number of vertices in $\mathrm{ASk}(C)$, which is close to the minimum number in an optimal tree for a given approximation error by factor 2, see Theorem 8. The second (geometric) guarantee about a near linear time for building $\mathrm{ASk}(C)$ is the number of points n in C, see Corollary 9.

To automatically characterize branching shapes of micelles (clusters of molecules in colloids), an Approximate Skeleton $\mathrm{ASk}(C)$ allows us to compute

- the topological type of any unorganized cloud C, e.g. count all *non-trivial* vertices of $\mathrm{ASk}(C) \subset \mathbb{R}^m$ whose degree is 1 (endpoints) or more than 2 (branching);
- the geometric characteristics of C, e.g. edge-lengths of $\mathrm{ASk}(C)$;
- the error of approximating a cloud C by its skeleton $\mathrm{ASk}(C)$, see Table 1.

Here is the pipeline of the Approximate Skeleton $\mathrm{ASk}(C)$.
Stage 1 in Sect. 3: for a cloud $C \subset \mathbb{R}^m$, we build an initial tree $\mathrm{core}(C)$, which has a small number of branching vertices within a Minimum Spanning Tree of C.
Stage 2 in Sect. 4: replace polygonal paths of $\mathrm{core}(C)$ by approximate paths with much fewer vertices to get $\mathrm{ASk}(C)$ in a near linear time within a given error (Fig. 2).
The key novelty and contributions to the data skeletonization are the following.

- Theorem 8 guarantees a small number of vertices in the Approximate Skeleton $\mathrm{ASk}(C)$ close to the minimum by factor 2 in an optimal tree within a given error.
- Corollary 9 guarantees a near linear time to compute $\mathrm{ASk}(C)$ within an error.

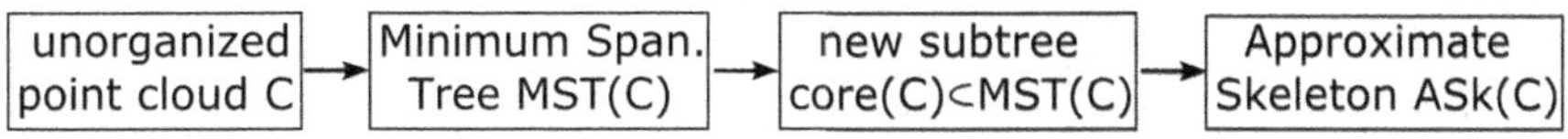

Fig. 2 Pipeline to compute an Approximate Skeleton ASk(C): MST(C) is classical, the new subtree core$(C) \subset$ MST(C) is introduced in Definition 6 in Sect. 3, final ASk(C) is built in Sect. 4

2 Basic Definitions and a Review of the Related Past Work

Definition 1 (a straight-line graph, ε-approximation) *A* straight-line *graph $G \subset \mathbb{R}^m$ (briefly, a* graph*) consists of vertices at points $q_1, \dots, q_k \in \mathbb{R}^m$ and undirected straight-line edges connecting pairs q_i, q_j, $i \neq j$, in such a way that any edges meet only at their common vertex. Let d be the Euclidean distance. For $\varepsilon > 0$, a cloud $C \subset \mathbb{R}^m$ is* ε-approximated *by a graph G if C is within the* ε-offset *that is the union of ε-balls at all points of G, i.e. $G^\varepsilon = \{p \in \mathbb{R}^m \mid d(p, q) \leq \varepsilon$ for some $q \in G\}$.*

Past Algorithms Without Guarantees. Singh et al. [27] approximated a cloud $C \subset \mathbb{R}^m$ by a subgraph of a Delaunay triangulation, which requires $O(n^{\lceil m/2 \rceil})$ time for n points of C and the three thresholds: a minimum number K of edges in a cycle and $\delta_{min}, \delta_{max}$ for inserting/merging 2nd order Voronoi regions. Similar parameters are need for *principal curves* [15], which were later extended to iteratively computed *elastic maps* [13]. Since it is often hard to estimate a rate of convergence for iterative algorithms, we discuss below non-iterative methods with theoretical guarantees.

The metric graph reconstruction (MGR) takes as an input a large metric graph Y, which is an abstract graph with weighted edges and outputs a smaller abstract metric graph $\hat{X}$. The distance between any points of a metric graph is defined as the length of a shortest path these points. If Y is a good ε-approximation to an unknown graph X, then Aanjaneya et al. [1, Theorem 5] proved the existence of a homeomorphism $X \to \hat{X}$ that distorts the metrics on X and $\hat{X}$ with a multiplicative factor $1 + c\varepsilon$ for $c > \frac{30}{b}$, where $b > 14.5\varepsilon$ is the length of a shortest edge of X.

The authors of the Reeb graph skeletonization [11, page 3] have checked that for the MGR algorithm from [1] "it is often hard to find suitable parameters in practice, and such local decisions tend to be less reliable when the input data are not as nice (such as a 'fat' junction region)", see this junction in the 2nd picture of Fig. 1.

Definition 2 (a Reeb graph) *Given a topological space $K \subset \mathbb{R}^m$ (or) with a function $f : K \to \mathbb{R}$, the* Reeb *graph $R_f(K)$ is obtained from K by collapsing each connected components of every level set of f to a single point, so the* Reeb *graph $R_f(K)$ is the quotient of K by the equivalence relation $a \sim b$ if and only if $f(a) = t = f(b)$ and the points $a, b \in K$ are in the same connected component of $f^{-1}(t) \subset K$.*

Skeletonization via Reeb-Type Graphs. The *Vietoris-Rips* complex VR$(C; \alpha)$ on a cloud C consists of all simplices spanned by points whose pairwise distances are at most α. Starting from a noisy sample C of an unknown graph G, Ge et al. [11, Theorem 3.1] proved that the Reeb graph of VR$(C; \alpha)$ has a correct homotopy type if there is a triangulated space K with a continuous deformation $h : K \to G$ that ε-approximates the metrics of K, G. The *homotopy type* of a graph is its equivalence class under continuous deformations when any edge can be collapsed to a point.

The Graph Reconstruction by Discrete Morse Theory (DMT). Dey et al. [6] substantially improved the discrete Morse-based framework and proved new homotopy guarantees when an input is a density function $\rho : K \to \mathbb{R}$, which 'concentrates' around a hidden geometric graph G. The key advantage of this approach is the unbounded noise model that allows outliers far away from an underlying graph G.

Since the molecules of a micelle form an unorganized cloud of points (with large bounded noise) around hidden tree structures, the Tree Reconstruction problem in Sect. 1 essentially differs from the above approaches. An initial unorganized cloud of points is not an abstract metric graph (as in the metric graph reconstruction problem) and not a simplicial complex with scalar values at vertices (as in the discrete Morse theory approach), so extra pre-processing was needed in Sect. 5.

The α-Reeb graph G by Chazal et al. [5] solves the metric graph reconstruction problem, where the input is not an unorganized cloud, but a large metric graph X that should be approximated by a smaller graph $\hat{X}$. For a base point $p \in X$, the image of the distance function $d(p, *) : X \to \mathbb{R}$ is covered by intervals I_j having a length α and 50% overlap. Every connected component of $f^{-1}(I_j) \subset X$ defines a node in the α-Reeb graph G. Two nodes are linked if the corresponding components overlap. Informally, α controls the size of a subset of X that maps to a single vertex of G. Theorem 4.9 in [5] says that if X is ε-close to an unknown graph with edges of minimum length 8ε, the output G is $34(\beta(G) + 1)\varepsilon$-close to X in the Gromov-Hausdorff distance between spaces, not within one space, where $\beta(G)$ is the first Betti number of G. The algorithm has the fast time $O(n \log n)$ for n points in X. Similarly to Reeb graphs, α-Reeb graphs are abstract without an intrinsic embedding into the space of the cloud C and can have self-intersections even for $X \subset \mathbb{R}^2$.

The Mapper [26] extends any clustering algorithm and outputs a network of interlinked clusters and needs a user-defined function $f : C \to \mathbb{R}$, which helps to link different clusters of a cloud C. Another parameter is a covering of the image of f by a given number k of intervals I_j (often with 50% overlap). Each of k subclouds $f^{-1}(I_j) \subset C$ is clustered. Every cluster defines a node in the Mapper graph. Two nodes are linked if the corresponding clusters overlap. M. Carriére et al. [4] have proved first theoretical guarantees for the Mapper output.

More recent persistence-based algorithms for graph reconstruction [14, 19, 20, 28] and image segmentation [8, 9, 21, 22] essentially find most persistent cycles hidden in a cloud, hence go beyond the tree reconstruction problem in Sect. 1.

Straightening polygonal curves is a key ingredient in many skeletonization algorithms. Douglas-Peucker's heuristic [7] approximates a long zigzag line by a simpler line with fewer vertices, see Sect. 4. The elegant algorithm by P. Agarwal et al. [2] guarantees a near linear time and a small number of vertices in a final polygonal approximation when used with the Frechet distance between curves in $\mathbb{R}^2$. For the Hausdorff distance and higher dimensions, there is no near linear time straightening with guarantees on the size of a skeleton to our best knowledge.

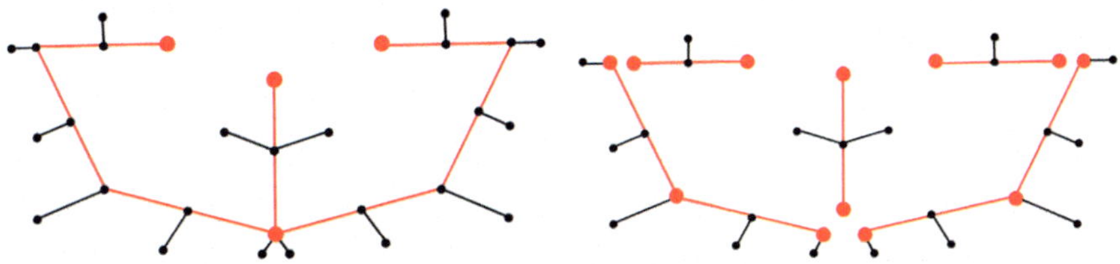

Fig. 3 **Left**. One vertex (large red dot at the bottom) of MST(C) has a high depth by Definition 5 and is connected by longest paths to 3 vertices of degree 1. The other vertices have at most 2 disjoint long paths within MST(C). **Right**. The paths of core(C) and subclouds are shown disjointly

Definition 3 (MST(C)) *For a cloud* $C \subset \mathbb{R}^m$, *a* Minimum Spanning Tree MST(C) *is a connected graph that has (1) the vertex set C, (2) no cycles, and (3) a minimum total length, where lengths of edges are measured in the Euclidean distance.*

If all distances between points of C are distinct, then MST(C) is unique. We write a Minimum Spanning Tree, similarly an Approximate Skeleton, to cover all cases.

Theorem 4 *[24, Theorem 5.1] For any cloud* $C \subset \mathbb{R}^m$ *of* n *points, a Minimum Spanning Tree* MST(C) *can be computed in time* $O(\max\{c^6, c_p^2 c_l^2\}c^{10} n \log n\, \alpha(n))$, *where* $\alpha(n)$ *is the inverse Ackermann function;* c, c_p, c_l *are defined in [24].*

3 A New Tree core(C) Defined for Any Point Cloud $C \subset \mathbb{R}^m$

This section introduces an important subtree core(C) $\subset$ MST(C), which has many fewer non-trivial vertices than a usually 'hairy' MST(C) from Definition 3.

A tree core(C) might still have too many zigzags and will be replaced by a better tree ASk(C) with fewer vertices in Sect. 4. A vertex of a degree $k \neq 2$ is called (topologically) *non-trivial*, because any vertex of degree 2 can be potentially removed by straightening algorithms in Sect. 4. Since MST(C) contains many non-trivial vertices, the next hard step is to identify those few vertices of MST(C) that represent 'true' vertices of a tree T, which we try to reconstruct from C.

Definition 5 introduces the depth characterizing how deep a vertex sits within MST(C). At a deep vertex of a degree $k \geq 3$ at least 3 sufficiently long paths (without common edges) should meet, see the 3 red long paths in Fig. 3. The previous procedural approach by M. Aanjaneya et al. [1, Fig. 1b] to detect branching points in a shape of C used more parameters than a single branching factor β below.

Definition 5 **(deep vertices)** *For a cloud* $C \subset \mathbb{R}^m$ *and a vertex* $v \in$ MST(C) *of a degree* $k \geq 3$, *let* $B_1, \ldots, B_k \subset$ MST(C) *be the branches (subtrees) joined at the vertex* v. *Let* l_i *be the length of a longest path within the branch* B_i *from* v *to another vertex,* $i = 1, \ldots, k$. *Assuming that* $l_1 \geq l_2 \geq \ldots$, *set* depth(v) $= \min\{l_1, l_2, l_3\}$. *Let*

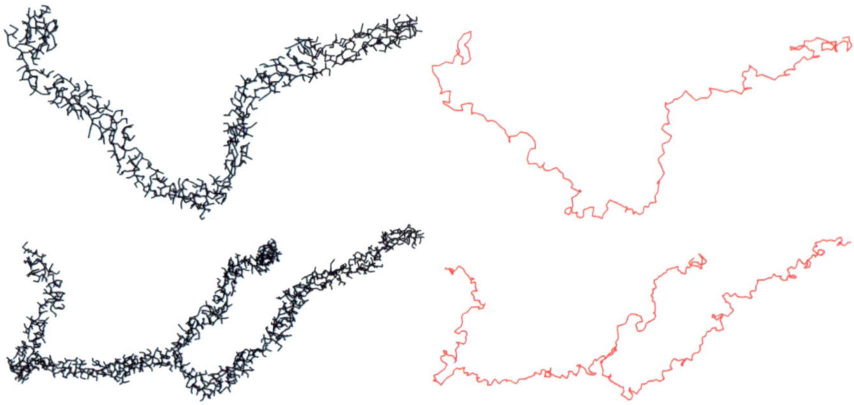

Fig. 4 **Left**: black MST(C) for the two clouds C in Fig. 1. **Right**: red core(C) in Definition 6

$l(C)$ be the average edge-length of MST(C). *For a branching factor $\beta > 0$, the vertices of* MST(C) *whose depths are larger than $\beta l(C)$ are called* deep.

Taking the minimum depth(v) $= \min\{l_1, l_2, l_3\}$ above guarantees that vertices in any short branches of MST(C) are not deep, hence deep vertices can not form small cliques. The experiments on real micelles in Sect. 5 justify that Definition 5 separates deep vertices from other shallow vertices for a long range of the factor β (Fig. 4).

Definition 6 introduces a subtree core(C), which non-essentially depends on the branching factor β and better approximates a cloud C than MST(C), see Fig. 3.

Definition 6 (core(C)) *In Definition 5, if we remove all deep vertices $v_1, \dots, v_m$,* MST(C) *splits into several subtrees. If the closure of such a subtree S has two deep vertices v_i, v_j, they are joined by a unique path $P_{ij} \subset S$. If S has one deep vertex v_i, take a longest path $P_i \subset S$ from v_i to another vertex $v_i' \in S$. We ignore P_i if its length is less than $\beta l(C)$, where β is the branching factor from Definition 5. All the vertices v_i, v_i' and the paths P_{ij}, P_i between them form the subtree* core(C) $\subset$ MST(C).

If the closure of a subtree S above has $k \geq 3$ deep vertices v_1, v_2, v_3, then S contains a vertex v with at least 3 paths to v_1, v_2, v_3. Then depth(v) $>$ depth(v_i), $i = 1, 2, 3$, so v is also deep and S should be split by removing v. Hence $k \leq 2$.

In Fig. 3 the two black edges at the red deep vertex v of degree 5 are too short, hence ignored in Definition 6. The tree core(C) consists of only 3 red long paths meeting at v. Here are the steps of Stage 1 for the Approximate Skeleton ASk(C).
Step 1a. If needed, split a cloud C in clusters to approximate them below.
Step 1b. If C is one cluster, find MST(C) by the fast algorithm from Theorem 4.
Step 1c. Find the depths of vertices in MST(C) by Algorithm 1 in Appendix A.
Step 1d. Identify all deep vertices of MST(C) by their depth(v) $= \min\{l_1, l_2, l_3\}$.
Step 1e. The subtree core(C) $\subset$ MST(C) is formed by all the paths P_i and P_{ij} from Definition 6 that have lengths more than $\beta l(C)$, where β is a given branching factor.

4 ASk(C): Approximate Skeleton of a Cloud $C \subset \mathbb{R}^m$

The tree core(C) from Definition 6 has only few non-trivial vertices, but contains noisy zigzags with too many *trivial* vertices of degree 2. This section discusses how to straighten these zigzags and decrease the total number of vertices.

We have tried Douglas-Peucker's heuristic [7], which was rather unstable and produced large zigzags on curved micelles in Fig. 1. The worst complexity is $O(n^2)$ in the number n of points for $d > 2$. A final approximation can have a size $\Omega(n)$ even in $\mathbb{R}^2$. Another problem with [7] are potential self-intersections even in $\mathbb{R}^2$, which are caused by large zigzags that approximate non-monotone curves [29].

The problem of straightening polygonal paths in a tree core(C) is harder than the curve simplification, because the input is a cloud of unorganized points. So a final approximation should take into account the points of a cloud C outside core(C).

Definition 7 *Let $L \subset \mathbb{R}^m$ be a straight line. An ordered cloud $C = \{p_1, \dots, p_n\} \subset \mathbb{R}^m$ is called* monotone *with respect to L if the order of points is preserved by the orthogonal projection of C to L.*

Since there are many paths of core(C) to straighten, we split the cloud C into monotone subclouds as formalized in Algorithm 2 in Appendix A. Since monotone subpaths can be quickly found only in $\mathbb{R}^2$ [25], Theorem 8 below will assume that each subpath P between non-trivial vertices of core(C) is monotone by Definition 7 with respect to the straight line connecting the endpoints of P.

All results in this section are proved in Appendix A. Here are the Stage 2 steps.

Step 2a. Split every polygonal path between non-trivial vertices (of degrees $k \neq 2$) in the subtree core(C) $\subset$ MST(C) into monotone subpaths by Algorithm 2.
Step 2b. Each monotone subpath of core(C) with endpoints (say) p_1, p_n has the subcloud C' approximated by a polygonal path via points of C' by Steps 2c–2f.
Step 2c. For each subcloud $C' = \langle p_1, \dots, p_n \rangle$ of points ordered by their orthogonal projections to $[p_1, p_n]$, start from ind(1) $= 1$ and find the next index ind(i) for $i = 2, \dots, m$ by repeating Steps 2d–2e, which is possible by Lemma 12 in Appendix A.
Step 2d (exponential). Find the smallest index j such that $d([p_{\text{ind}(i-1)} p_l], C') > \varepsilon$ for $l = \text{ind}(i-1) + 2^{j+1}$, $j = 0, 1, 2 \dots$ For every index l, compute the distance $d([p_{\text{ind}(i-1)} p_l], C')$ orthogonally to the line segment $[p_1 p_n]$ as in Definition 11.
Step 2e (binary). Search for the maximum ind(i) between $\text{ind}(i-1) + 2^j$ and $\text{ind}(i-1) + 2^{j+1}$ such that $d([p_{\text{ind}(i-1)} p_{\text{ind}(i)}], C') \leq \varepsilon$ by dividing the range in 2 halves.
Step 2f. The found indices ind(i) specify a polygonal path ε-approximating each monotone subcloud from Step 2b. Combine all these paths into a full skeleton.
Step 2g. Any edges of a length more than $\beta l(C)$ from Definition 5 are temporarily removed from the skeleton. Each remaining connected component with only short edges is collapsed to its center of mass. The resulting vertices are connected according to the temporarily removed edges to get the Approximate Skeleton ASk(C).

For a cloud $C \subset \mathbb{R}^m$, mark the endpoints of all monotone subpaths in core(C) obtained by Algorithm 2. Consider all skeletons $S \subset \mathbb{R}^m$ that have fixed vertices at the marked points of C such that any polygonal path between fixed vertices (say u, v) is monotone under the orthogonal projection to the line segment $[u, v]$.
The approximation problem for an error $\varepsilon > 0$ is to minimize the total number of vertices in a straight-line graph $S \subset \mathbb{R}^m$ whose each monotone path should ε-approximate the corresponding subcloud of C by the distance in Definition 11.

Theorem 8 *Let k be the minimum number of vertices over all graphs ε-approximating a given cloud $C \subset \mathbb{R}^m$. Then* ASk(C) *lies in $C^{2\varepsilon}$ and has at most k vertices.*

Theorem 8 estimates the number of vertices of ASk(C) when the geometric error is 2ε. In practice, the tree core(C) has an initial approximation error for a given cloud C, because many points of C may not be vertices of core$(C) \subset$ MST(C).

We measure the initial error $d($core$(C), C)$ by Definition 11 and take the maximum of $d([v_i v_j], C)$ over monotone paths of core(C) computed in Algorithm 2. Stage 2 approximates C by a graph simpler than core(C), but keeps the approximation error small. The error ε in Corollary 9 is $\gamma \times d($core$(C), C)$, where γ is an error factor that takes values in the interval $[1.1, 1.5]$ for the experiments in Sect. 5.

Corollary 9 *For any n points $C \subset \mathbb{R}^m$ and any error factor $\gamma > 1$, an Approximate Skeleton* ASk$(C) \subset \mathbb{R}^m$ *within the $\gamma d($core$(C), C)$-offset of the cloud C (as in Definition 1) can be computed in time $O(\max\{c^6, c_p^2 c_l^2\} c^{10} n \log n \, \alpha(n))$, where $\alpha(n)$ is the inverse Ackermann function, the constants c, c_p, c_l are defined in Appendix A.*

5 Comparisons of Five Algorithms on Real and Synthetic Data

This section experimentally compares the Approximate Skeleton ASk(C) with those four skeletonization algorithms from Sect. 2 that have theoretical guarantees and accept any cloud C of points: Mapper [26], Metric Graph Reconstruction MGR [1], α-Reeb graphs [5] and most recent discrete Morse theory (DMT) algorithm [6].

The Mapper [26] is very flexible in the sense that its parameters might be manually tuned for given data over numerous clustering algorithms. Having tried several possibilities, we have settled on the following choices from the original work [26].

1) Convert a cloud C into a connected neighborhood graph $N(C)$ with Euclidean edge-lengths by using a distance threshold. The filter function is the distance function in $N(C)$ from a root that is the furthest point from a random point in C.
2) The image of the filter function is covered by 10 intervals with the 50% overlap so that C splits into 10 subclouds when filter values are in one of the 10 intervals.

3) Each subcloud C' is clustered by the single-linkage clustering with the threshold $\tau\times$the average edge-length of MST(C'), where values of the factor τ are given in Table 1. The final Mapper graph has a single node representing each cluster.

The authors of the DMT algorithm by Dey et al. [6] have kindly made their code available at https://github.com/wangjiayuan007/graph_recon_DM. Starting from an unorganized cloud of points, e.g. centers of molecules of a micelle, we generated scalar values at nodes of a regular grid required for the DMT algorithm.

1) We subdivide the axis-aligned bounding box of a cloud $C \subset \mathbb{R}^3$ into small boxes: minimum 20 rectangular boxes (as close to cubic as possible) along each side.
2) The scalar values are found from the Kernel Density Estimate $KDE(p) = \sum\limits_{q\in C} \exp(-d(p,q))$ at every grid node p. The computed values are passed to the DMT with a parameter δ that regulates how small density values are replaced by 0.

The MGR algorithm has required much more efforts, because the original code was lost as confirmed by the main author of [1]. Since the algorithm was well-explained, we have implemented MGR ourselves and confirmed the earlier claim that "it is often hard to find suitable parameters" [11, page 3]. Trying many values of the key parameter r gave the zero success rate on the homeomorphism type.

Hence we have improved MGR by splitting this parameter into two: the first $r_1 = 15$ (values used in all experiments) was used for detecting vertex points, the second r_2 (three values $1, 1.5, 2$ in Table 1 experiments) was used for clustering points of different types. Only after using these different values, we have managed to push the success rates of MGR closer to 50% on the homeomorphism type.

The α-Reeb graph has the essential parameter α whose values $20, 25, 30$ were tried in all experiments. ASk(C) has little dependence on the branching factor β, e.g. all values $[20, 50]$ produced almost identical results in Table 1 and Fig. 9.

Since three output graphs (Mapper, α-Reeb and MGR) are abstract, to compute any geometric error of approximation, we map them to $\mathbb{R}^3$ by sending every node v of G to the average point (center of mass) of the cluster (for Mapper and MGR) or subgraph (for α-Reeb) corresponding to v. Each link between nodes is mapped as a line segment between the corresponding points in $\mathbb{R}^3$.

Figures 5, 6, 7 and 8 show clouds and outputs of 5 algorithms. Since real micelles have irregular shapes in $\mathbb{R}^3$, their 2D projections may contain intersections of edges.

Table 1 shows the average results of the three algorithms on the dataset of more than 100 real micelles (clouds of about 300 molecules) whose endpoints and homeomorphism types were manually detected. A *homeomorphism* is a 1-1 continuous map with a continuous inverse, so a homeomorphism type is a stronger shape descriptor than a homotopy type, which counts only linearly independent cycles.

The *most important error* measure for the tree reconstruction problem in Sect. 1 is the success rate for detecting a correct homeomorphism type. Indeed, an incorrect graph can be perfect on other errors, e.g. MST(C) is extremely fast, has the zero geometric error (for many distances between a cloud and a reconstructed graph) and even has a correct homotopy type (no cycles) for any underlying tree T.

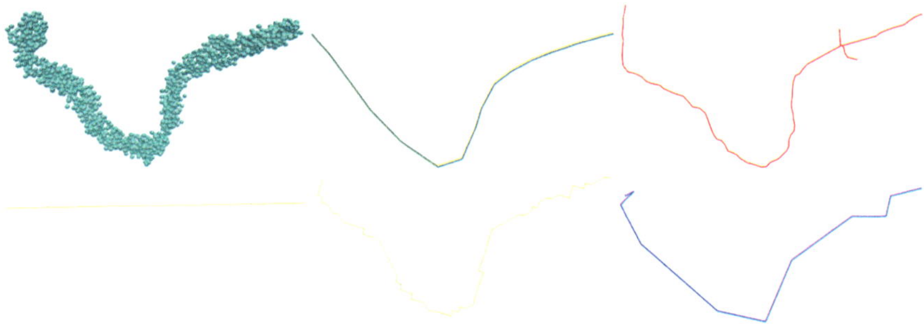

Fig. 5 **1st**: a cylindrical micelle with no branching vertices. **2nd**: Mapper, **3rd**: α-Reeb, **4th**: MGR, **5th**: DMT, **6th**: new ASk(C)

Fig. 6 **1st**: a branched micelle with exactly one degree 3 vertex, **2nd**: Mapper, **3rd**: α-Reeb, **4th**: MGR, **5th**: DMT, **6th**: new ASk(C)

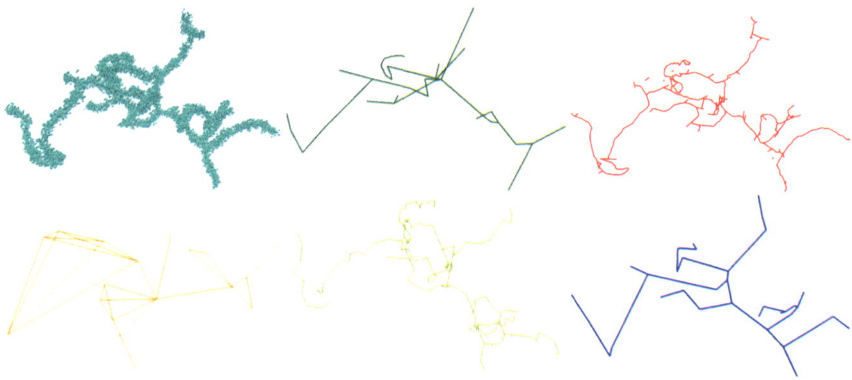

Fig. 7 **1st**: 'Christmas tree' micelle with several degree 3 vertices). **2nd**: Mapper, **3rd**: α-Reeb, **4th**: MGR, **5th**: DMT, **6th**: new ASk(C). All intersections come only from planar projections

Fig. 8 **1st**: a random point sample around an 8-star in $\mathbb{R}^3$, **2nd**: Mapper, **3rd**: α-Reeb, **4th**: MGR, **5th**: DMT, **6th**: new $\mathrm{ASk}(C)$

Hence the key results are in the middle column of Table 1 and the top right picture of Fig. 9. We included the success rate on the number of endpoints (degree 1 vertices) as a weaker topological error. As $\mathrm{MST}(C)$ shows above, only if an algorithm performs well on a topological reconstruction, it makes sense to evaluate the performance on other measures such as geometric distances and time.

Table 1 shows that the Mapper, MGR and DMT essentially depend on their parameters, because the success rates, run time and distance error significantly vary when the parameters are only slightly changed. The α-Reeb and ASk were stable, because Table 1 contains almost identical success rates for different parameters.

Both algorithms achieved best results on the most important measure of the homeomorphism success rate, though the top right picture in Fig. 9 highlights $\mathrm{ASk}(C)$ and MGR as the best for homeomorphism. In comparison with α-Reeb and MGR, the Approximate Skeleton $\mathrm{ASk}(C)$ is much faster and achieves similar distance errors, see the relevant results in both Table 1 and Fig. 9.

In addition to the comparison on more than 100 micelles, we have tested the algorithms on the much larger dataset of synthetic clouds generated as follows.

1) An N-*star* in $\mathbb{R}^3$ has one vertex at $0 \in \mathbb{R}^3$ and straight edges of length 100 to N *endpoints*in random directions with a minimum angle $\dfrac{\pi}{4}$ between edges.

Table 1 Columns 3–4 contain success rates for detecting the correct number of endpoints and a homeomorphism type of a graph over more than 100 real micelles. Column 5 contains the maximum Euclidean distance from points of a given cloud C to the reconstructed graph $G \subset \mathbb{R}^3$

Algorithm	Parameters	Endpoints success	Homeomorphism success	Time, ms	Max distance from C
Mapper	$\tau = 1.25$	54.21%	54.21%	**18**	4.59
Mapper	$\tau = 1.75$	66.36%	66.36%	**18**	4.62
Mapper	$\tau = 2.25$	68.20%	68.20%	19	4.67
MGR	$r_2 = 1$	48.60%	45.79%	25010	5.95
MGR	$r_2 = 1.5$	40.19%	40.19%	17410	5.51
MGR	$r_2 = 2$	29.91%	29.91%	25480	**3.46**
α-Reeb	$\alpha = 20$	**98.13%**	**98.13%**	367	10.49
α-Reeb	$\alpha = 25$	97.20%	97.20%	375	12.50
α-Reeb	$\alpha = 30$	**98.13%**	**98.13%**	373	14.19
DMT	$\delta = 0.1$	48.60%	45.79%	6290	5.95
DMT	$\delta = 0.2$	40.19%	40.19%	6192	5.51
DMT	$\delta = 0.3$	29.91%	29.91%	6410	**3.46**
ASk(C)	$\beta = 20$	**98.13%**	**98.13%**	42	5.16
ASk(C)	$\beta = 30$	**98.13%**	**98.13%**	42	5.16
ASk(C)	$\beta = 40$	97.20%	97.20%	42	5.31

2) For $N = 3, \ldots, 8$ and every of 100 random N-stars T, we found a minimum axis-aligned box containing T, enlarged this box by the noise bound of 10%.
3) We uniformly chose a random point p in the resulting box and checked if p is at a distance at most 10 (=10% of edge-lengths) from T. If successful, $500N$ such points form a noisy sample of the ground truth N-star $T \subset \mathbb{R}^3$.

Figure 9 shows 4 plots for the 4 error measures of 5 algorithms, which were averaged over 3 values of essential parameters as in Table 1. The Mapper threshold factor for single-edge clustering was $\tau \in \{1.25, 1.75, 2.25\}$. The α-Reeb scale was $\alpha \in \{20, 25, 30\}$. The branching factor of ASk(C) was $\beta \in \{20, 30, 40\}$.

For the correct number of endpoints, the new skeleton ASk(C) achieves 100% results on the synthetic clouds, because Definition 5 provides a very stable concept of a deep vertex not critically depending on a branching factor β. For the homeomorphism type, the minimum success of ASk(C) is 96%, because all short branches of MST(C) are removed to get core(C) homeomorphic to an underlying tree.

For the random point sample of the 8-star graph in Fig. 8, the 2nd, 3rd and 5th graphs have several branched vertices instead of one. The 5th graph has several zigzags, which would be straightened in core(C). The 4th graph has a triangular cycle because of incorrectly detected overlaps of clusters corresponding to vertices.

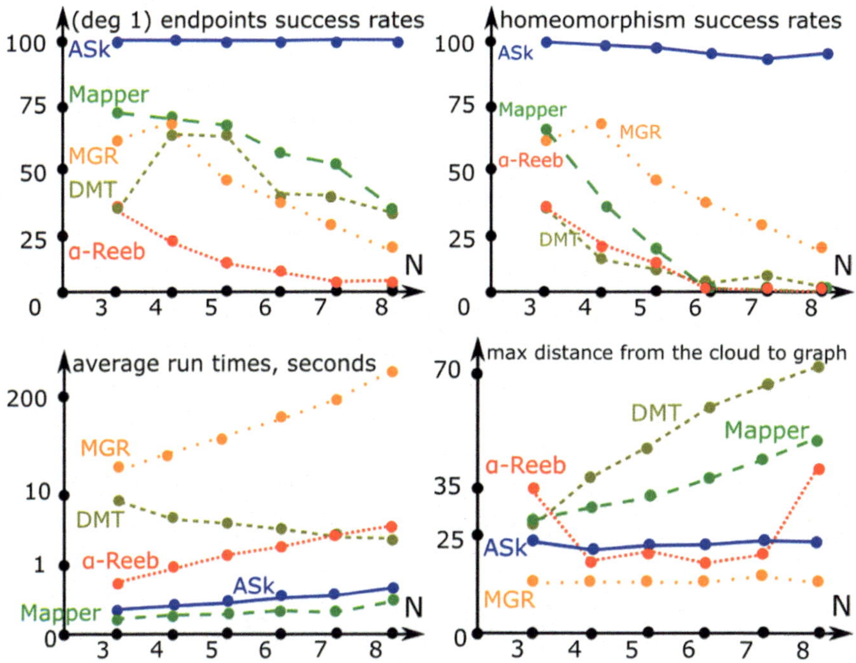

Fig. 9 For $N = 3, \ldots, 8$, each dot represents the average over 100 noisy samples around a random N-star graph in $\mathbb{R}^3$ and 3 parameters as in Table 1. Mapper: green and long-dashed, MGR: orange and sparsely dotted, α-Reeb: red and densely dotted, DMT: olive and short-dashed, ASk(C): blue and solid. **Top left**: the success rate in percentages for detecting a correct number of endpoints. **Top right**: the success rate in percentages for detecting the homeomorphism type. **Bottom left** (logarithmic scale): average run times in milliseconds. **Bottom right**: the max distance from a cloud C to reconstructed graphs. The exact numbers are in the txt files in the supplementary materials

6 Conclusions and a Discussion of the Approximate Skeleton

Though the current implementation was tested in $\mathbb{R}^3$, all steps and results work in any $\mathbb{R}^m$. Here is the summary of the key contributions to data skeletonization.

- The detection of deep (branched) vertices in Definition 5 uses a global structure of longest paths within MST(C), hence is more stable under a change of parameters.
- To improve the Metric Graph Reconstruction by M. Aanjaneya et al. [1], we have split one parameter r (used for detecting vertex points and also for clustering later) into two separate parameters (with default values) $r_1 = 15$, $r_2 \in [1, 2]$, which led to more successful (20–40% rates instead of 0%) reconstructions in Table 1.
- Theorem 8 proves the first size guarantees (on a small number of vertices) for the Approximate Skeleton ASk(C), while all past methods from Sect. 2 considered topological (mostly homotopy type) or metric properties of reconstructed graphs.

- Corollary 9 says that the Approximate Skeleton $\text{ASk}(C)$ can be quickly computed within a given error as required in the Tree Reconstruction Problem from Sect. 1.

Because of the page limit the last author couldn't include one more result on $\text{ASk}(C)$ with realistic conditions on an underlying tree $T \subset \mathbb{R}^m$ and its noisy sample C to guarantee that $\text{MST}(C)$ and $\text{ASk}(C)$ are homeomorphic to T. This is the first advance after Giesen's guarantees for shortest paths through sample points [12] in 1999. The C++ code of $\text{ASk}(C)$ is at https://github.com/YuryUoL/AsKAlgorithm.

In comparison with the past methods in Sect. 2, $\text{ASk}(C)$ starts from the most challenging input (an unorganized cloud of points $C \subset \mathbb{R}^m$ without any extra structure), outputs an embedded graph in $\mathbb{R}^m$ and provides two guarantees in Theorem 8 and Corollary 9. We plan to extend $\text{ASk}(C)$ to graphs with cycles, which can be still visualized [16, 17, 23] and encoded [18] in a 3-page book (a product of a line and the tree with 3 edges joined at one central vertex).

Acknowledgement This research was supported by the EPSRC grant EP/R018472/1. We thank all reviewers for their helpful suggestions.

Appendix A: Proofs of all the Statements from Sect. 4

The proof of Corollary 9 below uses Algorithm 1 for depths of vertices in $\text{MST}(C)$ from Definition 5. The depth is trivial (equal to 0) for any degree 1 vertex. For any other vertex v, the depth can be recursively computed from lengths of edges at v and depths of neighbors of v. Imagine a water flow simultaneously starting from all degree 1 vertices of $\text{MST}(C)$ and moving towards internal vertices inside $\text{MST}(C)$. At every vertex v of degree $k \geq 3$, the flow waits until v is reached from $k-1$ directions (edges at v), then the flow moves further in the remaining k-th direction.

Definition 11 introduces a new distance between a cloud $C \subset \mathbb{R}^m$ and a polygonal line. Recall that d is the Euclidean distance in $\mathbb{R}^m$. We assume that the points $C = \langle p_1, \ldots, p_n \rangle \subset \mathbb{R}^m$ are ordered by their orthogonal projections to the line $[p_1 p_n]$.

Definition 11 *For $1 \leq i < s < j \leq n$, let $H(p_s) \subset \mathbb{R}^m$ be the hyperspace that is orthogonal to $[p_1 p_n]$ and passes through p_s. The* distance *between p_s and $[p_i p_j]$ is measured orthogonally to $[p_1 p_n]$ as $d(p_s, [p_i p_j]) = d(p_s, H(p_s) \cap [p_i p_j])$, see Fig. 10. Consider the* distance $d([p_i p_j], C) = \max_{i<s<j} d(p_s, H(p_s) \cap [p_i p_j])$. *For $1 \leq \text{ind}(1) < \cdots < \text{ind}(k) \leq n$, the* distance *between C and the polygonal line $P = \langle p_{\text{ind}(1)}, \ldots, p_{\text{ind}(k)} \rangle$ is defined as $d(P, C) = \max_{2 \leq i \leq k} d([p_{\text{ind}(i-1),\text{ind}(i)}], C)$.*

Lemma 12 below justifies the steps of Stage 2 in Sect. 4, which outputs the Approximate Skeleton $\text{ASk}(C)$ starting from $\text{core}(C)$ obtained in Stage 1 at the end of Sect. 3.

Lemma 12 *Let $C = \langle p_1, \ldots, p_n \rangle$ be points ordered according to their orthogonal projections to $[p_1, p_n]$. For $\varepsilon > 0$, one can find indices $1 = \text{ind}(1) < \cdots < \text{ind}(m) =$*

Fig. 10 The distance from p_m to $[p_i p_j]$ in Definition 11 is measured orthogonally to $[p_1 p_n]$

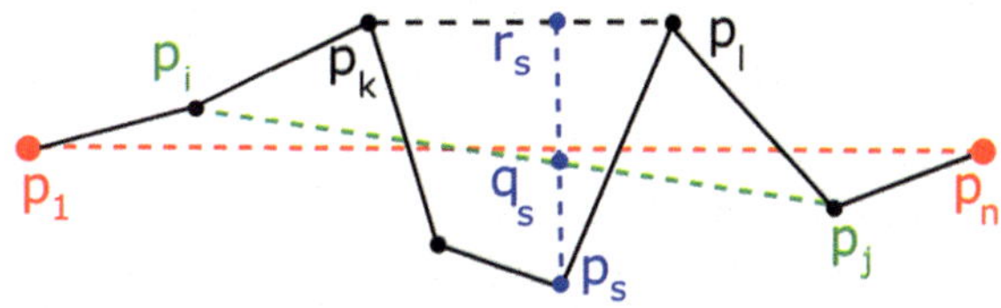

n in time $O(n \log n)$ *so that the estimates below for the distances in Definition 11 hold:*
(a) $d([p_{\mathrm{ind}(i-1)} p_k], C) \le \varepsilon$ *for* $\mathrm{ind}(i-1) < k \le \mathrm{ind}(i)$*,*
(b) $d([p_{\mathrm{ind}(i-1)} p_{\mathrm{ind}(i)+1}], C) > \varepsilon$ *for any* $1 < i < m$ *.*

The following lemma is needed for Theorem 8 and is conveniently illustrated in Fig. 10 below. Recall that the distances from Definition 11 are computed orthogonally to the straight segment $[p_1, p_n]$ passing through the endpoints of a monotone point cloud C.

Lemma 13 *Let* $C = \langle p_1, \ldots, p_n \rangle$ *be points ordered according to their orthogonal projections to* $[p_1 p_n]$*. Then* $d([p_k p_l], C) \le 2d([p_i p_j], C)$ *for any indices* $i \le k < l \le j$*.*

Input: the initial tree T = MST(C)
Initialize Minimal Binary Heap H of (vertex,depth)
For all deg 1 vertices $v \in T$, add $(v, 0)$ to H;
while H is not empty **do**
  (v, d) = H.pop(); // take the vertex v of a min depth
  set u = the only neighbor of v in T;
  $d_{new} = d +$ edge-length of (u, v);
  Remove the edge uv from T, but keep $u, v \in T$;
  Add (v, d_{new}) to front of the list Neighbors(u);
  if deg(u) = 0 in T **then**
    Set $flag[u] = true$;
  else if deg(v) = 1 **then**
    Add the pair (u, d_{new}) to the heap H
  end if
end while
Initialize a vector[] depths; // a future output
for all deg(v) > 2 vertices $v \in$ MST(C) **do**
  if $flag[v] = true$ **then**
    Set depths[v] = 3rd element of Neighbors(v);
  else
    Set depths[v] = 2nd element of Neighbors(v);
  end if
end for

Algorithm 1: Computing depths of vertices from Definition 5 in Step 1c by 'simultaneous flows' moving from endpoints.

Input: unordered set of points s and line l
Output: Sequence of vertices that form monotone paths k and t containing ordering of points s.
Define t to be ordering of s obtained from projecting s to l orthogonaly.
Define q to be the map from points s to their indices in t.
Define k to be queue and add first point of s to k.
Define a to be the first point of s.
while Point a is not the last point of t **do**
  Let point b be the next point from a in ordering t
  while Point b is not the last point of t and $k[a] < k[b]$ **do**
    Set a to be b.
    Set b to be the next point from b in ordering t.
  end while
  Add b to k.
  if b is the last point in ordering t **then**
    Exit the program.
  end if
  while Point b is not the last point of t and $k[b] < k[a]$ **do**
    Set a to be b.
    Set b to be the next point from b in ordering t.
  end while
  Add b to k.
end while

Algorithm 2: *monotone* subclouds of C. For each point p in a given cloud $C \subset \mathbb{R}^m$, find approximately its closest edge of core(C). Then any edge $e \subset$ core(C) has the *edge-cloud* $C(e) \subset C$ of points that are closer to e than to other edges of core(C). For every polygonal path $v_1, \ldots, v_k$ between non-trivial vertices $v_1, v_k \in$ core(C) define cloud $Y = \cup_{i=1}^{k-1} C((v_i, v_{i+1}))$ and straight line L spanned by points v_1 and v_k. Run the algorithm above with parameters (Y, L).

Proof. For any $k < m < l$, let $H(p_m) \subset \mathbb{R}^m$ be the hyperspace that is orthogonal to $[p_1 p_n]$ and passes through p_m. Consider the intersection points $q_m = H(p_m) \cap [p_i p_j]$ and $r_m = H(p_m) \cap [p_k p_l]$. Let $\varepsilon = d([p_i p_j], C)$, then $d(p_m, q_m) = d(p_m, [p_i p_j]) \leq \varepsilon$. Since the points p_k and p_l are ε-close to the segment $[p_i p_j]$, the intermediate point $r_m \in [p_k p_l]$ is also ε-close to $[p_i p_j]$, i.e. $d(r_m, q_m) = d(r_m, [p_i p_j]) \leq \varepsilon$. The triangle inequality implies that $d(p_m, r_m) \leq d(p_m, q_m) + d(q_m, r_m) \leq 2\varepsilon$. Taking the maximum over $k < m < l$, we get $d([p_k p_l], C) \leq 2\varepsilon$. □

Proof of Lemma 12. Assuming that indices $1 = \text{ind}(1) < \cdots < \text{ind}(i-1)$ were found, we search for the next index ind(i) as follows. Search exponentially by trying indices $k = \text{ind}(i-1) + 2^j$ for $j = 0, 1, \ldots$ while $d([p_{\text{ind}(i-1)} p_k], C) \leq \varepsilon$.

Each evaluation of the distance $d([p_{\text{ind}(i-1)} p_k], C)$ requires $O(k - \text{ind}(i-1))$ time, because we need to compare $k - \text{ind}(i-1) - 1$ distances to $[p_{\text{ind}(i-1)} p_k]$ (orthogonally to $[p_1 p_n]$) from every point of C between $p_{\text{ind}(i-1)}$ and p_k.

After finding $k = \text{ind}(i-1) + 2^j$ and $l = \text{ind}(i-1) + 2^{j+1}$ such that $d([p_{\text{ind}(i-1)} p_k], C) \leq \varepsilon$ and $d([p_{\text{ind}(i-1)} p_l], C) > \varepsilon$, we start a binary search for ind(i) in the range $[k, l)$ each time choosing one half of the current range until both conditions (a)-(b) hold.

Finding the next index $\text{ind}(i)$ requires $O(\log n)$ computations for the distance $d([p_k p_l], C)$, where $l - k \le \text{ind}(i) - \text{ind}(i-1)$, hence $O((\text{ind}(i) - \text{ind}(i-1)) \log n)$ time overall. Taking the sum over all $i = 2, \ldots, m$, the total time is $O(n \log n)$. □

Proof of Theorem 8. Since endpoints of all monotone polygonal paths of $\text{core}(C)$ are fixed in minimization problem before Theorem 8, we separately consider every corresponding monotone subcloud C' of points (say) $p_1, \ldots, p_n$ ordered by their orthogonal projections to the line through the line segment $[p_1 p_n]$. Let $1 = \text{opt}(1) < \cdots < \text{opt}(k) = n$ be indices of an optimal ε-approximation (polygonal path) Q to C'. In the notations of Lemma 12 for the approximation error 2ε we will prove below that $\text{opt}(i) \le \text{ind}(i)$ by induction on i. Then $n = \text{opt}(k) \le \text{ind}(k)$ and the size m of the list $1 = \text{ind}(1) < \cdots < \text{ind}(m) = n$, which is found in Lemma 12, is at most k as required. Taking the sum of upper bounds over all monotone paths of $\text{core}(C)$, we conclude that the 2ε-Approximate Skeleton $\text{ASk}(C)$ has the total number of vertices not greater than that that number for an ε-optimal skeleton S.

The base $i = 1$ means that $\text{opt}(1) = 1 = \text{ind}(1)$, i.e. both paths start from the point p_1. In the inductive step assume that $\text{opt}(i-1) \le \text{ind}(i-1)$. If $\text{opt}(i) \le \text{ind}(i-1)$, then $\text{opt}(i) \le \text{ind}(i)$ and the inductive step is complete. The remaining case is $\text{ind}(i-1) < \text{opt}(i)$. Since Q is an ε-approximation to C', we have $d([p_{\text{opt}(i-1)} p_{\text{opt}(i)}], C) \le \varepsilon$. Lemma 13 implies that $d([p_{\text{ind}(i-1)} p_l], C) \le 2\varepsilon$ for any index l such that $\text{ind}(i-1) < l \le \text{opt}(i)$. Lemma 12(b) for the approximation 2ε says that $d([p_{\text{ind}(i-1)} p_{\text{ind}(i)+1}], C) > 2\varepsilon$, hence $\text{opt}(i) \le \text{ind}(i)$. □

Definition 14 (expanion constants) *Let $C \subset \mathbb{R}^m$ be a cloud and $\bar{B}(p; r) = \{q \in \mathbb{R}^m \mid d(p, q) \le r\}$ be the closed ball with the center p and radius r. The* expansion *constant c_e is the smallest real number $c \ge 2$ such that $\forall x : |\bar{B}(x, 2r)| < c|\bar{B}(x, r)|$. Let c_s be the similarly defined constant for the metric space of line segments of* $\text{MST}(C)$, *then set $c = \max\{c_e, c_s\}$. Other constants c_p, c_l are similarly defined in [3].*

Proof of Corollary 9. The distance from Definition 11 measured orthogonally to the straight line through fixed endpoints $[p_1 p_n]$ is not smaller than the Hausdorff distance used for ε-offsets in Definition 1. Hence the algorithm from Lemma 12 produces required ε-approximations in the sense of Definition 1. The current implementation uses the single-edge clustering based on $\text{MST}(C)$, so Step 1a runs in $O(n)$ time. The total time is dominated by Step 1b computing $\text{MST}(C)$ in time $O(\max\{c_e^6, c_p^2 c_l^2\} c_e^{10} n \log n \, \alpha(n))$, where $\alpha(n)$ is the inverse Ackermann function.

Algorithm 1 in Step 1c has the pseudo-code above and maintains a binary tree on $O(n)$ vertices, which requires $O(n \log n)$ time. Selecting deep vertices in Step 1d and finding longest paths in Step 1e within subtrees of $\text{MST}(C)$ needs $O(n \log n)$ time by classical algorithms [10]. Step 2a to split C into subclouds is implemented by cover trees for line segments of $\text{core}(C) \subset \text{MST}(C)$ in time $O(c_s^{16} n \log n)$ as proved in [3]. By Lemma 12 Steps 2b-2g for approximating any subcloud of n_i points by a polygonal path runs in $O(n_i \log n_i)$ time. Hence the total time at Stage 2 for computing $\text{ASk}(C)$ over the cloud C of n points is $O(\max\{c^6, c_p^2 c_l^2\} c^{10} n \log n \, \alpha(n))$. □

Fig. 11 **Left**: ASk(C) for the branching factor $\gamma = 1.2$. **Middle**: $\gamma = 1.4$, **Right**: $\gamma = 1.6$

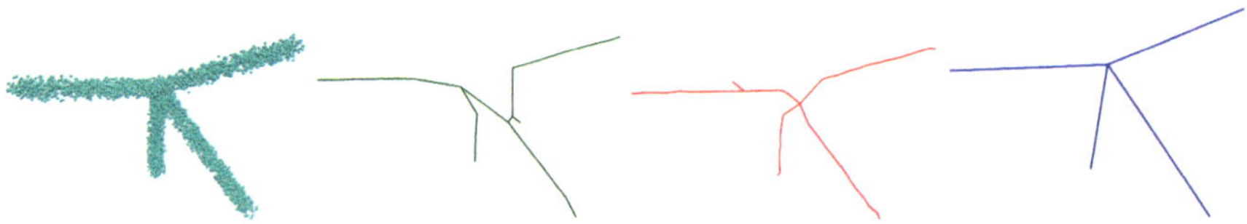

Fig. 12 **1st**: a sample around a 4-star in $\mathbb{R}^3$, **2nd**: Mapper, **3rd**: α-Reeb, **4th**: ASk(C)

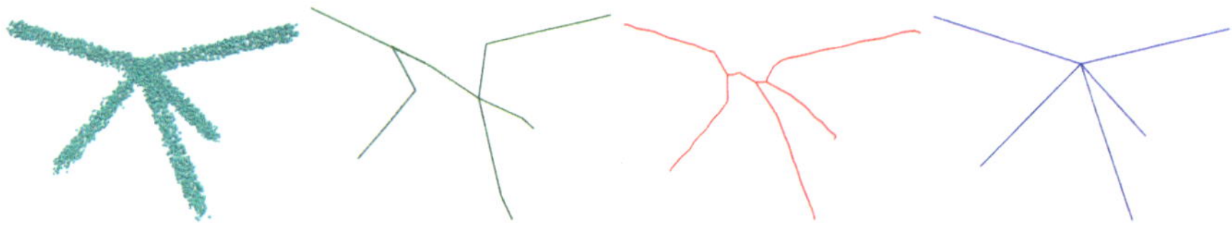

Fig. 13 **1st**: a sample around a 5-star in $\mathbb{R}^3$, **2nd**: Mapper, **3rd**: α-Reeb, **4th**: ASk(C)

Fig. 14 **1st**: a sample around a 6-star in $\mathbb{R}^3$, **2nd**: Mapper, **3rd**: α-Reeb, **4th**: ASk(C)

Appendix B: More Qualitative Comparisons of 3 Algorithms

Figures 12, 13, 14 and 15 show example outputs of 3 algorithms on real and randomly generated clouds in $\mathbb{R}^3$. In almost all cases the Mapper and α-Reeb graphs contain superfluous short edges, which affect the homeomorphism types.

The error factor γ from Corollary 9 affects the quality of approximation. Figure 11 shows that higher values of γ lead to more straightened curves (Fig. 16).

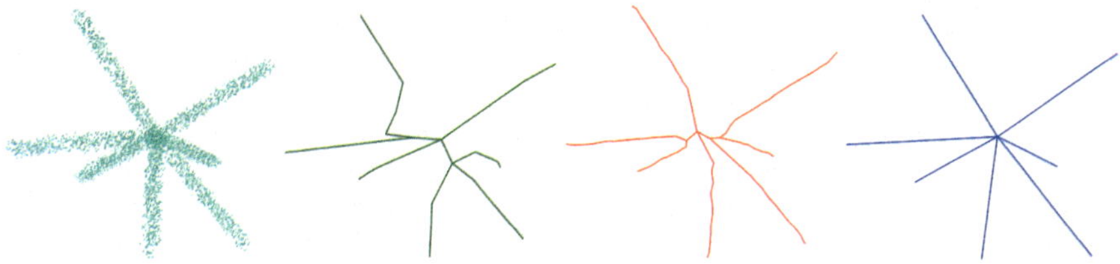

Fig. 15 **1st**: a sample around a 7-star in $\mathbb{R}^3$, **2nd**: Mapper, **3rd**: α-Reeb, **4th**: ASk(C)

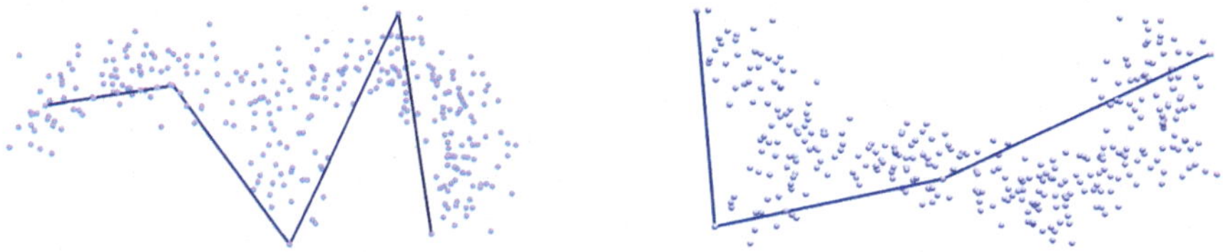

Fig. 16 Explaining 1.87% failures of ASk(C) in Table 1: two micelles C with short edges

Appendix C: Conditions on a Sample C of T so that MST$(C) \approx T$

For a noisy point cloud $C \subset \mathbb{R}^m$, a Minimum Spanning Tree MST(C) often looks as a hairy bunch of caterpillars with many short branches, which is far from an ideal skeleton approximating C. Theorem 15 gives the first ever (to the best of our knowledge) sufficient conditions on a noisy point cloud $C \subset \mathbb{R}^m$ when MST(C) is homeomorphic to an underlying tree T from which C was sampled.

Theorem 15 *Let C be a noisy sample of a straight-line tree $T \subset \mathbb{R}^m$. Conditions (15a)–(15g) on C and T below are sufficient for all 3 trees* MST(C), core(C) *and* ASk(C) *to be homeomorphic to T.*
(15a) C is geometrically close to T, i.e. $C \subset T^{\varepsilon}$ for some $\varepsilon > 0$;
(15b) the vertices $V(T)$ of T are approximated by C: $V(T) \subset C^{\varepsilon}$;
(15c) C is sufficiently sparse in the sense that the minimum distance δ between points of C has the lower bound $\frac{4}{\sqrt{3}}\varepsilon$;
(15d) the tree T is approximated by C, i.e. $T \subset C^{\rho}$ for some parameter $\rho > \frac{\delta}{2}$, whose upper bound is restricted by (15f) below;
(15e) the non-adjacent edges of the tree T are sufficiently away from each other, i.e. the distance between them is at least $2(\rho + \varepsilon)$.
(15f) the tree T has no small angles, i.e. the minimum angle γ between any adjacent edges of T satisfies $\sin\frac{\gamma}{2} > \frac{\rho+\varepsilon}{\delta-\varepsilon}$.
(15g) T has no short edges: the minimum edge-length is $2(\beta\rho + \varepsilon)$, where β is the branching factor from Definition 5.

Most of conditions (15a)–(15g) are trivially necessary for MST$(C) \approx T$. For example, T should be well-approximated by C in (15a)–(15b) with only small

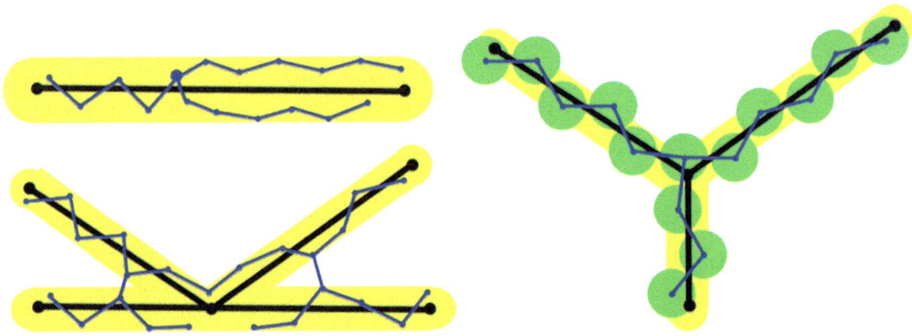

Fig. 17 A tree T is black, an ε-offset T^{ε} is yellow, C and MST(C) are blue. **Left**: hard cases when MST(C) $\not\approx T$ (too dense C or small angles of T). **Right**: a homeomorphism MST(C) $\approx T$ is guaranteed by Theorem 15, disjoint $\frac{\delta}{2}$-balls at the points of C are green

bounded noise, otherwise MST(C) will have long branches to outliers in C. The left picture in Fig. 17 shows challenging cases when ε-samples C of simple trees T can easily led to MST(C) $\not\approx T$, so Theorem 15 is really non-trivial.

Condition (15c) says that the points of C should be closer to T than to each other, though the minimal ratio $\delta/\varepsilon \approx 2.3$ isn't large. This sparsity guarantees that any edge of T will be approximated by a polygonal line of MST(C) without branching vertices. These polygonal lines of MST(C) can't have large zigzags between non-adjacent edges of T by (15e). Conditions (15d) and (15f) can be combined into the following 2-sided inequality
(15h) $\frac{\delta}{2} < \rho < (\delta - \varepsilon)\sin\frac{\gamma}{2} - \varepsilon$, which implies $\sin\frac{\gamma}{2} > \frac{1}{2}$, $\gamma > \frac{\pi}{3}$.

To guarantee the existence of ρ by (15h), we check when the left hand side is less than the right hand side: $\frac{\delta}{2} < (\delta - \varepsilon)\sin\frac{\gamma}{2} - \varepsilon$. After dividing both sides by ε, we get $(\frac{\delta}{\varepsilon} - 1)\sin\frac{\gamma}{2} - 1 > \frac{\delta}{2\varepsilon}$. Then the *sparsity* ratio satisfies the inequality

$$\frac{\delta}{\varepsilon} > \frac{\sin\frac{\gamma}{2} + 1}{\sin\frac{\gamma}{2} - \frac{1}{2}} \qquad (15i).$$

Inequality (15i) can be satisfied for any $\gamma > \frac{\pi}{3}$ when $\sin\frac{\gamma}{2} - \frac{1}{2} > 0$. For example, if $\gamma = \frac{\pi}{2}$, then $\frac{\delta}{\varepsilon} > \frac{\sqrt{2}+2}{\sqrt{2}-1} \approx 8.2$. If we set $\delta = 10\varepsilon$, (15i) gives $5\varepsilon < \rho < (\frac{9}{\sqrt{2}} - 1)\varepsilon \approx 5.36\varepsilon$. For the larger minimum angle $\gamma = \frac{2\pi}{3}$, inequality (15i) allows smaller ratios $\frac{\delta}{\varepsilon} > \frac{\sqrt{3}+2}{\sqrt{3}-1} \approx 5.1$. For $\delta = 6\varepsilon$ the range of ρ is $3\varepsilon < \rho < (5\frac{\sqrt{3}}{2} - 1)\varepsilon \approx 3.33\varepsilon$.

The homeomorphism guarantee from Aanjaneya et al. [1, Theorem 5] is incomparable with Theorem 15, because these results are stated for different inputs and metrics on graphs. For a rough comparison, [1, Theorem 5] says that, for a correct topological reconstruction from an (ε, R)-approximation, the shortest edge-length of a hidden graph is between 120ε and $4R$.

Conditions (15a)–(15g) may not be necessary for core(C), ASk(C) $\approx T$, because removing short branches from MST(C) produces core(C) often homeomorphic to T as justified by the experiments in Sect. 5. The key idea in the proof of Theorem 15 is to view the points of C as non-overlapping $\frac{\delta}{2}$-balls that sit ε-close to the tree T.

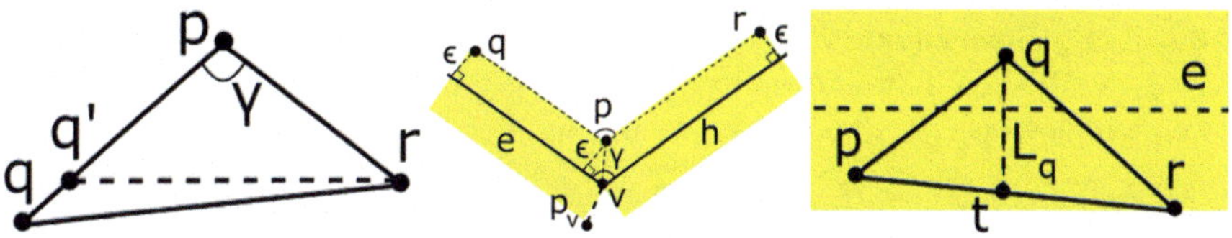

Fig. 18 **Left**: a proof of $d(q,r) \geq 2l \sin\frac{\gamma}{2}$ in Lemma 18. **Middle**: a proof of $d(q,r) > 2\rho$ in Lemma 20. **Right**: $\angle qpr < \frac{\pi}{3}$ in Lemma 22

If these balls are large in comparison with ε, their centres are connected following the shape of T without confusion. So connecting centers of the balls in a natural (optimal as in MST) way gives a tree homeomorphic to T.

The claims below are stated under the conditions of Theorem 15 and are needed to prove Proposition 25 to explicitly describe a structure of MST(C).

Lemma 16 *Any vertex v of T has a unique point $p_v \in C$ in the ε-ball v^ε at v.*

Proof. Condition (15b) implies that C has a point p_v the ε-ball v^ε at v. If there is another point $q \in C \cap v^\varepsilon$, then $d(p,q) \leq 2\varepsilon < \delta$, which contradicts (15c). □

Definition 17 *For an edge e of a tree $T \subset \mathbb{R}^m$, the* edge-cloud *$C_e \subset C$ consist of points that are in the ε-offset of e, but not in the ε-balls at the endpoints of e.*

Lemma 18 *For any non-collinear points $p, q, r \in \mathbb{R}^m$, if the angle $\angle qpr \geq \gamma$ and both Euclidean distances $d(p,q), d(p,r) \geq l$, then $d(q,r) \geq 2l \sin\frac{\gamma}{2}$.*

Proof. In the triangle $\triangle pqr$ if we keep the side lengths $d(p,q), d(p,r)$ and decrease the angle $\angle qpr$ to γ, then $d(q,r)$ can only decrease, so we can assume that $\angle qpr = \gamma$. Assuming that $d(p,q) \geq d(p,r)$, push q to the point $q' \in [pq]$ such that $d(p,q') = d(p,r)$, see the 2nd picture of Fig. 17. In the isosceles $\triangle pq'r$ the angle $\angle pq'r$ is acute. In $\triangle qq'r$ the obtuse angle $\angle qq'r$ is largest, hence the opposite side $[qr]$ is the longest, so $d(q,r) \geq d(q',r)$. The isosceles triangle $\triangle pq'r$ with $\angle q'pr = \gamma$ has $d(q',r) = 2d(p,r)\sin\frac{\gamma}{2} \geq 2l\sin\frac{\gamma}{2}$, so $d(q,r) \geq 2l\sin\frac{\gamma}{2}$. □

The ε-offset of any edge e in a straight-line tree $T \subset \mathbb{R}^m$ is the solid cylinder of the radius ε around e united with the ε-balls centered at the endpoints of e.

Lemma 19 *For any disjoint edges e, h of $T \subset \mathbb{R}^m$, the two subsets of C in $e^\varepsilon, h^\varepsilon$ (including the edge-clouds C_e and C_h) are more than 2ρ away from each other.*

Proof. For any points $p \in e^\varepsilon$ and $q \in h^\varepsilon$, their distances to the corresponding edges are $d(p,e) \leq \varepsilon$ and $d(q,h) \leq \varepsilon$. Since $d(e,h) > 2(\rho + \varepsilon)$ by condition (15e), the triangle inequality implies that $d(p,q) \geq d(e,h) - d(p,e) - d(q,h) > 2\rho$. □

Lemma 20 *For any adjacent edges e, h sharing a vertex v of a tree $T \subset \mathbb{R}^m$, the edge-clouds C_e and C_h are more than 2ρ away from each other.*

Proof. Let p_v be a unique vertex of C from Lemma 16. For any points $q \in C_e$ and $r \in C_h$, we'll show how inequality (15g) implies that $d(q, r) > 2\rho$ by using the lower bounds $d(p_v, q), d(p_v, r) \geq \delta$ from condition(15c).

Move the points $q \in C_e$ and $r \in C_h$ towards each other until they are on the boundaries of the ε-offsets e^ε, h^ε, respectively, so that $d(q, e) = \varepsilon = d(r, h)$.

Let p be the point in the bisector between the edges e, h meeting at v such that $[pq], [pr]$ are parallel to e, h, respectively, so that $d(p, e) = \varepsilon = d(p, h)$.

In the 1st picture of Fig. 18 in the right-angle triangle with the hypotenuse $[pv]$ and $\frac{1}{2}\angle(e, h) \geq \frac{\gamma}{2}$, we estimate $d(v, p) = \varepsilon / \sin \frac{1}{2}\angle(e, h) \leq \varepsilon / \sin \frac{\gamma}{2}$.

The triangle inequality implies that $d(p, p_v) \leq d(p_v, v) + d(v, p) \leq s\varepsilon$, where $s = 1 + 1/\sin \frac{\gamma}{2}$, and $d(p, q) \geq d(q, p_v) - d(p, p_v) \geq \delta - s\varepsilon$, also $d(p, r) \geq \delta - s\varepsilon$.

Lemma 18 for $l = \delta - s\varepsilon$ says that $\frac{1}{2}d(q, r) \geq (\delta - s\varepsilon) \sin \frac{\gamma}{2}$. After substituting $s = 1 + 1/\sin \frac{\gamma}{2}$, the last inequality becomes $\frac{1}{2}d(q, r) \geq (\delta - \varepsilon) \sin \frac{\gamma}{2} - \varepsilon > \rho$, which follows from (15g). Hence $d(q, r) > 2\rho$ as required. □

Lemmas 19 and 20 imply that the cloud C splits into the edge-clouds C_e and singletons $\{p_v\}$ over all edges e and vertices v of T.

Lemma 21 *All edges of* MST(C) *are not longer than* 2ρ *and not shorter than* δ, *which justifies condition (15d)* $\rho > \frac{\delta}{2}$.

Proof. Assume that the offset C^ρ is disconnected. If the tree T is in a single component of C^ρ, another component contains a point $p \in C$ that is more than $\rho > \frac{\delta}{2} > \varepsilon$ away from T, see (15cd), which contradicts condition (15a).

Since the connected tree T cannot overlap with more than one component of C^ρ, the offset C^ρ is connected. Then MST(C) is contained in the union C^ρ of ρ-balls with centers over all $p \in C$, so every edge of MST(C) is at most 2ρ

Condition (15c) means that the open $\frac{\delta}{2}$-balls centered at points of C are all disjoint, hence a shortest edge of MST(C) is at least δ. □

Lemma 22 *For any edge e of a straight-line tree $T \subset \mathbb{R}^m$, let $p, q, r \in e^\varepsilon$ be points ordered by their orthogonal projections to the straight line through e. If $d(p, q), d(q, r) \geq \delta > \frac{4}{\sqrt{3}}\varepsilon$, then the edge $[pr]$ is longest in the triangle $\triangle pqr$.*

Proof. The longest side is opposite to the largest angle. If both angles $\angle qpr$, $\angle qrp$ aren't greater than $\frac{\pi}{3}$, the remaining angle $\angle pqr \geq \frac{\pi}{3}$ should be the largest.

The 2nd picture of Fig. 18 shows the 2D section of the offset e^ε by the plane through the points p, q, r. In this 2D section the line L_q passing through the middle point q orthogonally to e meets the side $[pr]$ in a point t. The sine theorem in the triangle $\triangle pqt$ says that $d(p, q) \sin \angle qpr = d(q, t) \sin \angle ptq \leq d(q, t)$, which isn't greater than the width 2ε of the offset e^ε, so $\sin \angle qpr \leq \frac{2\varepsilon}{d(p,q)}$. The inequalities

$d(p, q) \geq \delta > \frac{4}{\sqrt{3}}\varepsilon$ imply that $\sin \angle qpr \leq \frac{\sqrt{3}}{2}$. Then $\angle qpr \leq \frac{\pi}{3}$ and similarly we get $\angle qrp \leq \frac{\pi}{3}$. □

Lemma 23 *For any edge* $e \subset T$, $\mathrm{MST}(C_e)$ *is a polygonal line going via points of the edge-cloud* C_e *according to their ordered orthogonal projections to* e.

Proof. By Lemma 22 two points $p, q \in C_e$ can be linked by an edge of $\mathrm{MST}(C_e)$ only if they have successive orthogonal projections to e (when there is no other point $r \in C_e$ whose projections separates the projections on p, q to e). □

Lemma 24 *For any edge* $e \in E(T)$ *with an endpoint* v, *let* $p_{ev} \in C_e$ *be the closest point from* C_e *to* p_v. *The whole edge-cloud* C_e *orthogonally projects to the infinite straight line through* e *on one side from the projection of* p_v.

Proof. The projection of p_v from the ε-ball v^{ε} to the line through e is the ε-neighborhood of the endpoint v. The edge-cloud C_e belongs to the solid ε-cylinder minus δ-ball around p_v (or minus the $(\delta - \varepsilon)$-ball around v), which projects to e more than ε away from v since $\delta > 2\varepsilon$ by (15c). □

Proposition 25 $\mathrm{MST}(C)$ *is the union* $\bigcup_{e \in E(T)} \mathrm{MST}(C_e) \bigcup_{v \in V(T)} [p_v, p_{ve}]$, *where* $p_v \in C$ *is a unique point from Lemma 16 for every vertex* $v \in V(T)$, *and* $p_{ev} \in C_e$ *is the closest point from the edge-cloud* C_e *to* p_v *for each edge* e *attached at* v.

Proof. By Lemma 21 any edge of $\mathrm{MST}(C)$ is at most 2ρ. Fix an edge e of T with endpoints u, v. By Lemmas 19 and 20 a point $p \in C_e$ can be linked by an edge of $\mathrm{MST}(C)$ only to points from the edge-cloud C_e or to the points p_u, p_v from Lemma 16, because all other points of C are more than 2ρ away from p. Hence all edges of $\mathrm{MST}(C)$ between points $p, q \in C_e$ depend only on the edge-cloud C_e and should belong to $\mathrm{MST}(C_e)$, so $\mathrm{MST}(C_e) \subset \mathrm{MST}(C)$.

For any edge e attached at a vertex v, the point p_v can be linked by an edge of $\mathrm{MST}(C)$ only to the point $p_{ev} \in C_e$ from Lemma 24. Indeed, let p_v be linked to two points $q, r \in C_e \subset e^{\varepsilon}$, Then q, r have orthogonal projections to the line through e on the same side from the projection of p_v by Lemma 24. Lemma 22 says that $[qr]$ is one of the shortest edges in the triangle $p_v qr$, hence $\mathrm{MST}(C)$ cannot include both edges $[p_v q]$ and $[p_v r]$.

For any edge e, the subtree $\mathrm{MST}(C_e)$ can be connected by edges of $\mathrm{MST}(C)$ only to the points p_u, p_v around the endpoints u, v of e. If one of these edges $[p_v p_{eu}]$ and $[p_v p_{ev}]$ is absent, then $\mathrm{MST}(C)$ splits into two components: one with C_e and one with p_v. Indeed, all other possible connections between singletons $\{p_v\}$ and $\mathrm{MST}(C_e)$ follow the combinatorial structure of T, which becomes disconnected after removing only one incidence of v and e. Hence $\mathrm{MST}(C)$ contains all expected edges $[p_v, p_{ve}]$, which join $\mathrm{MST}(C_e)$ over all $e \in E(T)$ into a connected tree by simulating the incidence of vertices and their edges in T. □

Proof of Theorem 15. In the notations of Proposition 25, MST(C) consists of polygonal lines that approximate all edges e of T and connect at points p_v close to vertices v of T. By Lemma 23, every polygonal line MST(C_e) is homeomorphic by the orthogonal projection to a subedge e' of the corresponding edge e.

These projections MST(C_e) $\to e'$ can be extended to the full homeomorphism MST(C) $\to T$ by mapping each point p_v from Lemma 16 to the corresponding vertex v, and the union of edges $[p_v, p_{ev}]$ around p_v to a small star in T with the center v and short arcs to the endpoints of subedges e' of edges at v.

Having proved that MST(C) $\approx T$, we show that every polygonal path P between non-trivial vertices of MST(C) is longer than $\beta l(C)$, where β is the branching factor from Definition 5 and $l(C)$ is the average edge-length of MST(C). Then Definitions 5–6 guarantee that P is included into core(C), hence core(C) = MST(C). Condition (15g) says that any edge e of T has a length at least $2(\beta\rho + \varepsilon)$. By condition (15b) the endpoints of the corresponding path $P \subset$ MST(C) are at most ε away from the vertices of e. So the length of P is at least $2\beta\rho \geq \beta l(C)$, because any edge of MST(C) is not longer than 2ρ by Lemma 21.

Having proved that $T \approx$ MST(C) = core(C), we conclude that $T \approx$ ASk(C), because the straightening core(C) $\to$ ASk(C) in Sect. 4 only replaces polygonal paths homeomorphic to [0, 1] by simpler paths also homemorphic to [0, 1]. □

References

1. Aanjaneya, M., Chazal, F., Chen, D., Glisse, M., Guibas, L., Morozov, D.: Metric graph reconstruction from noisy data. Int. J. Comp. Geometry Appl. **22**, 305–325 (2012)
2. Agarwal, P., Har-Peled, S., Mustafa, N., Wang, Y.: Near-linear time approximation algorithms for curve simplification. Algorithmica **42**(3–4), 203–219 (2005)
3. Beygelzimer, A., Kakade, S., Langford, J.: Cover trees for nearest neighbor. In: Proceedings of ICML, pp. 97–104 (2006)
4. Carriere, M., Oudot, S.: Structure and stability of the one-dimensional mapper. Found. Comput. Math. **18**, 1333–1396 (2018)
5. Chazal, F., Huang, R., Sun, J.: Gromov-Hausdorff approximation of filament structure using Reeb-type graph. Discrete Comput. Geom. **53**, 621–649 (2015). https://doi.org/10.1007/s00454-015-9674-1
6. Dey, T.K., Wang, J., Wang, Y.: Graph reconstruction by discrete Morse theory. arXiv:1803.05093 (2018)
7. Douglas, D., Peucker, T.: Algorithms for the reduction of the number of points required to represent a digitized line or its caricature. Cartographica **10**(2), 112–122 (1973)
8. Forsythe, J., Kurlin, V.: Convex constrained meshes for superpixel segmentations of images. J. Electron. Imaging **26**(6), 061609 (2017)
9. Forsythe, J., Kurlin, V., Fitzgibbon, A.: Resolution-independent superpixels based on convex constrained meshes without small angles. In: Bebis, G., et al. (eds.) Advances in Visual Computing. LNCS, Proceedings of ISVC 2016, vol. 10072, pp. 223–233. Springer, Cham (2016). https://doi.org/10.1007/978-3-319-50835-1_21
10. Fredman, M., Tarjan, R.: Fibonacci heaps and their uses in improved network optimization algorithms. J. ACM **34**(3), 596–615 (1987)
11. Ge, X., Safa, I., Belkin, M., Wang, Y.: Data Skeletonization via Reeb graphs. In: Proceedings of Neural Information Processing Systems, pp. 837–845 (2011)

12. Giesen, J.: Curve reconstruction, the traveling salesman problem and menger's theorem on length. In: Proceedings of SoCG, pp. 207–216 (1999)
13. Gorban, A., Zinovyev, A.: Principal graphs and manifolds. In: Handbook of Research on Machine Learning Applications and Trends, pp. 28–59 (2009)
14. Kalisnik, S., Kurlin, V., Lesnik, D.: A higher-dimensional homologically persistent skeleton. Adv. Appl. Math. **102**, 113–142 (2019)
15. Kégl, B., Krzyzak, A.: Piecewise linear skeletonization using principal curves. Trans. Pattern Anal. Mach. Intell. **24**, 59–74 (2002)
16. Kurlin, V.: 3-page embeddings of singular knots. Func. Anal. Appl. **38**, 14–27 (2004)
17. Kurlin, V.: 3-page diagrams of 3-valent graphs. Func. Anal. Appl. **35**, 230–233 (2001)
18. Kurlin, V.: Three-page encoding and complexity theory for spatial graphs. J. Knot Theory Ramifications **16**, 59–102 (2007)
19. Kurlin, V.: A homologically persistent skeleton is a fast and robust descriptor of interest points in 2D images. In: Proceedings of CAIP, vol. 9256, pp. 606–617 (2015)
20. Kurlin, V.: A one-dimensional homologically persistent skeleton of an unstructured point cloud in any metric space. Comput. Graph. Forum **34**(5), 253–262 (2015)
21. Kurlin, V., Harvey, D.: Superpixels optimized by color and shape. In: Pelillo, M., Hancock, E. (eds.) Energy Minimization Methods in Computer Vision and Pattern Recognition, vol. 10746, pp. 297–311. Springer, Cham (2018). https://doi.org/10.1007/978-3-319-78199-0_20
22. Kurlin, V., Muszynski, G.: Persistence-based resolution-independent meshes of superpixels. Pattern Recogn. Lett. **131**, 300–306 (2020)
23. Kurlin, V., Smithers, C.: A linear time algorithm for embedding arbitrary knotted graphs into a 3-page book. In: International Conference on Computer Vision, Imaging and Computer Graphics, pp. 99–122 (2015)
24. March, W., Ram, P., Gray, A.: Fast Euclidean minimum spanning tree: algorithm, analysis, and applications. In: Proceedings of SIGKDD, pp. 603–612 (2010)
25. Shin, H., Kim, D.S.: Optimal direction for monotone chain decomposition. In: International Conference on Computational Science and Its Applications, pp. 583–591 (2004)
26. Singh, G., Memoli, F., Carlsson, G.: Topological methods for the analysis of high dimensional data and 3D object recognition. In: Symposium Point-Based Graphics, pp. 91–100 (2007)
27. Singh, R., Cherkassky, V., Papanikolopoulos, N.: Self-organizing maps for the skeletonization of sparse shapes. IEEE Tran. Neural Networks **11**, 241–248 (2000)
28. Smith, P., Kurlin, V.: Skeletonisation algorithms with theoretical guarantees for unorganised point clouds with high levels of noise. Pattern Recogn. **115**, 107902 (2021)
29. Wu, S.T., da Silva, A., Márquez, M.: The Douglas-Peucker algorithm: sufficiency conditions for non-self-intersections. J. Braz. Comput. Soc. **9**(3), 67–84 (2004)

Topologically Robust B-spline Reconstruction of Fibers from 3D Images

Dennis Mosbach, Katja Schladitz, Bernd Hamann, and Hans Hagen

Abstract The micro-structure of wood-based insulation materials is analyzed to gain insight into how features on microscopic scales influence macroscopic thermal conductivity. Three-dimensional (3D) image data obtained by micro-computed tomography reveals a complex structure formed by cellulose fibers. To study the effect of geometry changes, simple B-spline representations of these fibers are highly desirable. A straightforward solution is to extract a triangulated isosurface from the 3D image and partition it into quadrilateral macro-cells with disk-like topology. For each cell, a B-spline surface is constructed by minimizing a least squares error term. However, the physical processing of the material affects the structure of the fibers. The resulting changes in surface topology cause difficulties for the quadrilateral partitioning. Image processing tools can solve these topological issues, but they also impact geometry. We present a novel approach that splits geometry and topology processing of the data. It allows for topological simplification while still preserving the geometry of a scanned object. Established B-spline approximation methods are used to create a model. The involved mathematical equations are described in detail with a focus on simple implementation. Our presented results demonstrate that smooth and accurate models can be created for challenging data.

D. Mosbach · H. Hagen
Computer Graphics and HCI Group, University of Kaiserslautern, 67663 Kaiserslautern, Germany
e-mail: hagen@cs.uni-kl.de

D. Mosbach (✉) · K. Schladitz
Image Processing Department, Fraunhofer ITWM, 67663 Kaiserslautern, Germany
e-mail: dennis.mosbach@itwm.fraunhofer.de

K. Schladitz
e-mail: katja.schladitz@itwm.fraunhofer.de

B. Hamann
Department of Computer Science, University of California, 95616 Davis, CA, USA
e-mail: hamann@cs.ucdavis.edu

I. Hotz et al. (eds.), *Topological Methods in Data Analysis and Visualization VI*,
Mathematics and Visualization, https://doi.org/10.1007/978-3-030-83500-2_14

1 Introduction

An important part in researching material properties is the generation of models representing microscopic structures which can be used for visualization and simulation purposes. How these models are obtained strongly depends on the size and complexity of the samples, the initial data acquisition, and the target application.

We are concerned with model generation of cellulose fibers. A cellulose fiber is understood as a hollow tube that is open on both ends. The empty space inside is called *lumen*. Models of such structures are helpful to understand and assess mechanical properties of a material on a macroscopic scale [1]. For example, this understanding allows researchers to establish a relation between the micro-structure and physical properties of paper [15]. Another example is the analysis of sound-dampening properties of wood-based acoustic insulation material with respect to its tortuosity, a parameter that relates physical, acoustic, and morphological properties of a material [20].

Our goal is to investigate heat conductivity properties of wood-based thermal insulation materials. The micro-structure of these materials consists of a complex system of cellulose fibers which occur in chunks, chips, and as individuals. To conduct simulations, B-spline surface models of such fibers are desired. These models should consist of a low number of continuously connected surfaces and include the inner and outer surfaces of the fiber wall. The main feature of interest is the shape of the fibers as well as the contained lumen. This means that small holes in the wall as well as roughness along the surface can be neglected and do not need to be present in the final model. B-spline surfaces are defined by a grid of control points. Storing just those control points requires far less memory than a high-resolution triangulation. Furthermore, B-splines are better suited to represent geometric detail than a down-sampled mesh. Changing control point locations directly affects the shape of the surface, which allows scientists to generate additional fibers for simulations from a small number of prototypes.

The initial data is a scan of a material sample given as 16-bit grayvalue image which is obtained by high-resolution synchrotron computed tomography. For denoising, a $5 \times 5 \times 5$ median filter is applied after which the image is binarized with a global gray value threshold. Individual fibers are then extracted from this image based on a local shape criterion [2].

However, the physical processing of the material causes damage to the micro-structure which leads to cuts and holes in the fiber walls as well as increased surface roughness in some areas. As Fig. 1 shows, these effects complicate the topological structure of the surface by adding numerous small tunnels, holes, and cavities.

A straightforward solution to create a B-spline model of those fibers is to extract the isosurface of such a fiber in form of a triangulated mesh. In order to compute a continuous B-spline model, this mesh needs to be partitioned into a collection of quadrilateral macro-cells. Each macro-cell can then be approximated by a B-spline surface. Due to the non-trivial topological situation on the fiber's surface, most approaches to compute such a quadrilateral decomposition either fail or produce a

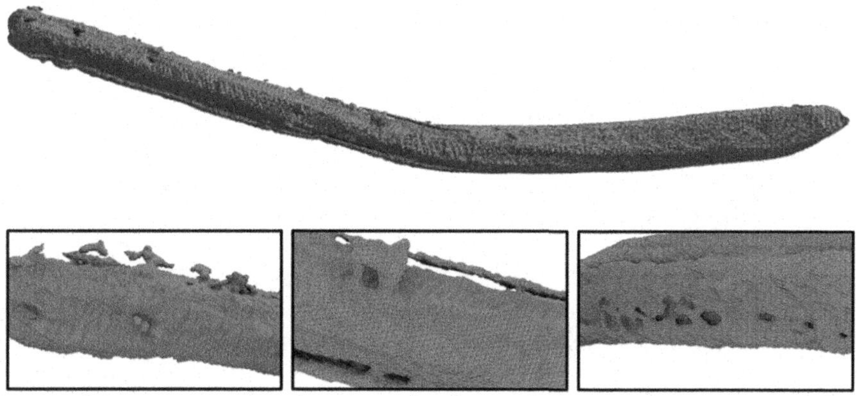

Fig. 1 Example of an extracted fiber. While the overall geometrical shape appears to be smooth, close-ups show artifacts affecting the local topology on the surface

very large amount of small macro cells. Standard image denoising operations can solve the topological problems, but might also introduce a geometrical error to the resulting surface.

Our contribution is a new approach that splits the data processing pipeline into geometry and topology processing. A strongly cleaned and repaired image is used to create the mesh of larger quadrilateral faces capturing the topology of the idealized cellulose fiber. This quad-mesh is then used to partition the original data, extracted from the unprocessed image, by assigning each data point to the nearest quad-face. Processing the data this way provides the necessary input for established B-spline approximation methods. The resulting B-spline models have fiber-like topology and closely approximate the geometry of the original data.

2 Related Work

While no methods exist to extract isosurfaces directly in form of B-splines, a variety of research has been published concerning the construction of B-spline approximations of discrete data given as point clouds or surface meshes. Such an explicit representation can easily be obtained from a 3D image by contouring methods like Marching Cubes.

Regular tensor product B-spline surfaces can only represent structures with disk-like topology. To construct a model for complex objects, the data needs to be decomposed into quadrilateral cells. The data points in each cell are then approximated by a single B-spline surface. The key challenge is to ensure certain continuity constraints along boundaries. While a watertight model, i.e., C^0-continuous transitions along shared boundaries, is almost always a requirement, it is often desired to also have at least tangent plane continuity, i.e., G^1-continuous transitions.

Eck and Hoppe [3] provide an in-depth description of the overall pipeline starting with a point cloud. Their first step is to establish a topological structure of the data by generating a fine surface mesh. Using a Voronoi-like tessellation and its dual Delaunay-like complex, the mesh is subdivided into a set of quadrilateral cells where each cell has a disk-like topology. Those cells are parameterized using harmonic maps. A B-spline approximation using a modification of Peters' scheme [18] is then computed to minimize the least squares error between the surfaces and the data points.

Gregorski et al. [8] propose a method to construct a set of B-spline surfaces approximating a given point cloud. They decompose their data into quadrangular cells based on a so-called strip tree which provides an adaptive subdivision of the data, similar to a quad-tree, into small boxes such that the points contained in each box can be approximated by B-spline surfaces with low error. Control points of neighboring surfaces are then adjusted during a post-processing step to ensure C^1-continuous transitions. Since they do not establish a topological structure of the surface, their approach works best on mostly flat objects.

Given a triangulated surface mesh and a manually defined subdivision into quadrilateral cells, Krishnamurthy and Levoy [13] do a remeshing of each cell to obtain a regular sampling of the data. This implicitly defines a parameterization and reduces the complexity of the B-spline approximation. However, to apply it to complex objects, an automatization of the quadrilateral decomposition is necessary.

Yoo [22] describes an approach to construct a B-spline model of human bones given as point cloud or sequence of computed tomography images. First, the input data is used to define an implicit surface on which a fine quad-mesh is constructed. Each quad and the normal vectors along its boundaries are then interpolated by a B-spline surface.

Yoshihara et al. [23] capture the topology of a given point cloud by constructing an implicit function and applying a level set method. At the cost of geometrical accuracy in noisy regions, this allows for a stable processing of difficult data. The object's surface is approximated by a Catmull-Clark subdivision surface and the corresponding control points are used to form a fine quad-mesh which is interpolated by B-spline surfaces.

Lin et al. [14] present a method to create a smooth B-spline model to approximate a mesh. The quadrilateral decomposition is done manually. After constructing a curve network representing the boundaries, the data points in each cell are approximated by a bi-quintic Bézier surface that interpolates the boundary curves. The resulting surface model is made G^1-continuous by also interpolating pre-computed normal vectors along the boundaries.

Zhao et al. [24] introduce an iterative approach, allowing individual surfaces to use differing knot vectors. Given a point cloud and a partition into quads, their approach constructs an initial set of B-spline surfaces to closely approximate the data and then ensures approximate G^1-continuity in a numerical post-processing step.

Based on their previous work, Peters and Fan [19] provide an in-depth theoretical analysis of G^1-continuous B-spline surface constructions. They state necessary constraints that B-splines based on a quad-mesh with arbitrary topology need to satisfy in order to obtain tangent plane continuity everywhere. A key statement is that in a general setting, a G^1-continuous B-spline surface model can only be constructed when the used knot vectors have at least two interior double knots.

Satisfying these requirements, Fan and Peters [4] provide a construction scheme for bi-cubic B-splines with exactly two double inner knots. Equations for all control points are explicitly given as linear combinations of the surrounding vertices of the quad-mesh, but the method can also be modified to approximate a set of discrete data points.

While a large variety of methods for many scenarios exists, they are primarily designed for structures that allow for a decomposition into a relatively small number of quadrilateral macro cells. However, this paper deals with rough fiber structures with complicated surface topology. A straightforward decomposition of the data into disk-like quadrilateral cells would require a very high number of small cells. Automatically removing these effects during a pre-processing step strongly affects the geometry of the resulting B-splines. Hence, an alternative solution is introduced here.

3 Pipeline

The key aspect of our method is splitting geometry and topology processing, see Fig. 2. After the data has been processed this way, standard methods for B-spline approximation can be applied.

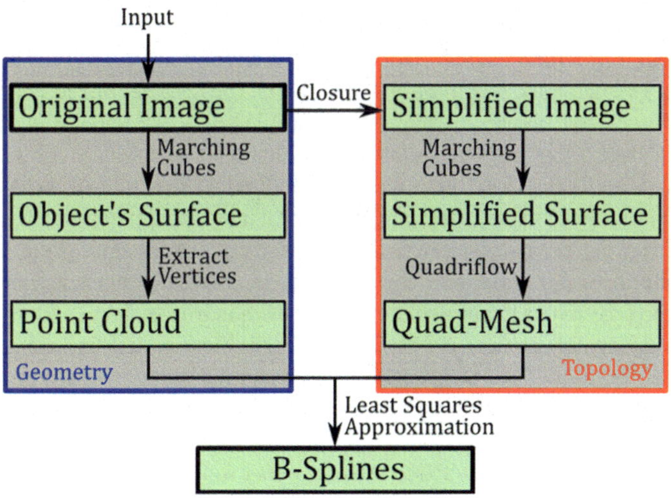

Fig. 2 Pipeline separating geometry and topology processing. To partition the data into quadrilateral cells, a morphological closure operation is applied to the input image. The isosurface of the image is extracted and re-meshed into a quad-mesh defining the topological structure of the B-spline model. The approximation is computed with respect to the data points extracted from the isosurface of the original mesh

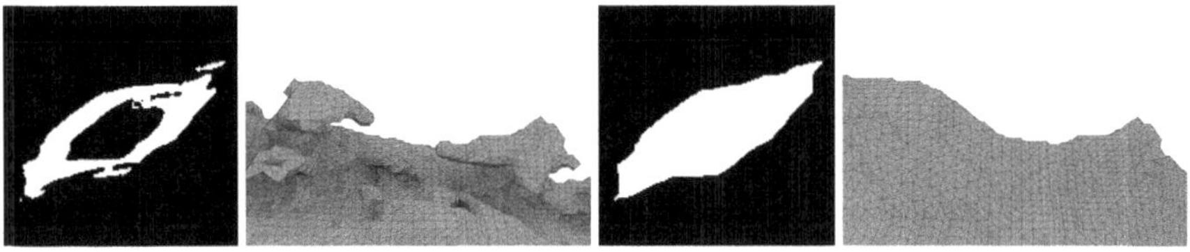

(a) (b) (c) (d)

Fig. 3 Performing a morphological closure operation simplifies the topology of the object. A slice of the 3D volume data **a** and a close-up of the corresponding region in the isosurface **b** highlight regions hard to handle for subsequent processing. Applying the closure operation simplifies the surface structure, as seen in **c** and **d**.

To compensate for small holes and the rough surface structure in the data, a topological simplification is employed. We apply a morphological closure to the volume image, i.e., we perform a sequential application of dilation and erosion on the object represented by the image's foreground [17]. The size of the filter mask needs to be chosen sufficiently large to fill the entire structure. Efficient implementations for large filter masks are available [16]. This operation preserves the original outer boundary in smooth regions and absorbs any outside material into the fiber's surface, see Fig. 3. The inside of the fiber and holes in its wall are closed. The resulting object has the topology of a cylinder. Hence, applying Marching Cubes leads to a surface mesh with a topological structure that is well-suited to be partitioned into large and evenly sized macro-cells. A re-meshing into a small number of quadrilateral cells is performed on this mesh, using a freely available implementation of the Quadriflow algorithm [11]. The resulting quad-mesh is a "rough geometrical approximation" of the data, but it provides a suitable topological structure for a B-spline model.

For the geometry-focused part of the pipeline, the original image is considered. As with the closed image, a triangulated isosurface is extracted using Marching Cubes. The vertices of this triangulation provide the geometry data for the B-spline approximation. To associate them with the topological structure of the quad-mesh, each point is assigned to the closest quad-face. An initial parameterization is obtained by projecting the points onto their associated faces. By assigning the corners of a quadrilateral to the corners of the unit square, a correspondence is established between the position of the projected point in the quadrilateral and a tuple in (u, v)-parameter space, where $(u, v) \in [0, 1]^2$, see Fig. 4. This initial parameterization is not optimal, but it suffices to generate a proper initial surface approximation. The parameterization is optimized during the iterative surface construction approach as described in Subsect. 4.2.1.

Cellulose fibers are hollow. To reconstruct a fiber wall and the contained lumen, the inside and outside of the isosurface representing the wall are considered separately. The topology is the same in both cases. Hence, the same quad-mesh can be used for both approximations. However, each data point needs to be classified as belonging to either the interior or exterior surface of the fiber wall. This classification can be done by applying a Euclidean Distance Transform to the closed image, assigning to each

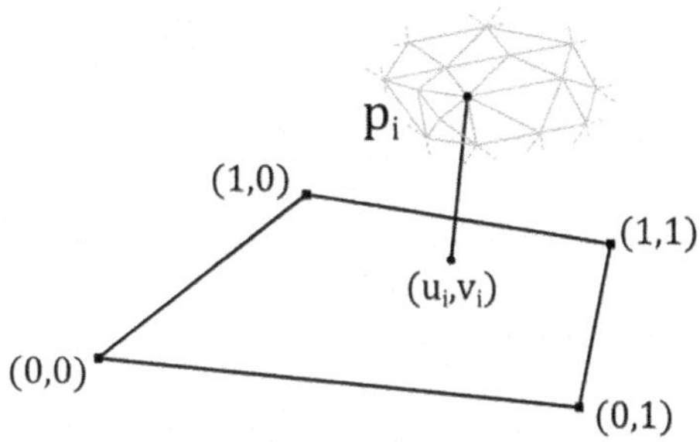

Fig. 4 Initial parameterization. Each vertex of the original isosurface is orthogonally projected onto the closest face of the quad-mesh. The corners of the quad correspond to the corners of the unit square $[0, 1]^2$. The relative position of the projected point in the quad defines the parameterization

voxel the distance to the closest point of the isosurface [7]. As the morphological closure still mostly preserves the outer surface, data points with low distances (in our case ≤ 2 voxels) are considered to belong to the outside surface of the wall, while data points with larger distance values are considered to belong to the inside. Once the classification is performed, separate surface models are constructed, one for each set of data points.

The quad-mesh obtained from the simplified image is closed. Since the layout produced by the Quadriflow algorithm is sensitive to sharp features, the actual open ends of the fibers align well with individual quad-faces. To reproduce the open structure of fibers, these quads are removed and the open space between inside and outside B-spline surfaces is closed via linear interpolation between the boundary curves.

Image processing operations, including closure, Euclidean Distance Transform and isosurface extraction with Marching Cubes, are performed with the implementations in MAVI [6]. The projection of the data points onto the quad-mesh is performed with the freely available software package libigl [12].

4 B-Spline Approximation

This Section reviews the construction of B-spline surfaces to approximate discrete data. The concepts are based on literature [5] and the methods mentioned in Sect. 2. Equations that need to be solved are stated in a uniform notation with a focus on simple implementation.

As input, a set of data points and their parameterization is required. Optionally, they can also have weights assigned to them, which is useful when considering a non-uniform distribution of data points. For complex objects that cannot be modeled with a single surface, a partition into quadrilateral cells is required. The data of each cell is then approximated by an individual B-spline surface that has to satisfy certain continuity constraints with neighboring surfaces.

After introducing the basic notation in Subsect. 4.1, Subsect. 4.2 discusses the construction of a single surface. Finally, a global system including the constraints necessary for a continuous model consisting of multiple B-spline surfaces is introduced in Subsect. 4.3.

4.1 B-Spline Surface Notation

A B-spline surface is defined by an order $k \in \mathbb{N}$, a knot vector for each parameter direction, and a rectangular grid of control points. To keep the construction of a model with multiple continuously connected surfaces as simple as possible, the B-splines considered here are restricted to quadratic grids of $n_c \times n_c$ control points and use the same knot vector

$$\tau = (\underbrace{0, \ldots, 0}_{k+1 \text{ times}}, \tau_1, \ldots, \tau_{n_c-k-1}, \underbrace{1, \ldots, 1}_{k+1 \text{ times}}) \in [0, 1]^{n_c+k+1}$$

with knots $\tau_i \leq \tau_{i+1}$ in u and v direction. The piecewise polynomial basis functions of order k defined by knot vector τ are denoted by $N_i(\cdot) := N_{i,k,\tau}(\cdot)$.

The B-spline surface is given by

$$S : [0, 1]^2 \to \mathbb{R}^3$$
$$S(u, v) = \sum_{i=1}^{n_c} \sum_{j=1}^{n_c} N_i(u) N_j(v) b_{i,j}.$$

4.2 Single B-Spline Surface Approximation

First, the approximation of a set of data points by a single B-spline surface is discussed. Given a set of n_p points $p_i \in \mathbb{R}^3$ with assigned parameter values $(u_i, v_i) \in [0, 1]^2$ and a weight $w_i \in \mathbb{R}$. The least squares error of the surface with respect to the data is denoted by

$$E_{\text{LS}} = \frac{1}{2} \sum_{i=1}^{n_p} w_i \left\| S(u_i, v_i) - p_i \right\|^2. \tag{1}$$

An equivalent notation can be used to simplify this term. Let $b \in \mathbb{R}^{n_c^2 \times 3}$ be a one dimensional list containing the B-spline control points and let $b_x, b_y, b_z \in \mathbb{R}^{n_c^2}$ be the vectors containing their x, y, and z coordinates. A correlation between the list index $\bar{i}$ and the index i, j in the original grid can be established with a bidirectional mapping, e.g., $\bar{i} = i n_c + j$. Using the same index mapping, let

$$N(u, v) = [N_1(u)N_1(v), \ \ldots, N_{n_c}(u)N_{n_c}(v)] \in \mathbb{R}^{1 \times n_c^2} \tag{2}$$

be a row vector containing the associated tensor products of basis functions. Then, Eq. (1) can be written as

$$E_{\text{LS}} = \frac{1}{2} \sum_{i=1}^{n_p} w_i \left\| N(u_i, v_i) b - p_i^T \right\|^2 .$$

Furthermore, let $N \in \mathbb{R}^{n_p \times n_c^2}$ be the matrix containing the basis function coefficients for the parameters u_i, v_i associated to each data point p_i and let a list of all data points be denoted by p, i.e.,

$$N = \begin{bmatrix} N(u_1, v_1) \\ \vdots \\ N(u_{n_p}, v_{n_p}) \end{bmatrix}, p = \begin{bmatrix} p_1^T \\ \vdots \\ p_{n_p}^T \end{bmatrix}.$$

The least squares error can then be expressed as

$$E_{\text{LS}} = \frac{1}{2} \sum_{\sigma \in \{x,y,z\}} (N b_\sigma - p_\sigma)^T W (N b_\sigma - p_\sigma)$$

with the weights as diagonal matrix $W = \text{diag}(w_1, ..., w_{n_p})$ and gradient

$$\nabla_\sigma E_{LS} = N^T W N b_\sigma - N^T W p_\sigma .$$

Minimizing the least squares error is equivalent to solving $\nabla_\sigma E_{\text{LS}} = 0$ for all components $\sigma \in \{x, y, z\}$.

Due to the discrete nature of the data, an approximation based on only the least squares error often includes unwanted wiggles. This is compensated by penalizing the deviation from a smooth surface by a fairing term. A commonly used term is the *thin-plate-energy* functional [3, 10]

$$E_{\text{TP}} = \frac{1}{2} \int_0^1 \int_0^1 \|\partial_{uu} S(u, v)\|^2 + 2 \|\partial_{uv} S(u, v)\|^2 + \|\partial_{vv} S(u, v)\|^2 \, du dv$$

which, for B-splines, can be rewritten as

$$E_{\text{TP}} = \frac{1}{2} \sum_{\sigma \in \{x,y,z\}} b_\sigma^T M b_\sigma$$

with gradient

$$\nabla_\sigma E_{\text{TP}} = M b_\sigma$$

where $M \in \mathbb{R}^{n_c^2 \times n_c^2}$ is a matrix containing the integrals of the basis functions given by

$$\begin{aligned} M = \int_0^1 \int_0^1 & (\partial_{uu} N(u, v))^T (\partial_{uu} N(u, v)) \\ & + 2(\partial_{uv} N(u, v))^T (\partial_{uv} N(u, v)) \\ & + (\partial_{vv} N(u, v))^T (\partial_{vv} N(u, v)) \mathrm{d}u \mathrm{d}v. \end{aligned}$$

The B-spline basis functions are piecewise polynomials. Hence, their integrals can either be determined analytically or by numeric integration, e.g., applying Gaussian quadrature to each knot interval $[\tau_i, \tau_{i+1})$ with a high enough degree to be exact for the given case.

To compute a smooth surface that approximates the data points, a linear combination of both terms is minimized, i.e.,

$$\min. \; (1 - \lambda) E_{\mathrm{LS}} + \lambda E_{\mathrm{TP}} \tag{3}$$

with the parameter $\lambda \in [0, 1)$ controlling the impact of the fairing term to the resulting surface. Since (3) is a system of quadratic equations with respect to the B-spline control points, its optimal solution can be found by solving

$$\begin{aligned} 0 &= (1 - \lambda) \nabla_\sigma E_{\mathrm{LS}} + \lambda \nabla_\sigma E_{\mathrm{TP}} \\ &= (1 - \lambda)(N^T W N b_\sigma - N^T W p_\sigma) + \lambda M b_\sigma \end{aligned}$$

for each coordinate $\sigma \in \{x, y, z\}$. It can be written as

$$((1 - \lambda)(N^T W N) + \lambda M) b = (1 - \lambda) N^T W p. \tag{4}$$

This is a linear system with respect to the control points b. It can be solved by a variety of freely available code packages, e.g., Eigen [9].

4.2.1 Iterative Approximation

The described approach aims on minimizing the distance between data points p_i and the points on the surface $S(u_i, v_i)$, associated to them through their parameterization. However, these are not necessarily the smallest distances between the data points and the surface. It is common practice to improve the quality of the approximation by employing an iterative approach alternating between the construction of the surface and an update to the parameter values.

After a surface has been constructed, the parameter values of each data point are updated, by minimizing

$$(u_i', v_i') = \operatorname{argmin}_{u, v \in [0,1]} \| S(u, v) - p_i \|.$$

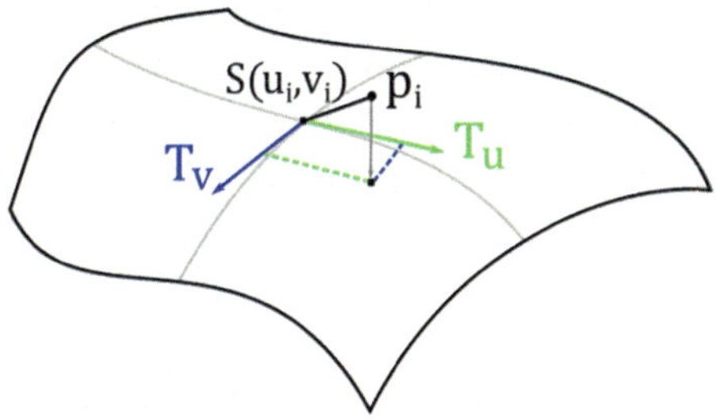

Fig. 5 Parameter optimization using first-order Taylor series. The error vector between each data point p_i and its corresponding point on the surface $S(u_i, v_i)$ is projected onto the tangent plane. The difference between the projected point and $S(u_i, v_i)$ is proportionally applied to the parameterization to obtain an improved parameter tuple (u_i', v_i')

which can either be done exactly using computationally expensive nonlinear optimization or approximately by considering low-degree Taylor series [21]. An example using a first-order Taylor series is shown in Fig. 5. The new parameter values are restricted to the unit square. To ensure stable behavior of the method, data points that are close to surface boundaries are not assigned to neighboring surfaces, even when the distance to them might be smaller.

4.3 Constructing a Continuous Surface Model

When constructing a model with n_s surfaces, the same principles than for a single surface apply, but some modifications are necessary to ensure continuous transitions between neighboring surfaces. Control points and data can be stored in a global list by concatenating the lists of each individual surface. The coefficient, weight, and smoothing matrices are combined to diagonal block matrices. Using

$$\widehat{b} = \begin{bmatrix} b^{(1)} \\ \vdots \\ b^{(n_s)} \end{bmatrix}, \widehat{p} = \begin{bmatrix} p^{(1)} \\ \vdots \\ p^{(n_s)} \end{bmatrix},$$

$$\widehat{N} = \begin{bmatrix} N^{(1)} & & 0 \\ & \ddots & \\ 0 & & N^{(n_s)} \end{bmatrix}, \widehat{W} = \begin{bmatrix} W^{(1)} & & 0 \\ & \ddots & \\ 0 & & W^{(n_s)} \end{bmatrix}, \widehat{M} = \begin{bmatrix} M & & 0 \\ & \ddots & \\ 0 & & M \end{bmatrix},$$

instead of the single surface terms in Eq. (4) leads to the same result as constructing each surface individually.

4.3.1 Achieving C^0-continuity by Adding Constraints. To obtain C^0-continuity, i.e., a watertight model, the row of control points along a shared boundary between any two neighboring surfaces needs to be identical. This can be expressed by a number of constraints of the form $\widehat{b}_i - \widehat{b}_j = 0$ with i, j being the indices in the global control point vector of the points that need to be identical. All constraints together can be expressed by a matrix vector product: $G\widehat{b} = 0$ where each row of matrix G ensures that two corresponding control points have to be identical.

With these constraints, the optimization problem (3) becomes

$$\begin{aligned} &\text{min. } (1-\lambda)\widehat{E}_{\text{LS}} + \lambda\widehat{E}_{\text{TP}} \\ &\text{s.t. } G\widehat{b} = 0. \end{aligned} \tag{5}$$

Using the method of Lagrange multipliers, an optimal solution of (5) can be found by solving

$$\left[\begin{array}{c|c} ((1-\lambda)\widehat{N}^T\widehat{W}\widehat{N} + \lambda\widehat{M} & G^T \\ \hline G & 0 \end{array}\right]\left[\begin{array}{c} \widehat{b} \\ \hline \Lambda \end{array}\right] = \left[\begin{array}{c} (1-\lambda)\widehat{N}^T\widehat{W}\widehat{p} \\ \hline 0 \end{array}\right] \tag{6}$$

with Lagrange multipliers Λ. This system can be implemented and solved in a straightforward way. However, adding the C^0-constraints in form of a matrix, increases the size of the overall system compared to the unconstrained case.

4.3.2 Achieving C^0-continuity by Reduction of Variables. An alternative is to implicitly enforce C^0-continuity, by using only one variable for each set of control points that are supposed to be identical. The matrix G is not needed and the overall number of unknowns is reduced, which allows better performance when solving the system. However, an index transformation is necessary to implement this system, i.e., each index $\hat{i}$ in the global (full) control point list $\hat{b}$ is assigned a new index $\tilde{i}$ in the reduced list of control points $\tilde{b}$. Using a matrix H with entries $H_{\hat{i},\tilde{i}} = 1$ and zero everywhere else, the original list of control points can be reconstructed as $\hat{b} = H\tilde{b}$. Plugging that in into the objective function (5) and solving for the gradient being equal zero leads to

$$\left((1-\lambda)H^T\widehat{N}^T\widehat{W}\widehat{N}H + \lambda H^T\widehat{M}H\right)\tilde{b} = (1-\lambda)H^T\widehat{N}^T\widehat{W}\widehat{p} \tag{7}$$

4.3.3 Achieving G^1-continuity The index transformation can be modified to also ensure G^1-continuity, i.e., transitions between surfaces with continuous tangent planes. Here, the method by Fan and Peters [4] is used. Given a quad mesh, they provide equations for a smooth surface model where all B-spline control points $\hat{b}_i$ are expressed as linear combinations of the quad mesh vertices q. Using these equations and an appropriate indexing, the overall B-spline control points can be expressed as $\hat{b} = Vq$.

By considering the vertex positions q as unknowns and using the matrix V instead of the index transformation H, (7) can be rewritten as

$$\left((1-\lambda)V^T\widehat{N}^T\widehat{W}\widehat{N}V + \lambda V^T\widehat{M}V\right)q = (1-\lambda)V^T\widehat{N}^T\widehat{W}\widehat{p}. \tag{8}$$

Solving this linear system corresponds to finding a quad mesh on which the scheme by Fan and Peters produces a set of B-spline surfaces minimizing the objective function (5). The resulting surfaces are G^1-continuous everywhere.

5 Results

We applied our method to a number of isolated fibers. Figure 6 shows an example, including intermediate steps. Further results are shown in Fig. 7. Even though many complicated topological situations arise in the data, our algorithm produces stable results, see Fig. 8. The overall shape and important characteristics, like length and bending, are well-preserved. Loose noise-like structures occurring in regions with high roughness are smoothed, while larger material parts are absorbed resulting in only a small distortion of the surface. Holes in the fiber walls are closed, and the

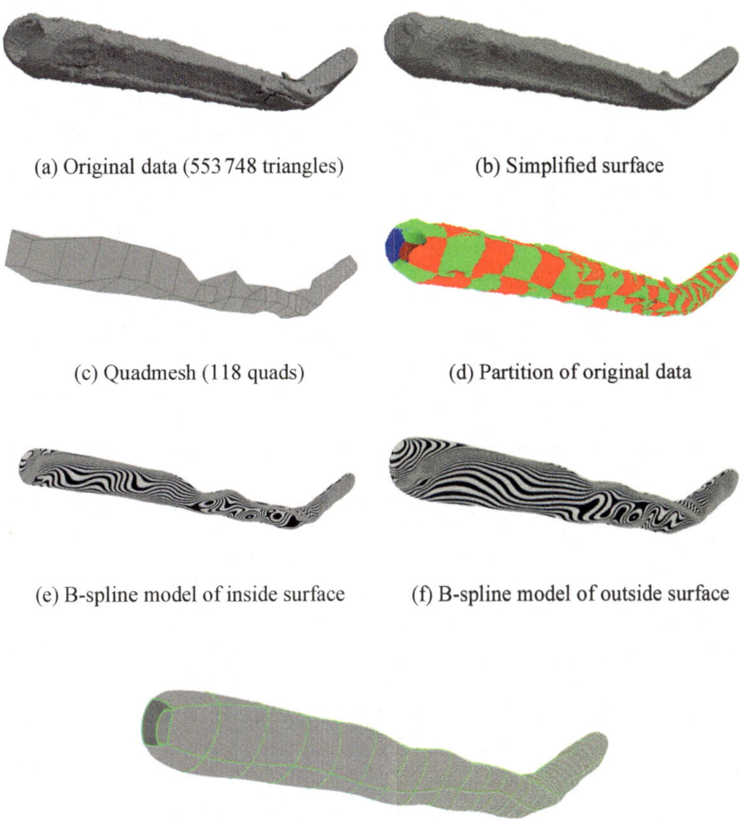

(a) Original data (553 748 triangles)

(b) Simplified surface

(c) Quadmesh (118 quads)

(d) Partition of original data

(e) B-spline model of inside surface

(f) B-spline model of outside surface

(g) Final B-spline model (236 B-spline surfaces)

Fig. 6 Main processing steps. **a** shows an isosurface of the original image. After applying a morphological closure operation, the topology is simplified **b**. A quad-mesh is computed for this surface **c**. The initial set of data points is partitioned by assigning each vertex to the nearest quad **d**. A smooth B-spline construction is applied to model the fiber wall's inside **e** and outside surfaces **f**. The final model is obtained by removing the B-spline surfaces at the end and linearly interpolating between the respective boundary curves **g**

(a) Isosurface (1 203 844 triangles)

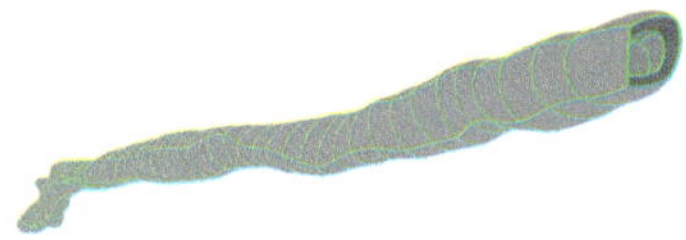

(b) B-spline model (676 B-spline surfaces)

(c) Isosurface (423 628 triangles)

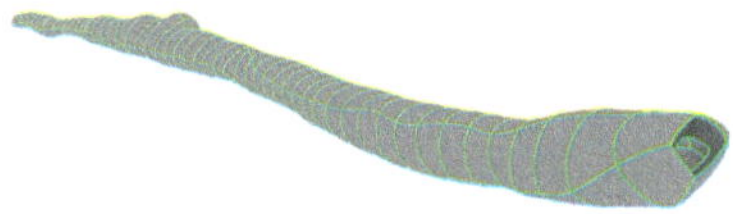

(d) B-spline model (432 B-spline surfaces)

(e) Isosurface (710 808 triangles)

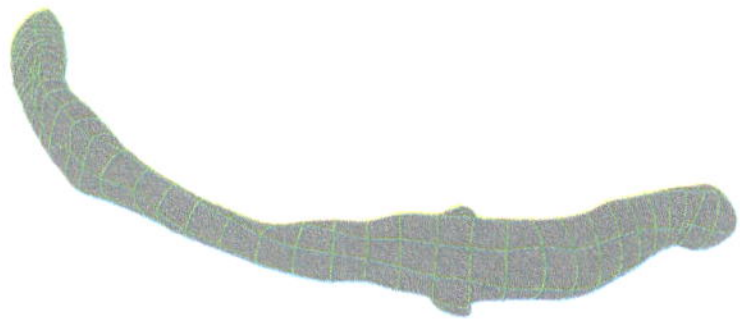

(f) B-spline model (580 B-spline surfaces)

(g) Isosurface (601 040 triangles)

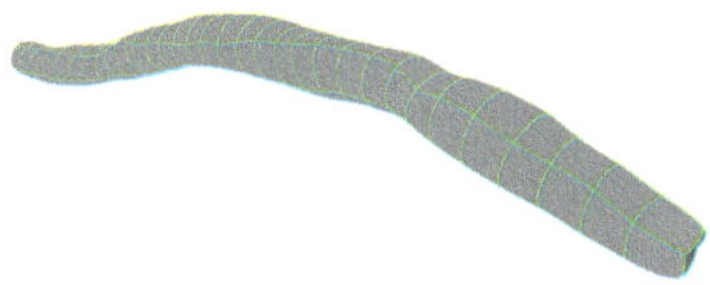

(h) B-spline model (268 B-spline surfaces)

Fig. 7 Additional results. Isosurfaces of the exact image data **a**, **c**, **e**, **g** and their B-spline approximations **b**, **d**, **f**, **h**

topological structure of the surface does not impact the stability of the quadrilateral partitioning.

Each resulting B-spline surface is of order $k = 3$ and is defined by 8×8 control points. The surface model is constructed with the G^1-continuous approach described in Subsection 4.3, which defines the knot vector as $\tau = (0, 0, 0, 0, \frac{1}{3}, \frac{1}{3}, \frac{2}{3}, \frac{2}{3}, 1, 1, 1, 1)$. The computation is done via three iterations of surface approximation and parameter optimization, with decreasing smoothness parameters $\lambda_1 = 0.9, \lambda_2 = 0.5, \lambda_3 = 0.1$.

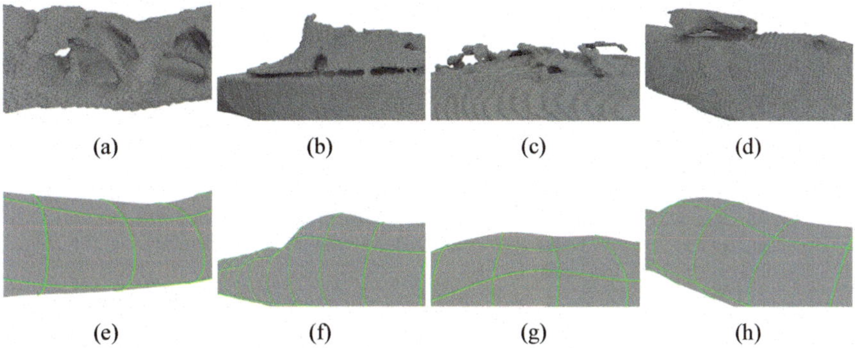

Fig. 8 Examples of data artifacts **a–d** and the B-spline approximations in those areas **e–h**

6 Conclusions

We have presented a new approach that reconstructs high-resolution cellulose fibers from 3D images with a low number of B-spline surfaces. Due to the production process of the material, the micro-structure contains complex topology that is problematic for established methods to handle properly. By splitting the processing pipeline into geometry and topology processing, we can deal with the topological difficulties and compute stable reconstructions of idealized structures.

After the data is processed with the pipeline, a B-spline surface model can be generated by using established B-spline methods. Our detailed description of the resulting linear system of equations focuses on an elegant implementation.

Our method can handle challenging situations in the image. For example, surface roughness has almost no impact. Even when larger material parts are present near the surface of a fiber, its wall is still reconstructed in a topologically correct way, and surplus material is treated as a local deformation of the surface. The triangulated meshes extracted from high-resolution images contain 423 628 to 1 203 844 triangles, and they are modeled with 236 to 676 B-spline surfaces.

We designed our method for fiber-like objects having simple overall topology to deal with local topological changes that arise in the presence of surface roughness or small holes and tunnels. Future work could address adaptations of our method to more complex shapes, e.g., chunks of connected fiber bundles.

Acknowledgements This research was funded by the Fraunhofer High Performance Center for Simulation- and Software-Based Innovation and supported by the Deutsche Forschungsgemeinschaft (DFG, German Research Foundation) - 252408385 - IRTG 2057, as well as a grant by the Deutsch-Französische Hochschule (DFH). The wood insulation investigation was funded by the German Federal Ministry of Food and Agriculture (BMEL) through its Agency for Renewable Resources, project LowLambda.

References

1. Badel, E., Delisee, C., Lux, J.: 3D structural characterisation, deformation measurements and assessment of low-density wood fibreboard under compression: The use of X-ray microtomography. Compos. Sci. Technol. **68**(7–8), 1654–1663 (2008)
2. Dobrovolskij, D., Engelhardt, M., Rack, A., Schladitz, K.: Shape classification for wood based insulation material. Presentation (2019). In: Contributed to ICTMS 2019, Cairns (Australia)
3. Eck, M., Hoppe, H.: Automatic reconstruction of B-spline surfaces of arbitrary topological type. In: Proceedings of the 23rd Annual Conference on Computer Graphics and Interactive Techniques, pp. 325–334. ACM (1996)
4. Fan, J., Peters, J.: Smooth bi-3 spline surfaces with fewest knots. Computer-Aid. Des. **43**(2), 180–187 (2011)
5. Farin, G.: Curves and Surfaces for CAGD: A Practical Guide, 5th edn. Morgan Kaufmann Publishers, San Francisco (2002)
6. Fraunhofer ITWM, Department of Image Processing: MAVI - modular algorithms for volume images. http://www.mavi-3d.de (2005). Accessed 23 Sept 2019
7. Godehardt, M., Mosbach, D., Roldan, D., Schladitz, K.: Efficient 3D erosion dilation analysis by sub-pixel EDT. In: International Symposium on Mathematical Morphology and Its Applications to Signal and Image Processing, pp. 243–255. Springer, Cham (2019)
8. Gregorski, B.F., Hamann, B., Joy, K.I.: Reconstruction of B-spline surfaces from scattered data points. In: Proceedings Computer Graphics International 2000, pp. 163–170. IEEE (2000)
9. Guennebaud, G., et al.: Eigen v3. http://eigen.tuxfamily.org (2010). Accessed 23 Sept 2019
10. Hagen, H., Schulze, G.: Automatic smoothing with geometric surface patches. Comput. Aid. Geom. Des. **4**(3), 231–235 (1987)
11. Huang, J., Zhou, Y., Niessner, M., Shewchuk, J.R., Guibas, L.J.: Quadriflow: A scalable and robust method for quadrangulation. Comput. Graph. Forum **37**, 147–160 (2018)
12. Jacobson, A., et al.: libigl: a simple C++ geometry processing library. https://libigl.github.io/ (2018). Accessed 23 Sept 2019
13. Krishnamurthy, V., Levoy, M.: Fitting smooth surfaces to dense polygon meshes. SIGGRAPH **96**, 313–324 (1996)
14. Lin, H., Chen, W., Bao, H.: Adaptive patch-based mesh fitting for reverse engineering. Computer-Aided Design **39**(12), 1134–1142 (2007)
15. Marulier, C., Dumont, P.J., Orgéas, L., du Roscoat, S.R., Caillerie, D.: 3D analysis of paper microstructures at the scale of fibres and bonds. Cellulose **22**(3), 1517–1539 (2015)
16. Mosbach, D., Hagen, H., Godehardt, M., Wirjadi, O.: Fast and memory-efficient quantile filter for data in three and higher dimensions. In: 2014 IEEE International Conference on Image Processing (ICIP), pp. 2928–2932. IEEE (2014)
17. Ohser, J., Schladitz, K.: 3D Images of Materials Structures: Processing and Analysis. John Wiley & Sons, New York (2009)
18. Peters, J.: Constructing C^1 surfaces of arbitrary topology using biquadratic and bicubic splines. In: Designing Fair Curves and Surfaces: Shape Quality in Geometric Modeling and Computer-Aided Design, pp. 277–293. SIAM (1994)
19. Peters, J., Fan, J.: On the complexity of smooth spline surfaces from quad meshes. Comput. Aid. Geom. Des. **27**(1), 96–105 (2010)
20. Peyrega, C., Jeulin, D., Delisée, C., Malvestio, J.: 3D morphological characterization of phonic insulation fibrous media. Adv. Eng. Mater. **13**(3), 156–164 (2011)
21. Rogers, D.F., Fog, N.: Constrained B-spline curve and surface fitting. Comput. Aid, Des. **21**(10), 641–648 (1989)
22. Yoo, D.J.: Three-dimensional surface reconstruction of human bone using a B-spline based interpolation approach. Computer-Aided Design **43**(8), 934–947 (2011)
23. Yoshihara, H., Yoshii, T., Shibutani, T., Maekawa, T.: Topologically robust B-spline surface reconstruction from point clouds using level set methods and iterative geometric fitting algorithms. Comput. Aid. Geom. Des. **29**(7), 422–434 (2012)
24. Zhao, X., Zhang, C., Xu, L., Yang, B., Feng, Z.: IGA-based point cloud fitting using B-spline surfaces for reverse engineering. Inf. Sci. **245**, 276–289 (2013)

Part IV: Overview Articles, Software and Viewpoints

Introduction to Vector Field Topology

Tobias Günther and Irene Baeza Rojo

Abstract Flow visualization is a research discipline that is concerned with the visual exploration and analysis of vector fields. An important class of methods are the topology-based techniques, which concentrate on individual structures in the domain that govern, constrain or guide the behavior of particles in the vector field. In this chapter, we give an overview of existing techniques for steady and unsteady vector fields in 2D and 3D. For time-dependent flows, we describe streamline-oriented and pathline-oriented approaches, eventually leading us to closely related feature-based visualization concepts such as reference frame invariance and Lagrangian coherent structures.

1 Introduction

Topology-based flow visualization concentrates on locations in the domain that govern the motion of the surrounding fluid. When flows are steady, this provides a compact description of asymptotic particle behavior. In time-dependent flows, there are still many open research problems despite decades of research [18]. One of the earliest topology-related visualization papers was published by Dallmann [23], who studied vortex separation. His work was not only influential in later topology-driven research, such as by Helman and Hesselink [55, 56] or Globus and Levit [34], but his visualization approaches were also source of inspiration for illustrative flow visualization methods [14]. The following manuscript is written for novice readers as an entry point to topology-based methods in flow visualization, primarily giving an

Present Address:
T. Günther (✉) · I. Baeza Rojo
Department of Computer Science, ETH Zürich, Zürich, Switzerland
e-mail: tobias.guenther@inf.ethz.ch

I. Baeza Rojo
e-mail: irene.baeza@inf.ethz.ch

T. Günther
FAU Erlangen-Nürnberg, Erlangen, Germany

I. Hotz et al. (eds.), *Topological Methods in Data Analysis and Visualization VI*,
Mathematics and Visualization, https://doi.org/10.1007/978-3-030-83500-2_15

overview of topological elements and related concepts from feature-based visualization. Over the course of this chapter, we concentrate mainly on differential methods, covering both steady as well as time-dependent vector fields. We refer the reader to existing surveys on topology-based methods, which cover older methods in more detail or also include scalar field and tensor field topology [18, 32, 53, 72, 90, 108, 130]. Topology-based methods are not only used for feature extraction, but also found applications in vector field modeling [117, 136], editing [20], simplification [73, 124, 125], smoothing [137], and compression [68, 75, 118].

The chapter is organized as follows. We begin with steady vector fields, covering the topological elements of 2D flows and afterwards 3D flows. When it comes to time-dependent vector fields, there are two classes of approaches: streamline-oriented and pathline-oriented topology. We summarize the streamline-oriented approaches first, covering the different transitions that topological elements undergo, and proceed with the current state on pathline-oriented approaches. Finally, we discuss ongoing research opportunities in the areas of reference frame extraction, high-dimensionality, uncertainty, and scalability.

Notation. In general, we denote scalar-valued quantities such as s with italic letters. Vector-valued quantities such as $\mathbf{v}$ are expressed in bold letters and matrices such as $\mathbf{J}$ are denoted in capital bold letters. When we discuss concepts in time-dependent flows, we use an overline symbol, e.g., $\overline{\mathbf{p}}$, to denote a vector in the space-time domain, which means that the time coordinate is added as an additional dimension.

2 Steady Vector Fields

Steady vector field topology is mainly interested in the asymptotic behavior of particles. That is, for each point in the domain, we would like to know where a particle will flow to in the limit and where it originally came from. In a steady vector field, the trajectory of a particle is commonly referred to as streamline. The limit behavior of particles is given by the topological skeleton, which consists of a number of points (critical points and special points on the boundary), which are connected with each other by streamlines, so-called separatrices. These separatrices divide the flow into regions with coherent behavior, as shown in Fig. 1. This means, every particle seeded from the same area will end up in the same sink or source, respectively, when tracing forward or backward. In the following, we introduce the elements of the topological skeleton for 2D and 3D flows. Note that most of these techniques only apply for steady vector fields. Time-dependent flows are covered in the subsequent section.

2.1 Two-Dimensional Flows

Given is a two-dimensional steady vector field $\mathbf{v}(\mathbf{x}) : \mathbb{R}^2 \to \mathbb{R}^2$:

Fig. 1 The topological skeleton of two steady 2D vector fields. Critical points (sources in red, sinks in blue, saddles in yellow) are connected by separatrices (black), which divide the flow into regions with the same asymptotic behavior. That is, particles released from any point inside such a region terminate in the same sink or source when tracing forward or backward, respectively

$$\mathbf{v}(\mathbf{x}) = \begin{pmatrix} u(x, y) \\ v(x, y) \end{pmatrix} \tag{1}$$

Since particles follow the flow tangentially, a particle trajectory $\mathbf{x}(\tau)$ can be computed by requesting that the tangent of the particle path is equal to the vector field direction at the particle position:

$$\frac{\mathrm{d}\mathbf{x}(\tau)}{\mathrm{d}\tau} = \mathbf{v}(\mathbf{x}(\tau)) \qquad \mathbf{x}(0) = \mathbf{x}_0 \tag{2}$$

This ordinary differential equation (ODE) is commonly solved with numerical integrators to compute the trajectory for a given initial position $\mathbf{x}_0$. We refer to the text book of Lapidus and Seinfeld [71] for a comprehensive introduction to the numerical integration of ordinary differential equations. In the following sections, we are also deeply interested in the spatial derivatives of the vector field, namely:

$$\frac{\partial \mathbf{v}(\mathbf{x})}{\partial \mathbf{x}} = \nabla \mathbf{v}(\mathbf{x}) = \mathbf{J}(\mathbf{x}) = \begin{pmatrix} \frac{\partial u(x,y)}{\partial x} & \frac{\partial u(x,y)}{\partial y} \\ \frac{\partial v(x,y)}{\partial x} & \frac{\partial v(x,y)}{\partial y} \end{pmatrix} \tag{3}$$

The above matrix is called the Jacobian $\mathbf{J}(\mathbf{x})$ of vector field $\mathbf{v}(\mathbf{x})$ and it captures the first-order flow behavior. In other words, it is a first-order estimate that tells us how particles behave in the immediate surrounding. We will later use this information to define and classify certain topological elements and flow features.

2.1.1 Critical Points

The key ingredient to steady vector field topology is the extraction of critical points. In general, critical points are locations $\mathbf{x}$ where the velocity vanishes:

$$\mathbf{v}(\mathbf{x}) = \mathbf{0} \tag{4}$$

The term critical point was coined in the flow visualization community by Helman and Hesselink [55, 56]. In other fields, these locations are also known as fixed points or stagnation points.

First-order Critical Points. A point $\mathbf{x}_0$ is a first-order critical point of $\mathbf{v}(\mathbf{x})$ if

1. $\mathbf{x}_0$ is a critical point of $\mathbf{v}(\mathbf{x})$
2. $\mathbf{v}(\mathbf{x})$ is differentiable at $\mathbf{x}_0$
3. $det(\mathbf{J}(\mathbf{x}_0)) \neq 0$.

The first condition is a minimal requirement. The second condition ensures that we can compute a Jacobian matrix, which is later relevant for the classification of first-order critical points. If the last condition is fulfilled, the critical point is said to be non-degenerate. In case of a first-order critical point, this means that the critical point is isolated, i.e., in an epsilon neighborhood there is no other critical point: $\mathbf{v}(\mathbf{x}_0 + \epsilon) \neq \mathbf{0}$. This protects us from trying to identify critical points inside of obstacles. The first-order critical points can be analyzed by an eigenanalysis of $\mathbf{J}(\mathbf{x}_0)$, as described in the following.

First-Order Classification. Let λ_1 and λ_2 be the possibly complex-valued eigenvalues of $\mathbf{J}(\mathbf{x}_0)$ and $\mathbf{c}_1$ and $\mathbf{c}_2$ be their corresponding eigenvectors, i.e.,

$$\mathbf{J}(\mathbf{x}_0) \cdot \mathbf{c}_i = \lambda_i \cdot \mathbf{c}_i \qquad \text{for} \qquad i \in \{1, 2\}. \tag{5}$$

By convention, we will assume that the eigenvalues are sorted in ascending order by their real parts, i.e., $Re(\lambda_1) \leq Re(\lambda_2)$. The eigenvalues characterize the behavior of the flow around the critical point and the corresponding eigenvectors indicate the direction of this behavior [55, 56]. For instance, if both eigenvalues have a positive real part, then we observe outgoing flow (source). If, on the other hand, both eigenvalues have a negative real part, then the flow is ingoing (sink). If one real part is positive and the other is negative, then a saddle is present. Finally, if both real parts are zero, then there is neither inflow nor outflow. The imaginary part of a pair of complex-conjugate eigenvalues denotes the swirling strength [145]. If this part is zero, then there is no rotation. Figure 2 illustrates the possible configurations of a steady 2D vector field.

Higher-Order Critical Points. When extending the classification of critical points beyond first-order, isolated critical points can become degenerate, i.e., $det(\mathbf{J}(\mathbf{x}_0)) = 0$. An example is shown in Fig. 3 (left), where the co-gradient vector field $\mathbf{v}(x, y) = (6xy, 3x^2 - 3y^2)^{\mathrm{T}}$ of a monkey saddle is visualized. Further, saddle critical points

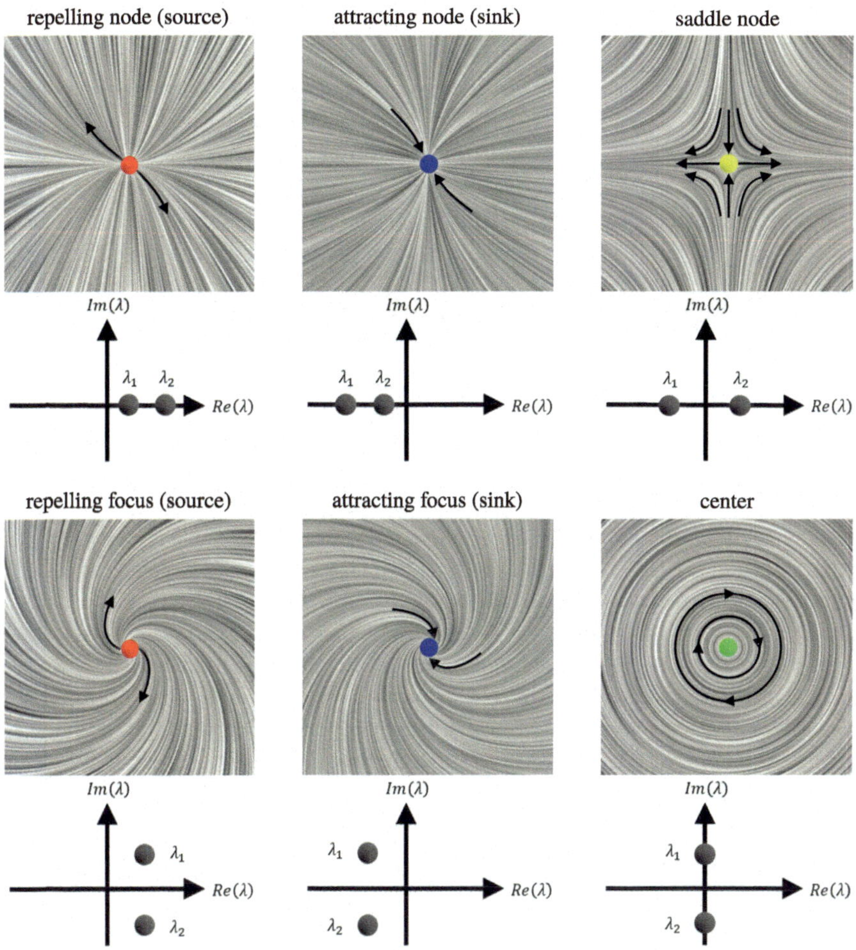

Fig. 2 Types of critical points in 2D steady vector fields (based on [55]). $Re(\lambda)$ denotes the real part of the eigenvalues λ_1 and λ_2, and $Im(\lambda)$ the imaginary part. The presence of imaginary parts (bottom) relates to vortical behavior

can have more than four sectors, which either exhibit parabolic, hyperbolic or elliptic behavior in their vicinity [107]. A parabolic sector contains streamlines that enter the critical point in one direction and leave it in the other direction. In an elliptic sector, streamlines enter the critical points in forward and backward direction. Finally in hyperbolic sectors, streamlines leave the critical point in both forward and backward direction. The latter is the kind of sector that is present around first-order saddles. Figure 3 shows examples of all three types of sectors around two higher-order critical points. Weinkauf et al. [136] visualized the different sectors around higher-order critical points in 3D, and Tricoche et al. [124] merged 2D critical points into higher-order critical points in order to simplify a given vector field.

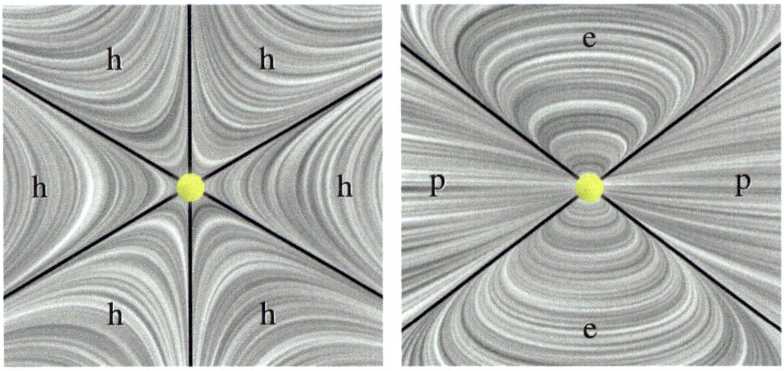

Fig. 3 Two examples of higher-order saddle critical points. Left, the co-gradient of a monkey saddle $\mathbf{v}(x, y) = (6xy, 3x^2 - 3y^2)$ is shown, which has six hyperbolic sectors. Right, the field $\mathbf{v}(x, y) = (2x^2 - 2y^2, xy)$ is visualized which has two parabolic and two elliptic sectors

Existence of Critical Points. Under certain circumstances, selected types of critical points cannot exist. For instance, in divergence-free flows, sources and sinks will never occur. Conservative vector fields, i.e., vector fields that are equivalent to the gradient of a scalar field, are always rotation-free, since their Jacobian is a symmetric matrix, having real eigenvalues only. Higher-dimensional flows that describe the motion of finite-sized objects in fluids do not contain sources [36, 40] and time-dependent flows do not contain any critical points in space-time, since particles always move forward in time. We address the topology of time-dependent flows in a later section.

Extraction of Critical Points. The numerical extraction of first-order critical points requires a root-finding in all components of the vector field. Equivalently, this can be seen as the intersection of the zero-level isolines of each flow component. For monotonic interpolation schemes, i.e., in bilinear (2D) and trilinear (3D) vectors fields, Globus et al. [34] discarded candidate cells by checking the signs of the components at the cell corners. If all signs are either all positive or all negative, then the cell cannot contain a critical point due to the mean value theorem. The position of the critical point can be located by recursive subdivision of the cell. Care must be taken, since this numerical scheme may result in duplicates, which have to be removed in a post-process. After a certain number of recursive subdivisions, exact locations may also be found by the application of multi-variate Newton-Raphson iterations [34]:

$$\mathbf{x}_{i+1} = \mathbf{x}_i - \nabla\mathbf{v}(\mathbf{x}_i)^{-1} \cdot \mathbf{v}(\mathbf{x}_i) \tag{6}$$

In practice, it is advisable to either use a QR factorization or the Moore-Penrose pseudoinverse of the Jacobian $\nabla\mathbf{v}(\mathbf{x})$ to avoid numerical issues in the matrix inversion. Using a singular value decomposition (SVD), $\nabla\mathbf{v}(\mathbf{x}_i) = \mathbf{U}\boldsymbol{\Sigma}\mathbf{V}^{\mathrm{T}}$, with $\mathbf{U}$ and $\mathbf{V}$ being orthonormal matrices and $\boldsymbol{\Sigma}$ being a diagonal matrix containing the singular values, the pseudoinverse is:

$$\nabla \mathbf{v}(\mathbf{x}_i)^+ = \mathbf{V}\boldsymbol{\Sigma}^+\mathbf{U}^\mathrm{T} \tag{7}$$

where $\boldsymbol{\Sigma}^+$ contains the reciprocals of the non-zero singular values. We refer the interest reader to the text book of Press et al. [92] (page 61) for a detailed explanation and discussion of the SVD. Note that the Newton method finds only one solution and requires an initial guess $\mathbf{x}_0$, for instance the center of a cell found by the aforementioned recursive subdivision. At most one critical point can exist per simplex, i.e., a triangle (2D) or tetrahedron (3D), and its location can be found analytically by inverting the barycentric interpolation.

2.1.2 Poincaré Index

The Poincaré index is a characteristic number of a closed curve γ in a 2D steady vector field $\mathbf{v} = (u, v)^\mathrm{T}$ [3]. It is computed by integrating the winding angle of the velocity vector (counterclockwise rotation) as we integrate along the closed curve γ in counterclockwise direction [125].

$$\mathrm{index}_\gamma = \frac{1}{2\pi}\oint_\gamma \mathrm{d}\alpha, \quad \text{with} \quad \alpha = \arctan\frac{v}{u} \tag{8}$$

The index of the curve is always an integer number. By placing a closed curve γ around a critical point such that no other critical point is inside the closed curve, the Poincaré index is extended to critical points. For first-order critical points, we have index +1 for sinks, sources and centers, and index -1 for saddles. If there are multiple critical points inside the closed curve γ, the indices of the interior critical points add up to coincide with the index of the curve γ. Thus, if no critical point is inside the area enclosed by the closed curve, the index is 0. For higher-order critical points, the Poincaré index is found by counting the number of elliptic sectors n_e and hyperbolic sectors n_h:

$$\mathrm{index}_{cp} = 1 + \frac{n_e - n_h}{2} \tag{9}$$

In linear vector fields, i.e., on triangular cells with barycentric interpolation, the index can be efficiently computed by accumulating the angle changes of the velocity vector along the three triangle edges [125], which can be used to efficiently test, whether a critical point exists inside the triangle. Scheuermann and Hagen [105] used this approach to check for critical points in neighboring triangles that may disappear after a diagonal flip.

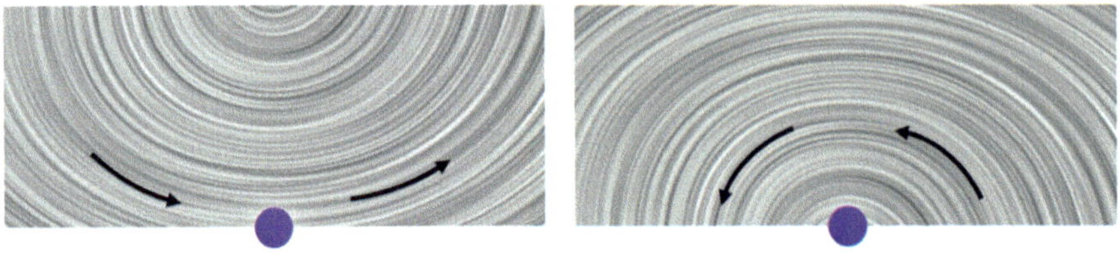

(a) In-bound boundary switch point (b) Out-bound boundary switch point

Fig. 4 Along open flow boundaries, the direction of in-flow and out-flow may change when walking along the boundary. These so-called boundary switch points are seeds of streamlines that are either in-bound (enter the domain) or out-bound (leave the domain)

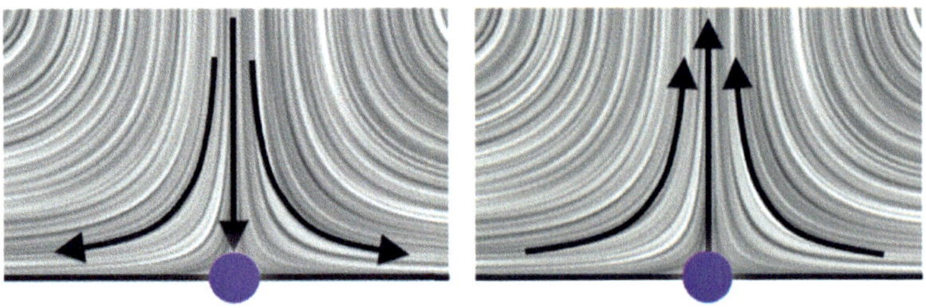

(a) Attachment point (b) Detachment point

Fig. 5 Examples of attachment and detachment points on closed boundaries

2.1.3 Boundary Switch Points

In case of open flow boundaries, i.e., if the fluid flow can enter or exit through a domain boundary, boundary switch points may exist [73]. When traveling along the domain boundary of a 2D flow, we either observe in-flow or out-flow, and the behavior may switch at certain locations. In practice, these locations can be determined as points at which the flow component that is normal to the boundary is zero and the flow vector is parallel to the domain boundary. Depending on whether the streamline that passes through the boundary switch point is staying in the domain or is always outside the domain, we refer to the boundary switch point as in-bound or out-bound, respectively. The type can be inferred from the direction in which the acceleration is pointing. For in-bound boundary switches, the acceleration points into the domain and for out-bound boundary switches it points outwards. An example for both cases is shown in Fig. 4. A topology-based flow visualization often views the entire domain. However, engineers might only want to study a region of interest, for example when the data becomes too large. To this end, Scheuermann et al. [106] found the structural changes of the streamlines in the region of interest by performing a topological analysis of the boundary.

(a) Four separatrices grow from a saddle critical point.

(b) Two separatrices grow from an in-bound boundary switch point.

(c) One separatrix grows from attachment or detachment point.

Fig. 6 Separatrices originate from the aforementioned topological elements

2.1.4 Attachment and Detachment Points

When a flow impinges directly on a wall it is forced to split left and right. The splitting point on the wall is called an attachment point [55], since in theory a single particle would get stuck on that location, whereas all other particles would slide left or right along the boundary. If the flow direction is reversed, this location is called a detachment point. See Fig. 5 for an example of both flow configurations. Both types of point are found in the same way. They are roots in the flow component that is tangential to the wall. Whenever a no-slip boundary condition is used, i.e., if the velocity on the wall is zero, the root finding test is taken an epsilon away from the boundary.

2.1.5 Separatrices

The above subsections introduced distinguished points in the domain that are of topological relevance. All the above points are connected by streamlines, which are called separatrices. At saddle critical points, separatrices emanate in direction of the eigenvectors. To numerically calculate them, the seed point is taken an epsilon away from the critical point, $\mathbf{x}_0 \pm \epsilon\, \mathbf{c}_i$, in direction of the eigenvectors $\mathbf{c}_i$, and the streamlines are traced in forward or backward direction, depending on the sign of the eigenvalue λ_i. For first-order critical points, four separatrices are connected to a saddle. In addition, we obtain two separatrices for in-bound boundary switch points and one separatrix for each attachment or detachment point. Figure 6 illustrates the separatrices that grow from saddle points, in-bound boundary switch points and attachment or detachment points. The net of separatrices spans the topological skeleton, which contains cells that are bounded by the separatrices. Each of the cells has the property that the origin and destination, i.e., the points reached in the limit by either forward or backward integration, are the same for all seed points within a cell. The topological skeleton was illustrated earlier in Fig. 1.

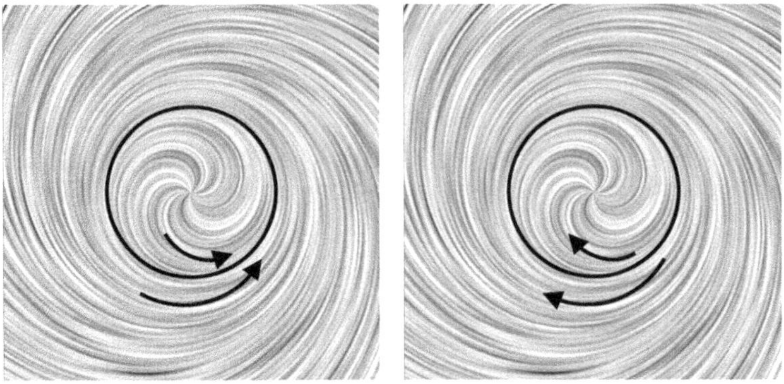

(a) Attracting closed streamline (b) Repelling closed streamline

Fig. 7 In 2D flows, isolated closed streamlines either act as sink or source on the surrounding vector field

2.1.6 Isolated Closed Streamlines

The last remaining topological element are closed streamlines, which are streamlines that connect to themselves. These structures are also known as periodic orbits. Similar to critical points, closed streamlines can be isolated or not. Isolated closed streamlines are either found by locating fixed points of the Poincaré map [142] by intersecting forward and backward integrated stream surfaces in space-time [121] or by Morse decomposition [20]. Due to the Poincaré index theorem, every 2D closed streamline contains at least one sink, source or center critical point, which can be used to guide the search for closed streamlines. In 2D flows, isolated closed streamlines are either acting as sink or source on the surrounding flow, as illustrated in Fig. 7, which means that they are either attracting or repelling. They can therefore not exist in divergence-free flows. In a divergence-free flow, we can find plenty of non-isolated closed streamlines.

2.2 Three-Dimensional Flows

The topological skeleton of steady 3D vector fields contains several more types of elements. The concepts, however, are very similar to the 2D case. Formally, we are given a three-dimensional steady vector field $\mathbf{v}(\mathbf{x}) : \mathbb{R}^3 \to \mathbb{R}^3$:

$$\mathbf{v}(\mathbf{x}) = \begin{pmatrix} u(x, y, z) \\ v(x, y, z) \\ w(x, y, z) \end{pmatrix} \tag{10}$$

with the spatial derivatives

$$\frac{\partial \mathbf{v}(\mathbf{x})}{\partial \mathbf{x}} = \nabla \mathbf{v}(\mathbf{x}) = \mathbf{J}(\mathbf{x}) = \begin{pmatrix} \frac{\partial u(x,y,z)}{\partial x} & \frac{\partial u(x,y,z)}{\partial y} & \frac{\partial u(x,y,z)}{\partial z} \\ \frac{\partial v(x,y,z)}{\partial x} & \frac{\partial v(x,y,z)}{\partial y} & \frac{\partial v(x,y,z)}{\partial z} \\ \frac{\partial w(x,y,z)}{\partial x} & \frac{\partial w(x,y,z)}{\partial y} & \frac{\partial w(x,y,z)}{\partial z} \end{pmatrix} \tag{11}$$

2.2.1 Critical Points

As in the 2D case, we call locations $\mathbf{x}_0$ at which the flow vanishes critical points of $\mathbf{v}(\mathbf{x})$, i.e., $\mathbf{v}(\mathbf{x}_0) = \mathbf{0}$. As before, a first-order classification is done by an eigenanalysis of the Jacobian $\mathbf{J}(\mathbf{x}) = \nabla \mathbf{v}(\mathbf{x})$. Let λ_i be the eigenvalues corresponding to the eigenvectors $\mathbf{c}_i$, i.e., $\mathbf{J}(\mathbf{x}) \cdot \mathbf{c}_i = \lambda_i \cdot \mathbf{c}_i$ for $i \in \{1, 2, 3\}$. We distinguish the following types of first-order critical points based on the real-part of the eigenvalues:

$$\text{source: } 0 < Re(\lambda_1) \leq Re(\lambda_2) \leq Re(\lambda_3) \tag{12}$$

$$\text{repelling saddle: } Re(\lambda_1) < 0 < Re(\lambda_2) \leq Re(\lambda_3) \tag{13}$$

$$\text{attracting saddle: } Re(\lambda_1) \leq Re(\lambda_2) < 0 < Re(\lambda_3) \tag{14}$$

$$\text{sink: } Re(\lambda_1) \leq Re(\lambda_2) \leq Re(\lambda_3) < 0 \tag{15}$$

Figure 8 illustrates the different types. Note that there are two types of saddles in 3D, which are categorized based on their dominant behavior into attracting or repelling saddles. Each of the four cases above can be further subdivided by considering the imaginary parts of the eigenvalues:

$$\text{Focus: } Im(\lambda_1) = 0 \quad \text{and} \quad Im(\lambda_2) = -Im(\lambda_3) \neq 0 \tag{16}$$

$$\text{Node: } Im(\lambda_1) = Im(\lambda_2) = Im(\lambda_3) = 0 \tag{17}$$

Without loss of generality we assumed λ_1 to be real-valued eigenvalue in the focus classification. Note that either none or two eigenvalues will have imaginary parts, since complex eigenvalues always appear in pairs of complex-conjugates. In case of complex eigenvalues, the rotation occurs in a plane, which is spanned by the real and imaginary parts of the corresponding complex-conjugate eigenvectors. This plane is also referred to as the swirling plane.

2.2.2 Boundary Switch Curves

In 3D domains, the boundary is represented by 2D surfaces. On these surfaces, the flow either enters or exits the domain. A line that separates these two behaviors is called a boundary switch curve [135]. It consists of all location $\mathbf{x}$ at which $\mathbf{v}(\mathbf{x})$ is parallel to the tangent plane of the domain boundary. Figure 9 shows the possible configurations.

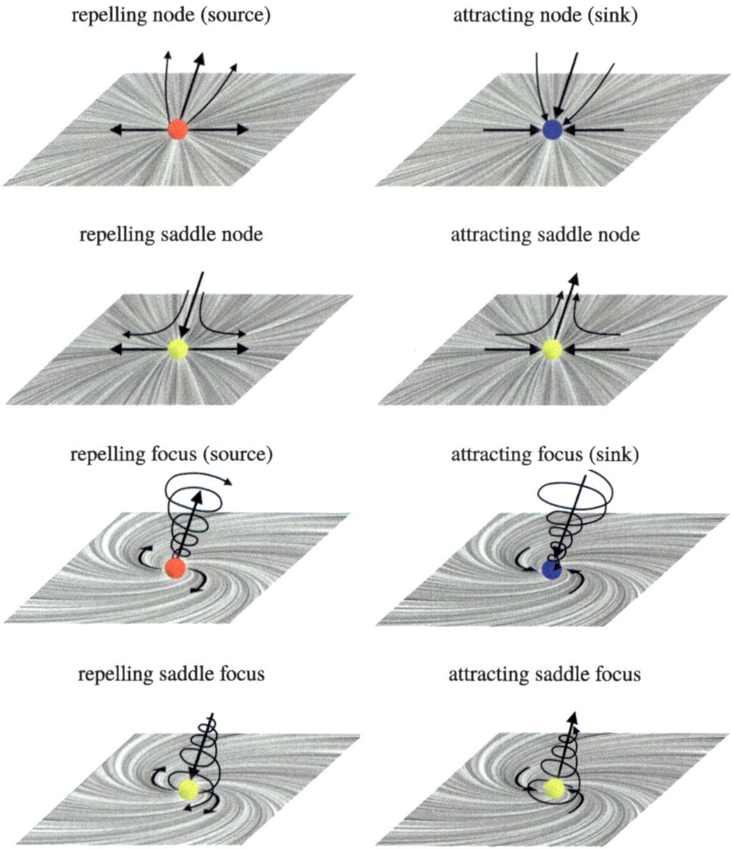

Fig. 8 Types of critical points in 3D steady vector fields

As we travel along a boundary switch curve, we can observe the behavior of streamlines passing through. Line sectors along which a passing streamline remains in the domain are called in-bound sectors. Equivalently, these are points at which the acceleration $\mathbf{a} = \mathbf{J}\mathbf{v}$ points into the domain. Conversely, if the streamline remains outside of the domain, the point is part of an out-bound sector. Here, the acceleration $\mathbf{a} = \mathbf{J}\mathbf{v}$ points out of the domain. Locations at which the behavior switches from in-bound to out-bound are called in-out points. In piecewise bilinear vector fields, i.e., if the vector field is given on a regular grid and is interpolated bilinearly, boundary switch curves are either straight line segments on the boundaries of the piecewise linear fields, or they are hyperbolas inside of the piecewise bilinear fields. Figure 9d gives an example.

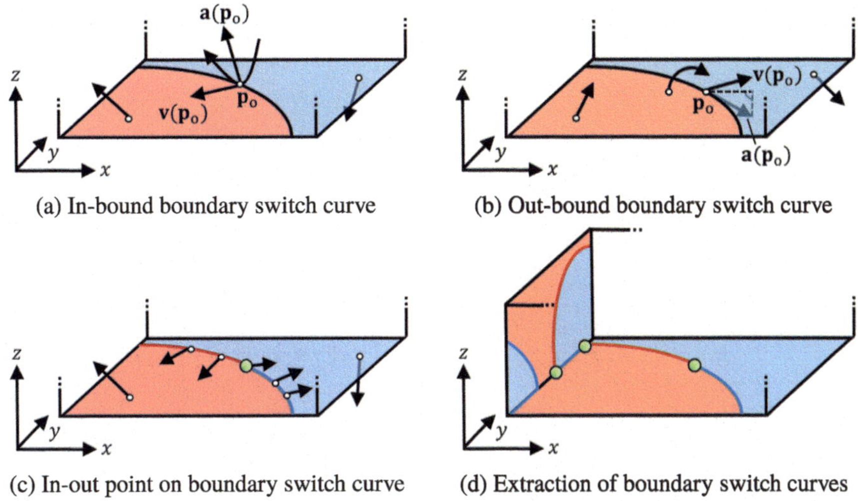

(a) In-bound boundary switch curve (b) Out-bound boundary switch curve

(c) In-out point on boundary switch curve (d) Extraction of boundary switch curves

Fig. 9 On open flow boundaries, there are curves that separate regions of in-flow and out-flow. Segments of the curves are classified based on the behavior of streamlines seeded from them. Curve points at which the behavior switches are called in-out points (green)

2.2.3 Attachment and Detachment Points

Similar to the 2D flow, attachment and detachment points occur on boundaries and obstacles. In the analysis of fluid flows, we are not only interested in the particular points, but also on the behavior of nearby particles. For instance, the attaching or detaching lines might be swirling and have more or less temporally-coherent behavior. Wiebel et al. [140] detected vortices that detach from a boundary by tracking critical points in the wall shear stress vector field and continuously releasing particles from the critical points, which assembles so-called generalized streaklines that show the swirling behavior. Nsonga et al. [80] developed an algorithm to extract the regions around attachment and detachment points, which are referred to as splats and antisplats. In the analysis of meteorological flows around mountains, scientists are interested in a characteristic number that distinguises whether a stratified air flow goes over or around a mountain. To decide this, meteorologists use the Froude number [54, 74], which was adapted from naval architecture.

2.2.4 Separatrices

In Sect. 2.2.1, we have seen that there are two types of saddles in 3D steady vector fields: repelling saddles (two positive real parts) and attracting saddles (two negative real parts). Starting an epsilon neighborhood away from the critical point, separatrices grow out in the direction of the eigenvectors. Two of the three eigenvalues have equal sign. Their corresponding eigenvectors span a surface from which a separating

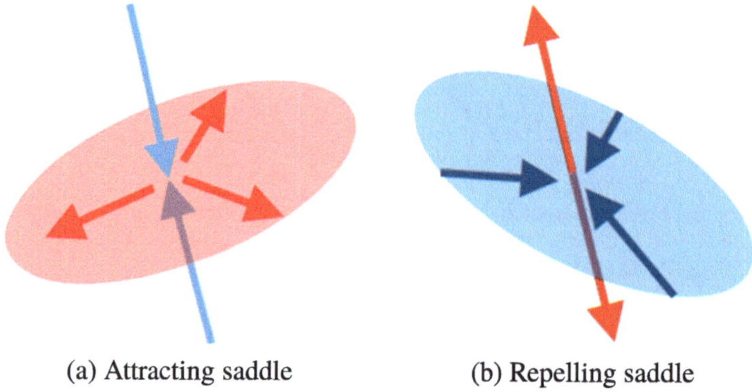

Fig. 10 Separating surfaces growing out from an attracting and a repelling saddle. The remaining eigenvector direction points in the direction of a virtual separatrix

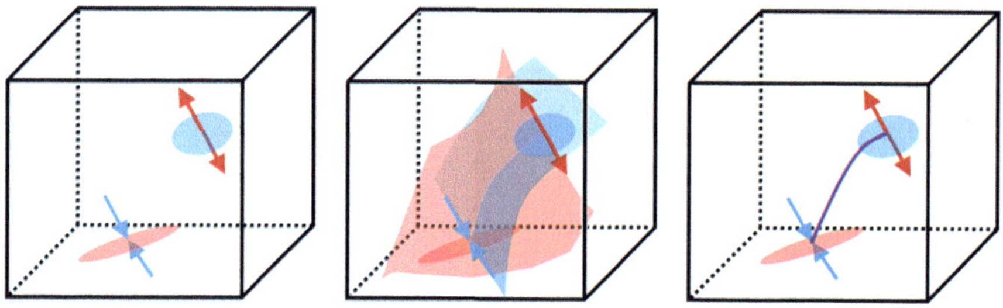

Fig. 11 Forward and backward-integrated separating surfaces can be intersected, giving saddle connectors [120]. Here, the saddles (left), their separatring surfaces (middle) and the intersection (right) are shown

surface is growing, as illustrated in Fig. 10. A stream surface can be thought of as the union of infinitely many streamlines that were seeded along a seeding curve. In our case, the seeding curve lies in the plane spanned by the eigenvectors. We refer to Schneider et al. [109] for a stream surface extraction algorithm that adapts the refinement to the presence of critical points. Alternatively, Wiebel et al. [139] investigated the extraction of separatrices using suitable cross sections of the flow. The remaining eigenvalue with opposite sign emanates a streamline in direction of the eigenvector, which is referred to as *virtual separatrix* [1], as it does not truly separate space. In addition, separating surfaces grow from in-bound boundary switch curves. Aside from the pure geometric extraction of separating surfaces, illustrative techniques have been investigated [14] to produce informative surface visualizations that were inspired by the early work of Dallmann [23].

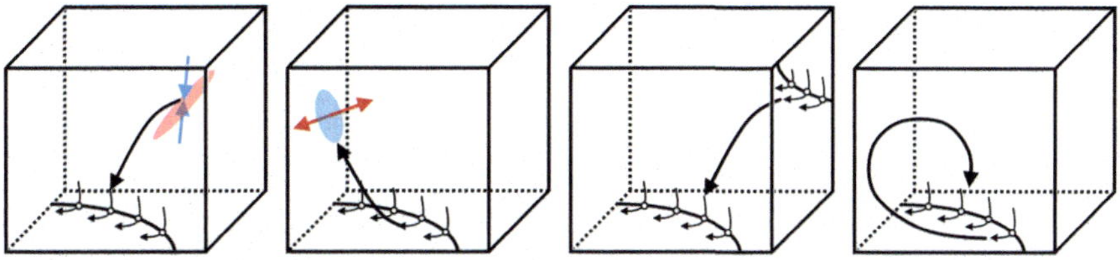

Fig. 12 Boundary switch connectors arise when separating surfaces of saddles and boundary switch curves are intersected [135]. Here, the different configurations are shown

2.2.5 Saddle Connectors and Boundary Switch Connectors

The visualization of separating surfaces becomes cluttered very quickly, since the separating surfaces may occlude each other on the screen. The picture can be simplified considerably by intersecting forward-integrated and backward-integrated separating surfaces, as shown in Fig. 11. The intersecting curve is called a saddle connector [120], which is a single streamline that connects the two saddles. Saddle connectors are found by intersecting the separating surfaces of saddles and also by intersecting the separating surfaces growing from boundary switch curves. The latter gives rise to boundary switch connectors [135], as shown in Fig. 12. Boundary switch connectors arise from connection with saddles, from connection with other boundary switch curves or even by self-connection to themselves.

2.2.6 Isolated Closed Streamlines

Isolated closed streamlines are theoretically well-understood and studied for general dynamical systems [143]. In 3D, closed streamlines can act as sink, source, center or saddle [63]. The notion of isolated closed streamlines that locally act as centers or saddles leads us to the feature curves that are described next.

All the above elements are considered topological elements, and are therefore part of the topological skeleton. In the following, we introduce two feature curves that are closely related to topology as they also order the flow.

2.2.7 Vortex Corelines

The first feature curve are vortex corelines. These are segments of streamlines that other streamlines rotate around, see Fig. 13a. Globus et al. [34] identified them as virtual separatrices growing out from focus saddle points, by tracing them in the direction of the eigenvector with corresponding real eigenvalue. For steady vector fields, Sujudi and Haimes [115] proposed the reduced velocity criterion:

$$\mathbf{v} - (\mathbf{v}^{\mathrm{T}}\mathbf{e})\mathbf{e} = \mathbf{0} \tag{18}$$

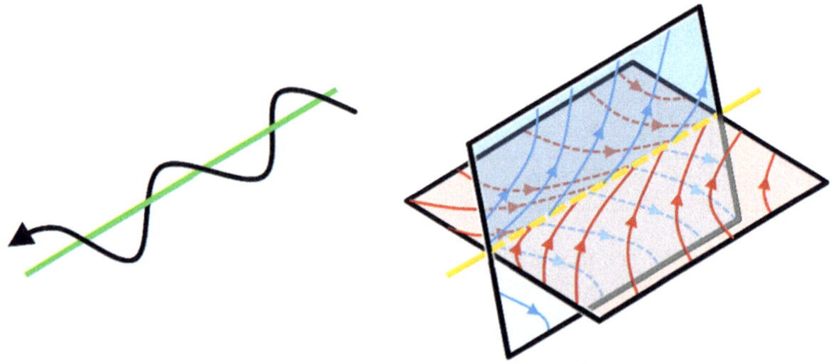

(a) Vortex coreline (b) Bifurcation line

Fig. 13 In 3D flows, two types of feature curves are often of interest: vortex corelines (green) and bifurcation lines (yellow)

This criterion identifies locations at which the flow $\mathbf{v}$ flows in the direction of the eigenvector $\mathbf{e}$ with corresponding real eigenvalue. Since the Jacobian is also requested to have complex eigenvalues, the criterion makes sure that the projection of the flow vector onto the swirling plane gives zero. The criterion was applied in practice in various situations [31, 65]. Peikert and Roth [85] reformulated this criterion into the parallel vectors form:

$$\mathbf{v} \parallel \mathbf{J}\mathbf{v} \quad \Leftrightarrow \quad \mathbf{v} \times \mathbf{J}\mathbf{v} = \mathbf{0} \tag{19}$$

with the condition that two eigenvalues of $\mathbf{J}$ are complex. Two vector fields are parallel if their cross product is zero, which turns the extraction into a component-wise root-finding problem. We refer to Peikert and Roth for more details on the numerical extraction of parallel vectors solutions [85]. As shown by Roth and Peikert [97], this criterion assumes that the curvature of the resulting coreline is zero. The curvature of a continuous 3D curve is calculated using curve tangent $\dot{\mathbf{x}}(t) = \frac{\mathrm{d}\mathbf{x}(t)}{\mathrm{d}t}$ and acceleration $\ddot{\mathbf{x}}(t) = \frac{\mathrm{d}^2\mathbf{x}(t)}{\mathrm{d}t^2}$:

$$\kappa(\mathbf{x}(t)) = \frac{\dot{\mathbf{x}}(t) \times \ddot{\mathbf{x}}(t)}{\|\dot{\mathbf{x}}(t)\|^3} \tag{20}$$

Since streamlines are tangent curves of the vector field $\mathbf{v}(\mathbf{x})$, we have $\dot{\mathbf{x}}(t) = \mathbf{v}(\mathbf{x}(t))$ and $\ddot{\mathbf{x}}(t) = \mathbf{J}(\mathbf{x}(t)) \cdot \mathbf{v}(\mathbf{x}(t))$. Thus, with Eq. (19), the curvature in Eq. (20) evaluates to zero.

In generalization, Roth and Peikert [97] introduced a criterion for bent vortex corelines as $\mathbf{v} \parallel (\nabla\mathbf{a})\mathbf{v}$, which assumes that corelines have zero torsion. We refer to Günther and Theisel [42] for an overview of vortex extraction methods, including density-based methods [138], extremum lines [102, 103], and integration-based methods [7].

2.2.8 Bifurcation Lines

Bifurcation lines differ from vortex corelines in one aspect: instead of complex eigenvalues, we need to have saddle-like behavior in the plane that is spanned by the two eigenvectors that are not parallel to the flow [87, 96]. In consequence, all eigenvalues are real-valued. Bifurcation lines of a 3D steady flow are illustrated in Fig. 13b. In the fluid dynamics literature, these feature curves are also known as hyperbolic trajectories [47]. We discuss the recent definitions and extraction algorithms for time-dependent flows later.

2.2.9 Invariant Manifolds

In an n-dimensional vector field $\mathbf{v}(\mathbf{x}) : \mathbb{R}^n \to \mathbb{R}^n$ sets of locations may exist that particles will never leave during advection. Formally, we can say that any point $\mathbf{x}_0 \in S$ of such an invariant set $S \subset \mathbb{R}^n$ remains in S, i.e., $\mathbf{x}_0 + \int_0^\tau \mathbf{v}(\mathbf{x}(t))\, dt \in S$ for any τ. If the set S is a manifold in the domain, the structure is called an *invariant manifold*. Invariant manifolds are an essential building block of the steady vector field topology. Every critical point, every closed orbit and every separatrix is an invariant manifold. Strictly speaking, every trajectory is an invariant manifold, too, which is why being an invariant manifold is only a necessary condition for a topological element. Under this definition, bifurcation lines and vortex corelines are not topological elements, since particles can flow out of a vortex coreline or a bifurcation line, especially when a specific criterion has to be fulfilled, such as the presence of swirling behavior or a sectional separation. For this reason, vortex corelines and bifurcation lines are usually considered to be feature curves.

2.3 Remarks

As the name suggests, differential vector field topology requires that the vector field is differentiable. In practice, however, a different vector field discretization, noise in the data or numerical integration errors during streamline integration will influence the result. To address these issues, alternatives have been explored, including discrete vector field topology [21, 22] and combinatorial vector field topology [26, 93, 94]. In this overview, we concentrated on the continuous approaches, i.e., differential vector field topology. However, we encourage the interested reader to explore the discrete and combinatorial approaches as well. All methods in the previous sections were designed for steady vector fields. In the following, we cover the topology of time-dependent flows.

3 Unsteady Flows

The previous section assumed that the vector field is not changing over time. In practice, however, vector fields are often time-dependent. As shown by Theisel et al. [122], the topology of time-dependent vector fields can be analyzed from two different angles: by studying streamlines or by studying pathlines. To start with, we look at the difference between a streamline and a pathline. For this, let $\mathbf{v}(x, y, t)$ be a time-dependent vector field:

$$\mathbf{v}(x, y, t) = \begin{pmatrix} u(x, y, t) \\ v(x, y, t) \end{pmatrix} \tag{21}$$

When freezing the vector field in time, i.e., if we only consider the trajectory of a particle in one single time slice at time t_0, we obtain a streamline $\mathbf{x}(\tau)$:

$$\text{streamline:} \qquad \frac{\mathrm{d}\mathbf{x}(\tau)}{\mathrm{d}\tau} = \mathbf{v}(\mathbf{x}(\tau), t_0) \qquad \mathbf{x}(0) = \mathbf{x}_0 \tag{22}$$

where $\mathbf{x}_0$ is the seed point. Such a trajectory marks the instantaneous path of a particle, which is more relevant for the study of magnetic field lines. In fluid flow analysis, we are concerned with the trajectory of a particle over a continuously advancing time:

$$\text{pathline:} \qquad \frac{\mathrm{d}\mathbf{x}(t)}{\mathrm{d}t} = \mathbf{v}(\mathbf{x}(t), t) \qquad \mathbf{x}(t_0) = \mathbf{x}_0 \tag{23}$$

Here, $\mathbf{x}_0$ and t_0 are the seed position and seed time, respectively. The latter ODE is not autonomous, since it depends on time t, which is not a state variable of the dynamical system. In other words, the numerical particle integrator needs additional information to sample the correct time slice. However, we can turn the definition of the tangent curves into autonomous first-order ODEs by making time an explicit state variable. To do so, we lift the vector field one dimension up for which we have two options:

$$\bar{\mathbf{s}}(x, y, t) = \begin{pmatrix} u(x, y, t) \\ v(x, y, t) \\ 0 \end{pmatrix} \qquad \bar{\mathbf{p}}(x, y, t) = \begin{pmatrix} u(x, y, t) \\ v(x, y, t) \\ 1 \end{pmatrix} \tag{24}$$

The vector field $\bar{\mathbf{s}}$ is called streamline vector field, since its tangent curves are streamlines of $\mathbf{v}(\mathbf{x}, t)$. Since the last component of $\bar{\mathbf{s}}$ is zero, the time will not change during particle integration, i.e., we obtain streamlines. On the other hand, the vector field $\bar{\mathbf{p}}$ is called the pathline vector field, since its tangent curves are pathlines of $\mathbf{v}(\mathbf{x}, t)$. Here, the last components is one, which causes the time to flow forward at the correct step size, as we numerically integrate the trajectory. On a side note, other characteristic curves such as streaklines and timelines can similarly be expressed as tangent curves of lifted vector fields [132, 134].

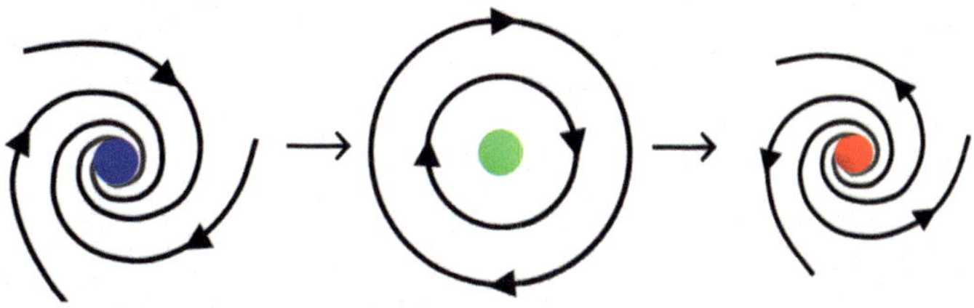

Fig. 14 When critical points change their type, a Hopf bifurcation is present. Here, an attracting focus (blue) turns into a repelling focus (red) by transitioning through a center (green)

Note that neither $\overline{\mathbf{s}}$ not $\overline{\mathbf{p}}$ have isolated critical points. The vector field $\overline{\mathbf{s}}$ contains critical lines, i.e., paths of critical points in space-time, and vector field $\overline{\mathbf{p}}$ contains no critical points at all, since the last component is always unequal to zero.

3.1 Streamline-Oriented Topology

Streamline-oriented topology is concerned with the changes in the asymptotic behavior of streamlines, when the vector field changes over time. Essentially, this means that we observe how the topological skeleton is evolving. In the following, we discuss the possible events in unsteady 2D vector fields. We refer to Tricoche et al. [126] and Theisel et al. [122] for more details. The last section will explain the differences for 3D unsteady vector fields.

3.1.1 Fold Bifurcations

Fold bifurcations occur when two critical points collapse or appear. This will always happen in pairs of critical points. Governed by the Poincaré index theorem, the only possibility is for saddles (index -1) to merge with either a sink, source or center (index $+1$). Conversely, two critical points will always appear together in pairs. In 2D flows, it is never possible for sinks to collapse with sources. Fold bifurcations can be found as critical points in the space-time domain of the vector field:

$$\begin{pmatrix} u(x,y,z) \\ v(x,y,z) \\ det(\mathbf{J}(x,y,z)) \end{pmatrix} = \mathbf{0} \tag{25}$$

The last component of this vector field contains the determinant of the Jacobian. In the event of two critical points merging, the two critical points are no longer isolated, due to the presence of the other point, i.e., the determinant briefly vanishes to zero. An example of a fold bifurcation is shown later in a space-time visualization in Fig. 16.

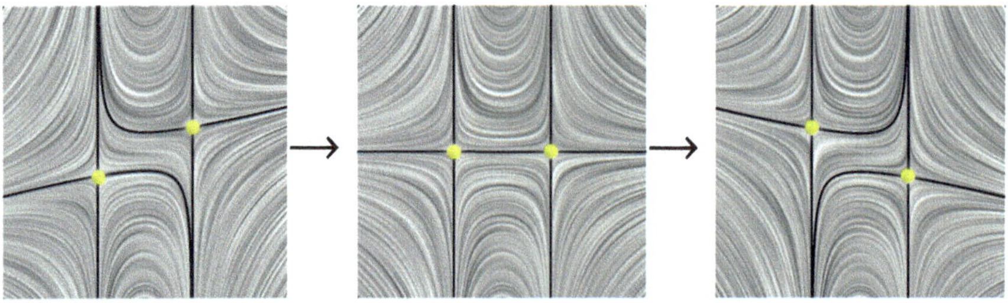

(a) Heteroclinic saddle connection (separatrices of two different saddles connect)

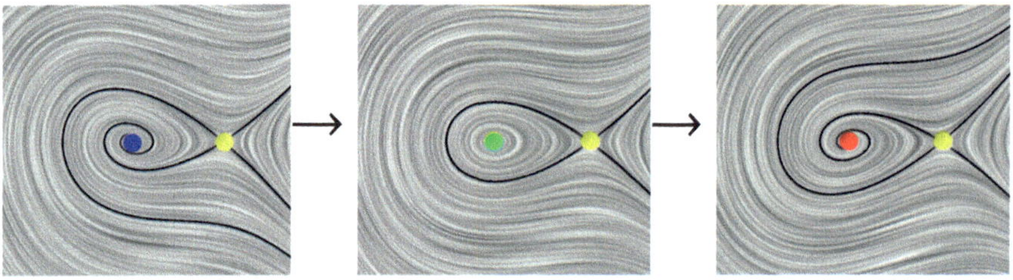

(b) Homoclinic saddle connection (separatrices of the same saddle connect)

Fig. 15 When separatrices connect saddle points, we obtain heteroclinic (two different saddles) or homoclinic (same saddle) saddle connectors. Homoclinic saddle connectors occur when the enclosed critical point undergoes a Hopf bifurcation. The illustrations above show three time steps of time-dependent vector fields that contain such bifurcations

3.1.2 Hopf Bifurcation

A Hopf bifurcation is the change in the type of a critical point. Thereby, a repelling focus may turn into an attracting focus via briefly transitioning through a center, as illustrated in Fig. 14. Alternatively, an attracting focus may turn into a repelling focus via a center. In the space-time domain, these locations are found as critical points of the vector field:

$$\begin{pmatrix} u(x,y,z) \\ v(x,y,z) \\ u_x(x,y,t) + v_y(x,y,t) \end{pmatrix} = \mathbf{0} \tag{26}$$

where the last component is the divergence of the flow, which is zero for the center configuration that is briefly visited when transitioning from an attracting focus to a repelling focus or vice versa.

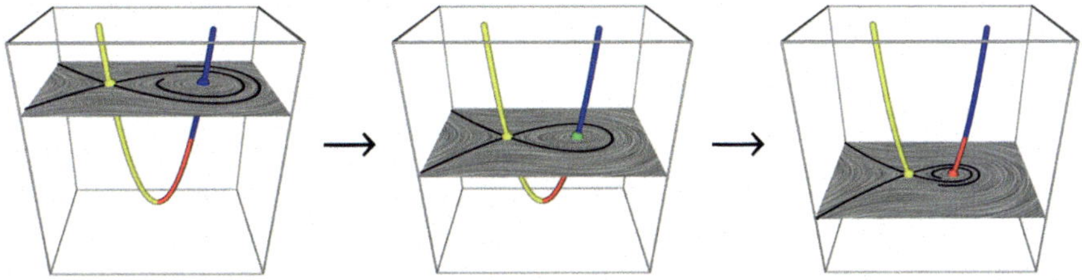

Fig. 16 Space-time visualization of a Hopf bifurcation (critical point changes type) and a fold bifurcation (two critical points merge). The paths of critical points are color-coded based on the type of the critical point (attracting is blue, repelling is red, saddle is yellow). The vertical axis of the 3D domain represents the time axis

3.1.3 Saddle Connection

Usually, the probability for a streamline to directly enter a saddle is zero. However, whenever two saddles slide past each other, there is a brief moment in time, when the separatrices of the two saddles meet, as illustrated in Fig. 15a. This brief connection is called a heteroclinic saddle connection. There is also a special case called a homoclinic saddle connection, which is sometimes also referred to as periodic blue sky bifurcation. In this case, the separatrix of a saddle connects back to the same saddle, as shown in Fig. 15b. This usually happens if the area that is enclosed by the self-connecting separatrix contains a critical point that undergoes a Hopf bifurcation. All types of saddle connections are momentary events, which can be found by intersecting the forward and backward integrated separating surfaces, as described by Theisel et al. [120].

3.1.4 Cyclic Fold Bifurcations

Previously, we have seen another type of topological structure: isolated closed streamlines. Two isolated closed streamlines may collapse onto each other, letting both of them disappear. This event is referred to as a cyclic fold bifurcation. Conversely, the isolated closed streamlines may be created together. Both kinds of events are either found by tracking closed streamlines through adjacent time slices [126] or by searching for adjacent curves in space-time [122].

3.1.5 Space-Time Visualizations

The changes of the topological skeleton in 2D time-dependent flows are best shown in 2D space-time by mapping time to the third dimension. The paths of critical points appear as curves. Hopf bifurcations are points on the curves, as shown in Fig. 16 and Fold bifurcations are the junctions at which curves meet. Starting from saddles, separating surfaces can be grown. We refer to Theisel et al. [122] for examples.

3.1.6 Streamline-Oriented Topology in 3D

Most of the previous bifurcation events carry over to the 3D case. Fold bifurcations are similar to the 2D case. Hopf bifurcations include transitions from sources to attracting saddles (and vice versa), or from sinks to repelling saddles (and vice versa). Saddle connections connect a single separatrix of one saddle to the separating surface of another saddle. Saddle connectors and boundary switch connectors can also collapse and disappear, which is a called a connector fold bifurcation. We refer the reader to the work of Weinkauf [131] for a more elaborate explanation of streamline-oriented topology in 3D.

3.2 *Pathline-Oriented Topology*

In the previous section, we visualized how streamlines are changing when a vector field is evolving over time. The observation of streamlines, however, is not particularly meaningful, when we want to assess the behavior of particles over time. In this case, we are rather interested in the observation of pathlines. There is, however, an inherent problem. While the streamline-oriented topology could trace particles in the time slice for an infinite amount of time, enabling an asymptotic observation of the flow, a pathline-oriented topology, is limited in the integration duration by the temporal domain of the data set. Unless the flow is periodic, we cannot study asymptotic behavior. In fact, not even critical points exist in the lifted vector field $\overline{\mathbf{p}}$ in Eq. (24), since the last component is always non-zero. In the absence of an asymptotic picture of the flow, we will fall back to a finite-time description of the behavior, which requires a formal definition of flow maps.

3.2.1 Flow Maps

In a time-dependent vector field $\mathbf{v}(\mathbf{x}, t)$, we use the *flow map* $\phi_{t_0}^{\tau}(\mathbf{x}_0)$, which maps a particle that was seeded at position $\mathbf{x}_0$ and at time t_0 to the location that it reaches after an integration in $\mathbf{v}(\mathbf{x}, t)$ for a given duration τ:

$$\phi_{t_0}^{\tau}(\mathbf{x}_0) = \mathbf{x}_0 + \int_{t_0}^{t_0+\tau} \mathbf{v}(\mathbf{x}(t), t)\, dt, \quad \text{with} \quad \mathbf{x}(t_0) = \mathbf{x}_0 \tag{27}$$

The flow map has a number of useful properties, such as:

(a) Hyperbolic LCS (b) Elliptic LCS (c) Parabolic LCS

Fig. 17 Overview of the different types of Lagrangian coherent structures (LCS) [49]. **a** shows vortices in the wake of Heard Island. The line structures that separate the vortices are the hyperbolic LCS. The image was released by NASA and was captured with the MODIS instrument aboard their Aqua satellite. **b** depicts ocean eddies, which carry plankton and pollutants across the ocean. These flow structures can persist for weeks. The image was released by the National Science Foundation and is credited to NASA. **c** shows a jet stream, which is a fast moving and narrow air current. In this image, it carries cirrus clouds vertically across the image. The cloud band has a distinct pattern that is created by the air current. This image is in public domain

$$\phi_t^0(\mathbf{x}) = \mathbf{x} \tag{28}$$

$$\frac{\mathrm{d}\phi_t^\tau(\mathbf{x})}{\mathrm{d}\tau} = \mathbf{v}(\phi_t^\tau(\mathbf{x}), t_0 + \tau) \tag{29}$$

$$\phi_{t_0}^{\tau_1+\tau_2}(\mathbf{x}) = \phi_{t_0+\tau_1}^{\tau_2}(\phi_{t_0}^{\tau_1}(\mathbf{x})) \tag{30}$$

$$\phi_{t+\tau}^{-\tau}(\phi_t^\tau(\mathbf{x})) = \mathbf{x} \tag{31}$$

Equation (28) expresses the identity flow map, where a particle is traced for duration $\tau = 0$, i.e., it stays at its seed point. Equation (29) states that the derivative of the flow map with respect to the integration duration is exactly the flow direction at the end point of the flow map, which is fulfilled, since the flow map follows the trajectory of a pathline. Equation (30) shows that two flow maps can be concatenated. In practice, the flow maps are discretized, which leads to discretization errors upon concatenation [19]. And finally, Eq. (31) means that the flow map is invertible. Flow maps are an essential building block of many feature definitions, including recirculations [141] and Lagrangian coherent structures (LCS) [49], which are described in the following.

3.2.2 Lagrangian Coherent Structures

Based on the flow map of the previous section, we can now express particle behavior that unfolds during a certain finite time range. Similar to separatrices in the previous section, we are interested in sets of particles that order the flow into regions of coherent behavior. This leads us to the definition of material lines. There are three types of material lines, which have been summarized recently by Haller [49]. We

(a) 2D Backward FTLE

(b) 3D Forward FTLE using [5], CC-BY 4.0.

Fig. 18 Hyperbolic LCS are frequently approximated by the finite-time Lyapunov exponent (FTLE). Forward FTLE approximates repelling structures and backward FTLE approximates attracting structures. Left, we see the fluid flow past a heated cylinder and right, we see a close-up of a cumulus cloud convection simulation

refer to Onu et al. [82] for a discussion of various LCS extraction techniques. An overview of the three types is given in Fig. 17.

Hyperbolic LCS are material lines that attract or repel locally the strongest. These lines act as transport barriers that particles do not cross. The lines are either attracting nearby particles or they are repelling them away. A common measure to approximate hyperbolic LCS is through the finite-time Lyapunov exponent (FTLE). FTLE linearly approximates the expansion of a virtual sphere by measuring the maximal separation of nearby-released particles. The expansion rate is captured by the right-hand Cauchy-Green tensor $\nabla\phi^{\mathrm{T}}\nabla\phi$, where ϕ is the aforementioned flow map. With τ being the integration duration, FTLE is defined as, c.f. Shadden [112]:

$$\mathrm{FTLE} = \frac{1}{|\tau|} \ln \sqrt{\lambda_{\max}(\nabla\phi^{\mathrm{T}}\nabla\phi)} \tag{32}$$

The flow map gradient is typically calculated by finite differences [52]. Alternatively, Kasten et al. [62] proposed localized FTLE, which linearizes the flow at each step to measure and concatenate the rate of expansion. To obtain hyperbolic LCS at subgrid accuracy and to avoid the numerical computation of flow map derivatives, Kuhn et al. [69] tracked timeline cells over time, which are adaptively refined whenever the timeline segments intersect. A benchmark comparison of multiple FTLE extraction algorithms was performed by Kuhn et al. [70]. To reduce the number of redundant particle integrations and to thereby improve performance, Brunton and Rowley [15] concatenated flow maps. Garth et al. [30] accelerated the computation and later limited the visualization of FTLE to boundaries in the flow to reduce clutter [33]. Sadlo et al. [98] and Barakat et al. [9] adaptively refined the flow maps. Barakat et al. [8] developed an interactive computation and rendering framework for unsteady 3D flows, in which the FTLE values are stored in a view-dependent and adaptively refined sparse grid. To avoid discretization artifacts entirely, Monte Carlo methods have been used [5, 38], which invoke exact FTLE calculations at each volume sample of a Monte Carlo renderer. Compared to the previous methods, this approach is far from interactive speed, but it is able to generate a ground truth image that can serve as baseline. Examples of FTLE visualizations are shown in Fig. 18. Haller introduced hyperbolic trajectories [47], which are trajectories that experience in one spatial direction repelling behavior and in the other attracting behavior in the locally strongest way. Based on this concept, Sadlo and Weiskopf [99] defined time-dependent saddles as the intersection of forward and backward FTLE ridges, which was later extended by Üffinger et al. [129] to 3D. This approach requires the computation of the finite-time Lyapunov exponent in the entire domain in order to find the ridge intersections, which is an expensive computation. For this reason, Hofmann and Sadlo [59] developed a refinement scheme that is applied to an initial guess that was computed locally [6, 78]. Thereby, the extraction time of distinguished hyperbolic trajectories [61], i.e., hyperbolic LCS, is greatly reduced. Bujack et al. [16] recently discussed different options to calculate the separating and repelling behavior in finite time windows. The above methods were developed for continuous vector fields. For scenarios, in which the velocity field is represented in a particle-based manner, Agranovsky et al. [2] and Shi et al. [113] developed FTLE extraction algorithms from particle data.

Elliptic LCS are lines that bound regions that rotate coherently [50] or do not stretch much during advection. The latter is expressed more formally as curves across which the averaged material stretching rate shows no leading-order variability [110]. In incompressible 2D flows, these lines preserve arc length and surface area. These structures are tightly related to vortex identification [42], in particular to vortex boundaries. For instance, Haller [49] selects the outermost nested elliptic LCS as boundary of a coherent vortex. More recently, Katsanoulis et al. [64] characterized elliptic LCS as lines that inhibit the diffusion of vorticity. For a summary of more vortex identification methods we refer to Günther and Theisel [42].

Parabolic LCS are transport barriers along which the material shearing is minimized, which corresponds to the cores of jets. Since these structures are embedded inside non-stretching structures, their stretching is also low. Farazmand and Haller [24] defined them as minimally hyperbolic, structurally stable chains of tensorlines that connect singularities of the Cauchy–Green strain tensor field. In the visualization community, jets in atmospheric flows have been identified as lines with maximal velocity magnitude [66] in a local coordinate frame that is aligned with the flow direction.

3.2.3 Coherent Sets and Almost-Invariant Sets

Dynamical systems are classified into autonomous or non-autonomous systems based on whether they depend on an independent variable, such as time. In autonomous systems, regions of the domain that resist mixing over a finite-time duration are referred to as almost-invariant sets. In time-dependent systems, these regions are known as coherent sets. Coherent sets can be seen as counterpart to LCS, since LCS divide the domain into regions of coherent transport behavior. Froyland and Padberg-Gehle [27] recently introduced tracking algorithms for those regions.

3.2.4 Local Approaches

All previous methods define the vector field topology of time-dependent flows by means of particle behavior over a finite time window. A number of methods investigated the reduction of the time-dependent flow to a steady flow by estimating a transport component. Weinkauf et al. [133] demonstrated that a constant flow vector can be subtracted from a von-Kármán vortex street to approximately eliminate a linear transport. This works well in most parts of the domain, aside from the region right behind the obstacle, in which the vortices accelerate. Fuchs et al. [28] located points at which the acceleration vanishes and removed the velocity at those locations. Similarly, Bujack et al. [17] used the determinant of the Jacobian to pin down the flow features. Bhatia et al. [12] removed a harmonic flow component using a Helmholtz-Hodge decomposition (HHD). The HHD splits the flow into a divergence-free and an irrotational part. In case the domain is bounded or not simply-connected, a harmonic component can appear, which is both divergence-free and irrotational. Bhatia et al. [12] proposed to model the ambient motion of the flow features by the harmonic component. Following up on research on reference frame optimization [37, 46] for vortex extraction, Baeza Rojo and Günther [6] observed topological elements in a spatially-varying reference frame, in which the flow becomes steady. The latter connects to a formal property that is highly relevant for flow feature extraction in time-dependent flow, i.e., reference frame invariance. Section 4 explains this concept in more detail.

3.2.5 Desirable Properties

Recently, Bujack et al. [18] collected the most commonly-used mathematical properties that an approach for a time-dependent vector field topology should enjoy. Reciting Bujack et al., these properties are:

Coincidence With the Steady Flow Topology. A method that was designed for time-dependent flow should contain the standard steady vector field topology as special case, when it is applied to a steady flow [90].

Induction of a Partition of the Domain. The unsteady counterpart to the topological skeleton should divide the domain into regions of coherent temporal behavior. This means, there are material boundaries that order the flow [90].

Lagrangian Invariance. A common measure for the physical meaningfulness of a time-dependent topology is the requirement that all topological structures are invariant manifolds of the flow [28, 49]. This means that the path of a time-dependent critical point becomes a pathline, and that separatrices become material lines or surfaces that are advected with the flow.

Reference Frame Invariance. The result should be the same, independent of the choice of the reference frame [45]. In the following section, we introduce reference frame transformations and the classes of invariance that are commonly desired.

Bujack et al. [18] reviewed the mathematical properties of many of the existing extraction algorithms, using carefully-crafted benchmarks and proofs. To this day, there is no algorithm that fulfills all desirable properties.

4 Concepts

In the following, we introduce a number of concepts that are important in the ongoing research on a time-dependent vector field topology.

4.1 Reference Frame Transformation

A desirable property for any feature definition is its invariance to the motion of the observer. First of all, this requires a formal definition of reference frame motion. Recently, Baeza Rojo and Günther [6] used displacement transformations to describe the motion of the observer, which originates from continuum mechanics [111]. They used a displacement vector field $\mathbf{F}(\mathbf{x}, t)$, which moves a space-time point $(\mathbf{x}, t)$ to its destination $(\mathbf{x}^*, t)$ via

$$\mathbf{x}^* = \mathbf{x} + \mathbf{F}(\mathbf{x}, t) \tag{33}$$

Thereby, $\mathbf{F}$ is an invertible transformation that maps between two differential spaces, i.e., $\mathbf{F}$ is a diffeomorphism. By differentiation with respect to time, we see that a given vector field $\mathbf{v}(\mathbf{x}, t)$ is transformed to $\mathbf{v}^*(\mathbf{x}^*, t)$ via:

$$\mathbf{v}^*(\mathbf{x}^*, t) = [\mathbf{I} + \nabla\mathbf{F}(\mathbf{x}, t)] \cdot \mathbf{v}(\mathbf{x}, t) + \mathbf{F}_t(\mathbf{x}, t) \tag{34}$$

It is interesting to note that other existing classes of reference frame transformations, such as Galilean transformations [133] (equal-speed translations), objective transformations [37, 48, 128] (smooth rotation and translation) and affine transformations [41] are all included as special cases.

4.2 Reference Frame Invariance

In flow visualization, reference frame invariance has first been studied in the context of flow feature extraction, namely for the detection of vortices. Whenever an observer sees a vector-valued property, this vector will change with the movement of the observer. An example of this reference frame dependence is shown in Fig. 19. Here, three different observers look at the same flow, each seeing different flow patterns. This is because the motion of the observer and the flow feature add up, as illustrated

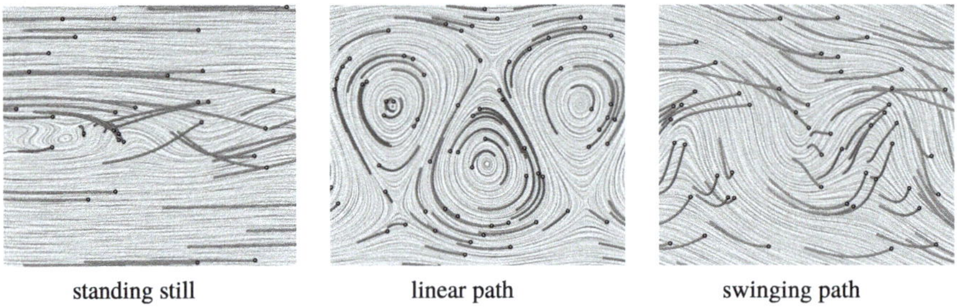

Fig. 19 Three observers see the same vector field from different reference frames. Here, a line integral convolution (time slice) and pathlines (black) are shown for three different reference frame movements: standing still, linearly translating and swinging along a sine curve–unfortunately, all give different results. Only the observer in the middle sees vortex structure. However, moving the observer a little bit faster of slower, would show the flow structures in a different location. Illustration from [37]

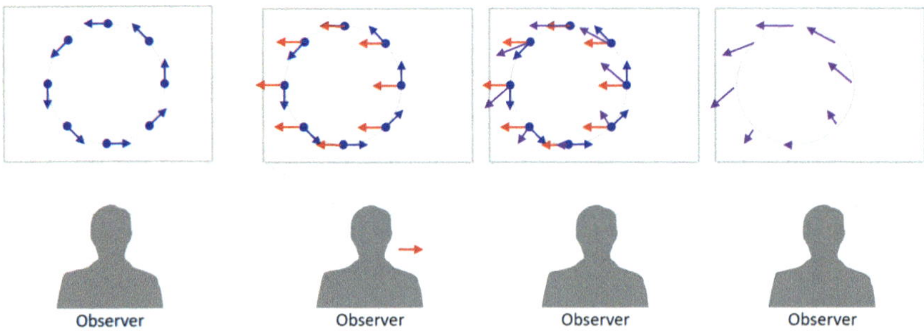

(a) Steady flow (b) Unsteady flow, since the observer is moving to the right.

Fig. 20 The observed vector field is the combination of the feature (blue) and its motion of the feature (red). In **a**, the flow is observed from a steady frame of reference. We see a closed streamline (blue). In **b**, the observer is moving to the right (red arrow), which creates an apparent opposite motion of the vortex. What the observer sees is the superposition of the red and blue vector, resulting in the purple vector. Note that the purple vectors no longer point along a closed streamline (right-most image). If we can estimate the ambient motion (red), it can be removed to reveal the original feature (blue)

in Fig. 20. Choosing the right reference, i.e., estimating the motion of the feature correctly, is quite important for the successful characterization of a vortex [77, 95]. Feature definitions can be classified based on the class of invariance they possess. The two most common reference frame invariances are the following.

Galilean invariance is the invariance of a measure under equal-speed translations of the reference frame of the form:

$$\mathbf{x}^* = \mathbf{x} + \mathbf{c}_0 + t\,\mathbf{c}_1, \quad t^* = t - a \tag{35}$$

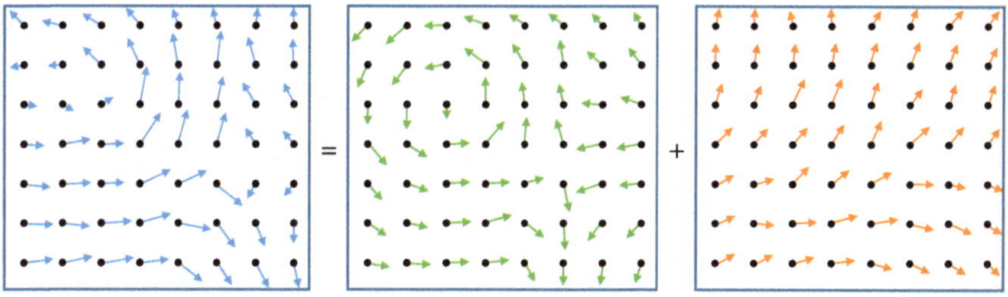

(a) Input flow $\mathbf{v}(\mathbf{x}, t)$ (b) Steady frame $\mathbf{w}(\mathbf{x}, t)$ (c) Ambient motion $\mathbf{f}(\mathbf{x}, t)$

Fig. 21 Example of a reference frame decomposition using the displacement optimization [6]. A given input vector field $\mathbf{v}(\mathbf{x}, t)$ is split into a flow in the steady reference frame $\mathbf{w}(\mathbf{x}, t)$ and the ambient motion $\mathbf{f}(\mathbf{x}, t)$

where $\mathbf{c}_0$ and $\mathbf{c}_1$ are constant vectors and a is a constant. Every measure that is derived from the Jacobian $\mathbf{J}$ of the vector field is Galilean invariant. There are only few measures that are Galilean invariant and include temporal derivatives, for instance the acceleration $\mathbf{a} = \mathbf{J}\mathbf{v} + \mathbf{v}_t$ [101] and the subtraction of the feature flow field $(\mathbf{v} - \mathbf{f})$ [119]. Due to the relativity of the observer and the feature we want to track, all Galilean invariant feature definitions are able to identify features that perform equal-speed translations. If a vortex performs any other type of movement, Galilean invariant methods will not produce the correct solution.

Objectivity refers to the invariance of a measure under a smooth rotation and translation of the reference frame, cf. [127]:

$$\mathbf{x}^* = \mathbf{Q}(t)\,\mathbf{x} + \mathbf{c}(t) \quad , \quad t^* = t - a \tag{36}$$

where $\mathbf{Q} \in SO(3)$ is a rotation matrix, $\mathbf{c}$ is a translation vector, and a is a constant. We assume $\mathbf{Q}$ and $\mathbf{c}$ to be smooth functions of t. The most recent feature defintions aim to be *objective* [37, 49, 50, 64]. Among the objective quantities are the divergence $\nabla \cdot \mathbf{v}$, the strain rate tensor $\mathbf{S}$, and the flow map gradient $\nabla\phi_{t_0}^{\tau}(\mathbf{x})$.

4.3 Topology in Steady Reference Frames

Recently, Baeza Rojo and Günther [6] optimized for steady reference frames by describing the reference frame motion as inhomogeneous displacement transformation. The motion of features in the steady reference frame is thereby modeled by more than just rotations and translations. The method thereby become invariant to even more classes of motion than with objectivity. Given the optimal reference frame, the flow $\mathbf{v}(\mathbf{x}, t)$ is decomposed into:

$$\underbrace{\mathbf{v}(\mathbf{x},t)}_{\text{inputfield}} = \underbrace{\mathbf{w}(\mathbf{x},t)}_{\text{steadyframe}} + \underbrace{\mathbf{f}(\mathbf{x},t)}_{\text{ambientmotion}} \tag{37}$$

While they used the optimal frame $\mathbf{w}(\mathbf{x}, t)$ to extract topological structures, the remaining ambient flow $\mathbf{f}(\mathbf{x}, t)$ can be used to track critical points over time. An example is given in Fig. 21. The numerical extraction of the reference frame still remains challenging and becomes harder the more degrees of freedoms are added. The linear optimization is under-constrained, which requires a regularizer that in turn places assumptions about the smoothness of the ambient motion and its derivatives. In case, the feature moves with constant speed in a constant direction, i.e., Galilean invariance is fulfilled, then $\mathbf{f}(\mathbf{x}, t) = -\mathbf{J}^{-1}\mathbf{v}_t$ is the normalized feature flow field [35, 41, 119]. This means, the approach of Weinkauf et al. [133] appears as special case. Reference frame optimizations have previously been done for vortex extraction, using local optimization [37, 41], global optimization [46] and deep learning [67].

4.4 High-Dimensional Flows

All sections above concentrated on 2D or 3D domains. Vector fields also arise in the description of dynamical systems [76], namely when describing how an arbitrarily high-dimensional state of a system is changing over time. This vector field is called the *phase flow*, which depends on the current state of the dynamical system. Typical fluid flows can be considered as first-order ODEs. Many processes not only depend on the position of an object, but also on its velocity, which leads us to second-order ODEs. Examples of such systems are oscillators, pendulums, n-body problems [10] and the motion of finite-sized objects in fluids [86, 114, 116].

Similar to the fluid flows above, such dynamical systems also contain topologically-relevant structures and features. An introduction to the topology of dynamical systems was given by Abraham and Shaw [1]. Hofmann et al. [57] extracted and visualized all types of critical points in a four dimensional vector field. Depending on the actual structure of the phase flow, the possible types of topological structures may be limited. Inertial particles for instance, i.e., finite-sized in fluids, exhibit an attraction by a globally attracting manifold [51, 79], which causes halve the eigenvalues of the phase flow's Jacobian to be negative, i.e., sources cannot exist [36, 40]. This attracting manifold was visualized by Baeza Rojo et al. [4] for varying particle sizes. Further, Günther et al. [40] visualized stable sets of inertial systems interactively using multi-dimensional stacking. FTLE has been calculated by Garaboa-Paz and Pérez-Muñuzuri [29] by measuring the expansion in full phase space, whereas Sagrista et al. [100] visualized the separation in the subspaces as well, using multi-dimensional stacking. Inertial vortex corelines have been extracted with the assumption of Galilean invariance [39] and objectivity [43]. The latter requires the search for parallel vectors in the high-dimensional space, for which Hofmann et al. [58] introduced the dependent vectors operator. Recently, Bartolovic [10] introduced an optimization-based dimensionality reduction method for high-dimensional trajectories that preserves geometric and topological properties.

4.5 Uncertainty

As the computational resources become more capable, the computation of multiple simulations with slightly different input conditions or parameters becomes more affordable. This allows domain scientists to quantify the uncertainty of simulation models or measurements, for instance in meteorological forecasting. These days, uncertainty visualization is considered to be among the top challenges in scientific visualization [13]. For uncertain vector field topology, integral curves and fixed points have been generalized to define an uncertain vector field topology by integrating particle density functions [83]. Petz et al. [88] used Monte Carlo sampling to extract probabilistic local features such as critical points from Gaussian distributed vector fields. Bhatia et al. [11] introduced edge maps to introduce a fuzzy topology that performs a topological decomposition based on growing of streamwaves. Hummel et al. [60] considered the variance among ensemble members to identify LCS that exist in many members. Obermaier and Joy [81] categorized ensemble visualizations into location-based methods that compare ensemble properties at a fixed location and feature-based methods that first extract, match and compare features. Guo et al. [44] used Monte Carlo sampling to determine the probability for LCS in uncertain vector fields. Uncertainty visualization remains a very active topic with plenty of research on different frontiers, including the modeling of correlations [89, 91] and the visualization of confidence [25, 84, 104, 144].

5 Outlook

While the topological elements of steady vector fields are fairly well understood, there is still an ongoing discussion on the definition of a time-dependent vector field topology. As recently shown by Bujack et al. [18], there is no approach yet that holds up in all unsteady flow scenarios. Meanwhile, the continuum mechanics and fluid dynamics community are pushing the frontiers of the Lagrangian flow analysis, which is tightly related to the classical view of vector field topology. In fact, we hope to see more synergies between the different research directions. Aside from new theoretical contributions on the feature definitions, we can also expect to see more work on uncertain data, since ensemble simulations and model variability are of high interest in the application domains. The analysis of high-dimensional flows and general dynamical systems will benefit more from synergies with dimensionality reduction research. As the data sizes are growing, global topology-based methods will see more parallelization across multiple compute nodes within in-situ environments. While scalar field topology algorithms now become more available through open-source packages such as the Topology ToolKit (TTK) [123], a similar platform for vector field topology is missing. We hope that such effort is taken soon. A general availability increases the adaption in practice, which in turn reveals new research challenges.

Acknowledgements This work was supported by the Swiss National Science Foundation (SNSF) Ambizione grant no. PZ00P2_180114 and by ETH Research Grant ETH-07 18-1.

References

1. Abraham, R.H., Shaw, C.D.: Dynamics - The Geometry of Behaviour. The Visual Mathematics Library. Aerial Press Incorporated, Santa Cruz (1984)
2. Agranovsky, A., Garth, C., Joy, K.: Extracting flow structures using sparse particles. In: Proceedings of the Vision, Modeling and Visualization, pp. 151–160 (2011)
3. Andronov, A.A.: Qualitative Theory of Second-order Dynamic Systems, vol. 22054. Halsted Press, Canberra (1973)
4. Baeza Rojo, I., Gross, M., Günther, T.: Visualizing the phase space of heterogeneous inertial particles in 2D flows. Comput. Graph. Forum (Proc. EuroVis) **37**(3), 289–300 (2018)
5. Baeza Rojo, I., Gross, M., Günther, T.: Accelerated Monte Carlo rendering of finite-time Lyapunov exponents. IEEE Trans. Vis. Comput. Graph. (Proc. IEEE Scientific Visualization 2019) (2020, to appear)
6. Baeza Rojo, I., Günther, T.: Vector field topology of time-dependent flows in a steady reference frame. IEEE Trans. Vis. Comput. Graph. (Proc. IEEE Scientific Visualization 2019) (2020, to appear)
7. Banks, D.C., Singer, B.A.: A predictor-corrector technique for visualizing unsteady flow. IEEE Trans. Visual Comput. Graphics **1**, 151–163 (1995)
8. Barakat, S.S., Garth, C., Tricoche, X.: Interactive computation and rendering of finite-time Lyapunov exponent fields. IEEE Trans. Visual Comput. Graphics **18**(8), 1368–1380 (2012)
9. Barakat, S.S., Tricoche, X.: Adaptive refinement of the flow map using sparse samples. IEEE TVCG (Proc. SciVis) **19**(12), 2753–2762 (2013)
10. Bartolovic, N., Gross, M., Günther, T.: Phase space projection of dynamical systems. Comput. Graph. Forum (Proc. EuroVis) (2020)
11. Bhatia, H., et al.: Flow visualization with quantified spatial and temporal errors using edge maps. IEEE TVCG **18**(9), 1383–1396 (2012)
12. Bhatia, H., Pascucci, V., Kirby, R.M., Bremer, P.T.: Extracting features from time-dependent vector fields using internal reference frames. Comput. Graph. Forum (Proc. EuroVis) **33**(3), 21–30 (2014)
13. Bonneau, G.P., et al.: Overview and State-of-the-Art of Uncertainty Visualization, pp. 3–27. Springer, London (2014)
14. Born, S., Wiebel, A., Friedrich, J., Scheuermann, G., Bartz, D.: Illustrative stream surfaces. IEEE Trans. Visual Comput. Graphics **16**(6), 1329–1338 (2010)
15. Brunton, S.L., Rowley, C.W.: Fast computation of FTLE fields for unsteady flows: a comparison of methods. Chaos **20**, 017503 (2010)
16. Bujack, R., Dutta, S., Baeza Rojo, I., Zhang, D., Günther, T.: Objective finite-time saddles and their connection to FTLE. In: Eurographics Conference on Visualization - Short Papers (2019)
17. Bujack, R., Hlawitschka, M., Joy, K.I.: Topology-inspired Galilean invariant vector field analysis. In: IEEE Pacific Visualization Symposium, pp. 72–79 (2016). https://doi.org/10.1109/PACIFICVIS.2016.7465253

18. Bujack, R., Yan, L., Hotz, I., Garth, C., Wang, B.: State of the art in time-dependent flow topology: interpreting physical meaningfulness through mathematical properties. Comput. Graph. Forum (Proc. Eurovis) (2020). https://doi.org/10.1111/cgf.14037
19. Chandler, J., Obermaier, H., Joy, K.I.: Interpolation-based pathline tracing in particle-based flow visualization. IEEE Trans. Visual Comput. Graphics **21**(1), 68–80 (2015)
20. Chen, G., Mischaikow, K., Laramee, R.S., Pilarczyk, P., Zhang, E.: Vector field editing and periodic orbit extraction using Morse decomposition. IEEE Trans. Visual Comput. Graphics **13**(4), 769–785 (2007)
21. Chen, G., Mischaikow, K., Laramee, R.S., Zhang, E.: Efficient Morse decompositions of vector fields. IEEE Trans. Vis. Comput. Graph. **14**(4), 848–862 (2008). https://doi.org/10.1109/TVCG.2008.33
22. Conley, C.C.: Isolated Invariant Sets and the Morse Index, vol. 38. American Mathematical Society, Providence (1978)
23. Dallmann, U.: Topological structures of three-dimensional vortex flow separation. In: 16th Fluid and Plasmadynamics Conference, p. 1735 (1983)
24. Farazmand, M., Blazevski, D., Haller, G.: Shearless transport barriers in unsteady two-dimensional flows and maps. Physica D **278**, 44–57 (2014)
25. Ferstl, F., Bürger, K., Westermann, R.: Streamline variability plots for characterizing the uncertainty in vector field ensembles. IEEE Trans. Vis. Comput. Graph. (Proc. IEEE Scientific Visualization 2015) **22**(1), 767–776 (2016)
26. Forman, R.: Combinatorial vector fields and dynamical systems. Math. Z. **228**(4), 629–681 (1998)
27. Froyland, G., Padberg-Gehle, K.: Almost-invariant and finite-time coherent sets: directionality, duration, and diffusion. In: Ergodic Theory, Open Dynamics, and Coherent Structures, pp. 171–216. Springer (2014)
28. Fuchs, R., et al.: Toward a Lagrangian vector field topology. Comput. Graph. Forum **29**(3), 1163–1172 (2010). https://doi.org/10.1111/j.1467-8659.2009.01686.x
29. Garaboa-Paz, D., Pérez-Muñuzuri, V.: A method to calculate finite-time Lyapunov exponents for inertial particles in incompressible flows. Nonlin. Proc. Geophys. **22**(5), 571–577 (2015)
30. Garth, C., Gerhardt, F., Tricoche, X., Hagen, H.: Efficient computation and visualization of coherent structures in fluid flow applications. IEEE Trans. Vis. Comput. Graph. (Proc. IEEE Visualization) **13**(6), 1464–1471 (2007)
31. Garth, C., Laramee, R.S., Tricoche, X., Schneider, J., Hagen, H.: Extraction and visualization of swirl and tumble motion from engine simulation data. In: Topology-Based Methods in Visualization. Visualization and Mathematics, pp. 121–135. Springer, Heidelberg (2007)
32. Garth, C., Tricoche, X.: Topology-and feature-based flow visualization: methods and applications. In: SIAM Conference on Geometric Design and Computing, pp. 25–46. IEEE Computer Society, Los Alamitos (2005)
33. Garth, C., Wiebel, A., Tricoche, X., Joy, K., Scheuermann, G.: Lagrangian visualization of flow-embedded surface structures. In: Computer Graphics Forum, vol. 27, pp. 1007–1014. Wiley Online Library (2008)
34. Globus, A., Levit, C., Lasinski, T.: A tool for visualizing the topology of three-dimensional vector fields. In: Proceedings of the IEEE Visualization, pp. 33–40 (1991)
35. Günther, T.: Opacity optimization and inertial particles in flow visualization. Ph.D. thesis, University of Magdeburg (2016)
36. Günther, T., Gross, M.: Flow-induced inertial steady vector field topology. Comput. Graph. Forum (Proc. Eurographics) **36**(2), 143–152 (2017)
37. Günther, T., Gross, M., Theisel, H.: Generic objective vortices for flow visualization. ACM Trans. Graph. (Proc. SIGGRAPH) **36**(4), 141:1–141:11 (2017)
38. Günther, T., Kuhn, A., Theisel, H.: MCFTLE: Monte Carlo rendering of finite-time Lyapunov exponent fields. Comput. Graph. Forum (Proc. EuroVis) **35**(3), 381–390 (2016)
39. Günther, T., Theisel, H.: Vortex cores of inertial particles. IEEE Trans. Vis. Comput. Graph. (Proc. IEEE SciVis) **20**(12), 2535–2544 (2014)

40. Günther, T., Theisel, H.: Inertial steady 2D vector field topology. Comput. Graph. Forum (Proc. Eurographics) **35**(2), 455–466 (2016)
41. Günther, T., Theisel, H.: Hyper-objective vortices. IEEE Trans. Vis. Comput. Graph. **26**, 1532–1547 (2018)
42. Günther, T., Theisel, H.: The state of the art in vortex extraction. Comput. Graph. Forum **37**(6), 149–173 (2018)
43. Günther, T., Theisel, H.: Objective vortex corelines of finite-sized objects in fluid flows. IEEE Trans. Vis. Comput. Graph. (Proc. IEEE SciVis) **25**(1) (2019). https://doi.org/10.1109/TVCG.2018.2864828
44. Guo, H., He, W., Peterka, T., Shen, H.W., Collis, S.M., Helmus, J.J.: Finite-time Lyapunov exponents and Lagrangian coherent structures in uncertain unsteady flows. IEEE Trans. Visual Comput. Graphics **22**(6), 1672–1682 (2016)
45. Hadjighasem, A., Farazmand, M., Blazevski, D., Froyland, G., Haller, G.: A critical comparison of Lagrangian methods for coherent structure detection. Chaos Interdisc. J. Nonlinear Sci. **27**(5), 053104 (2017)
46. Hadwiger, M., Mlejnek, M., Theußl, T., Rautek, P.: Time-dependent flow seen through approximate observer killing fields. IEEE Trans. Visual Comput. Graphics **25**(1), 1257–1266 (2019). https://doi.org/10.1109/TVCG.2018.2864839
47. Haller, G.: Finding finite-time invariant manifolds in two-dimensional velocity fields. Chaos Interdisc. J. Nonlinear Sci. **10**(1), 99–108 (2000)
48. Haller, G.: An objective definition of a vortex. J. Fluid Mech. **525**, 1–26 (2005)
49. Haller, G.: Lagrangian coherent structures. Annu. Rev. Fluid Mech. **47**, 137–162 (2015)
50. Haller, G., Hadjighasem, A., Farazmand, M., Huhn, F.: Defining coherent vortices objectively from the vorticity. J. Fluid Mech. **795**, 136–173 (2016)
51. Haller, G., Sapsis, T.: Where do inertial particles go in fluid flows? Physica D **237**, 573–583 (2008). https://doi.org/10.1016/j.physd.2007.09.027
52. Haller, G., Yuan, G.: Lagrangian coherent structures and mixing in two-dimensional turbulence. Physica D **147**(3–4), 352–370 (2000)
53. Heine, C., et al.: A survey of topology-based methods in visualization. Comput. Graph. Forum **35**(3), 643–667 (2016). https://doi.org/10.1111/cgf.12933
54. Heinze, R., Raasch, S., Etling, D.: The structure of kármán vortex streets in the atmospheric boundary layer derived from large eddy simulation. Meteorol. Z. **21**(3), 221–237 (2012)
55. Helman, J.L., Hesselink, L.: Representation and display of vector field topology in fluid flow data sets. Computer **22**(8), 27–36 (1989)
56. Helman, J.L., Hesselink, L.: Visualizing vector field topology in fluid flows. IEEE Comput. Graphics Appl. **11**, 36–46 (1991)
57. Hofmann, L., Rieck, B., Sadlo, F.: Visualization of 4D vector field topology. Comput. Graph. Forum **37**(3), 301–313 (2018)
58. Hofmann, L., Sadlo, F.: The dependent vectors operator. Comput. Graph. Forum **38**(3), 261–272 (2019)
59. Hofmann, L., Sadlo, F.: Extraction of distinguished hyperbolic trajectories for 2D time-dependent vector field topology. Comput. Graph. Forum (Proc. Eurovis) (2020). https://doi.org/10.1111/cgf.13982
60. Hummel, M., Obermaier, H., Garth, C., Joy, K.I.: Comparative visual analysis of Lagrangian transport in CFD ensembles. IEEE Trans. Visual Comput. Graphics **19**(12), 2743–2752 (2013)
61. Ide, K., Small, D., Wiggins, S.: Distinguished hyperbolic trajectories in time-dependent fluid flows: analytical and computational approach for velocity fields defined as data sets (2002)
62. Kasten, J., Petz, C., Hotz, I., Noack, B., Hege., H.C.: Localized finite-time Lyapunov exponent for unsteady flow analysis. In: Proceedings of Vision, Modeling and Visualization, pp. 265–274 (2009)
63. Kasten, J., Reininghaus, J., Reich, W., Scheuermann, G.: Toward the extraction of saddle periodic orbits. In: Topological Methods in Data Analysis and Visualization III, pp. 55–69. Springer (2014)

64. Katsanoulis, S., Farazmand, M., Serra, M., Haller, G.: Vortex boundaries as barriers to diffusive vorticity transport in two-dimensional flows. arXiv preprint arXiv:1910.07355 (2019)
65. Kenwright, D., Haimes, R.: Vortex identification-applications in aerodynamics: a case study. In: Proceedings. Visualization 1997 (Cat. No. 97CB36155), pp. 413–416. IEEE (1997)
66. Kern, M., Hewson, T., Sadlo, F., Westermann, R., Rautenhaus, M.: Robust detection and visualization of jet-stream core lines in atmospheric flow. IEEE Trans. Visual Comput. Graphics **24**(1), 893–902 (2017)
67. Kim, B., Günther, T.: Robust reference frame extraction from unsteady 2D vector fields with convolutional neural networks. Comput. Graph. Forum (Proc. EuroVis) **38**(3), 285–295 (2019)
68. Koch, S., Kasten, J., Wiebel, A., Scheuermann, G., Hlawitschka, M.: 2d vector field approximation using linear neighborhoods. Vis. Comput. **32**(12), 1563–1578 (2016)
69. Kuhn, A., Engelke, W., Rössl, C., Hadwiger, M., Theisel, H.: Time line cell tracking for the approximation of Lagrangian coherent structures with subgrid accuracy. Comput. Graph. Forum **33**(1), 222–234 (2014)
70. Kuhn, A., Rössl, C., Weinkauf, T., Theisel, H.: A benchmark for evaluating FTLE computations. In: Proceedings of 5th IEEE Pacific Visualization Symposium (PacificVis 2012), pp. 121–128, Songdo, Korea (2012)
71. Lapidus, L., Seinfeld, J.H.: Numerical Solution of Ordinary Differential Equations. Academic Press, New York (1971)
72. Laramee, R., Hauser, H., Zhao, L., Post, F.: Topology-based flow visualization, the state of the art. In: Topology-based Methods in Visualization. Mathematics and Visualization, pp. 1–19. Springer, Heidelberg (2007)
73. de Leeuw, W., van Liere, R.: Collapsing flow topology using area metrics. In: Proceedings of the Conference on Visualization, VIS 1999, pp. 349–354 (1999)
74. Leo, L.S., Thompson, M.Y., Di Sabatino, S., Fernando, H.J.: Stratified flow past a hill: dividing streamline concept revisited. Bound. Layer Meteorol. **159**(3), 611–634 (2016)
75. Lodha, S., Renteria, J., Roskin, K.: Topology preserving compression of 2D vector fields. In: Proceedings of the IEEE Visualization, pp. 343–350 (2000)
76. Löffelmann, H., Doleisch, H., Gröller, E.: Visualizing dynamical systems near critical points. In: Spring Conference on Computer Graphics and its Applications, pp. 175–184 (1998)
77. Lugt, H.J.: The dilemma of defining a vortex. In: Recent Developments in Theoretical and Experimental Fluid Mechanics, pp. 309–321. Springer (1979)
78. Machado, G.M., Boblest, S., Ertl, T., Sadlo, F.: Space-time bifurcation lines for extraction of 2D Lagrangian coherent structures. Comput. Graph. Forum (Proc. EuroVis) **35**(3), 91–100 (2016)
79. Mograbi, E., Bar-Ziv, E.: On the asymptotic solution of the Maxey-Riley equation. Phys. Fluids **18**(5), 051704 (2006)
80. Nsonga, B., Niemann, M., Fröhlich, J., Staib, J., Gumhold, S., Scheuermann, G.: Detection and visualization of splat and antisplat events in turbulent flows. IEEE Trans. Vis. Comput. Graph. **26**, 3147–3162 (2019)
81. Obermaier, H., Joy, K.I.: Future challenges for ensemble visualization. IEEE Comput. Graphics Appl. **34**(3), 8–11 (2014)
82. Onu, K., Huhn, F., Haller, G.: LCS tool: a computational platform for Lagrangian coherent structures. J. Comput. Sci. **7**, 26–36 (2015)
83. Otto, M., Theisel, H.: Vortex analysis in uncertain vector fields. Comput. Graph. Forum (Proc. EuroVis) **31**(3), 1035–1044 (2012). https://doi.org/10.1111/j.1467-8659.2012.03096.x
84. Pang, A.T., Wittenbrink, C.M., Lodha, S.K.: Approaches to uncertainty visualization. Vis. Comput. **13**(8), 370–390 (1997)
85. Peikert, R., Roth, M.: The "parallel vectors" operator - a vector field visualization primitive. In: Proceedings of the IEEE Visualization, pp. 263–270 (1999)
86. Peng, J., Dabiri, J.O.: Transport of inertial particles by Lagrangian coherent structures: application to predator-prey interaction in jellyfish feeding. J. Fluid Mech. **623**, 75–84 (2009). https://doi.org/10.1017/S0022112008005089

87. Perry, A.E., Chong, M.S.: A description of eddying motions and flow patterns using critical-point concepts. Annu. Rev. Fluid Mech. **19**(1), 125–155 (1987)
88. Petz, C., Pöthkow, K., Hege, H.C.: Probabilistic local features in uncertain vector fields with spatial correlation. In: Computer Graphics Forum, vol. 31, pp. 1045–1054. Wiley Online Library (2012)
89. Pfaffelmoser, T., Reitinger, M., Westermann, R.: Visualizing the positional and geometrical variability of isosurfaces in uncertain scalar fields. In: Computer Graphics Forum, vol. 30, pp. 951–960. Wiley Online Library (2011)
90. Pobitzer, A., et al.: The state of the art in topology-based visualization of unsteady flow. Comput. Graph. Forum **30**(6), 1789–1811 (2011)
91. Pöthkow, K., Weber, B., Hege, H.C.: Probabilistic marching cubes. In: Computer Graphics Forum, vol. 30, pp. 931–940. Wiley Online Library (2011)
92. Press, W.H., Teukolsky, S.A., Vetterling, W.T., Flannery, B.P.: Numerical Recipes in C, vol. 2. Cambridge University Press, Cambridge (1996)
93. Reich, W., Schneider, D., Heine, C., Wiebel, A., Chen, G., Scheuermann, G.: Combinatorial vector field topology in three dimensions. In: Topological Methods in Data Analysis and Visualization II, pp. 47–59. Springer (2012)
94. Reininghaus, J., Lowen, C., Hotz, I.: Fast combinatorial vector field topology. IEEE Trans. Visual Comput. Graphics **17**(10), 1433–1443 (2011). https://doi.org/10.1109/TVCG.2010.235
95. Robinson, S.K.: Coherent motions in the turbulent boundary layer. Annu. Rev. Fluid Mech. **23**(1), 601–639 (1991)
96. Roth, M.: Automatic extraction of vortex core lines and other line type features for scientific visualization, vol. 2. Ph.D. dissertation number 13673, ETH Zurich (2000)
97. Roth, M., Peikert, R.: A higher-order method for finding vortex core lines. In: Proceedings of the IEEE Visualization, pp. 143–150 (1998)
98. Sadlo, F., Peikert, R.: Efficient visualization of Lagrangian coherent structures by filtered AMR ridge extraction. IEEE Trans. Vis. Comput. Graph. (IEEE Visualization) **13**(6), 1456–1463 (2007)
99. Sadlo, F., Weiskopf, D.: Time-dependent 2-D vector field topology: an approach inspired by Lagrangian coherent structures. Comput. Graph. Forum **29**(1), 88–100 (2010). https://doi.org/10.1111/j.1467-8659.2009.01546.x. https://onlinelibrary.wiley.com/doi/abs/10.1111/j.1467-8659.2009.01546.x
100. Sagristà, A., Jordan, S., Just, A., Dias, F., Nonato, L.G., Sadlo, F.: Topological analysis of inertial dynamics. IEEE Trans. Vis. Comput. Graph. (Proc. IEEE SciVis 2016) **23**(1), 950–959 (2017)
101. Sahner, J.: Extraction of vortex structures in 3D flow fields. Ph.D. thesis, University of Magdeburg, Germany (2009)
102. Sahner, J., Weinkauf, T., Hege, H.C.: Galilean invariant extraction and iconic representation of vortex core lines. In: Proceedings of Eurographics/IEEE VGTC Symposium on Visualization (EuroVis), pp. 151–160 (2005)
103. Sahner, J., Weinkauf, T., Teuber, N., Hege, H.C.: Vortex and strain skeletons in Eulerian and Lagrangian frames. IEEE Trans. Visual Comput. Graphics **13**(5), 980–990 (2007)
104. Sanyal, J., Zhang, S., Dyer, J., Mercer, A., Amburn, P., Moorhead, R.: Noodles: a tool for visualization of numerical weather model ensemble uncertainty. IEEE Trans. Visual Comput. Graphics **16**(6), 1421–1430 (2010)
105. Scheuermann, G., Hagen, H.: A data dependent triangulation for vector fields. In: Proceedings of Computer Graphics International 1998, pp. 96–102. IEEE Computer Society Press, Los Alamitos (1998)
106. Scheuermann, G., Hamann, B., Joy, K., Kollmann, W.: Visualizing local vector field topology. J. Electr. Images **9**, 356–367 (2000)
107. Scheuermann, G., Kruger, H., Menzel, M., Rockwood, A.: Visualizing nonlinear vector field topology. IEEE Trans. Visual Comput. Graphics **4**(2), 109–116 (1998)

108. Scheuermann, G., Tricoche, X.: Topological methods for flow. The Visualization Handbook, p. 341 (2005)
109. Schneider, D., Reich, W., Wiebel, A., Scheuermann, G.: Topology aware stream surfaces. Comput. Graph. Forum (Proc. EuroVis) **29**(3), 1153–1161 (2010)
110. Serra, M., Haller, G.: Objective Eulerian coherent structures. Chaos Interdisc. J. Nonlinear Sci. **26**(5), 053110 (2016)
111. Shabana, A.A.: Computational Continuum Mechanics. Wiley, Hoboken (2018)
112. Shadden, S.C., Lekien, F., Marsden, J.E.: Definition and properties of Lagrangian coherent structures from finite-time Lyapunov exponents in two-dimensional aperiodic flows. Physica D **212**(3–4), 271–304 (2005). https://doi.org/10.1016/j.physd.2005.10.007
113. Shi, L., Zhang, L., Cao, W., Chen, G.: Analysis enhanced particle-based flow visualization. Electr. Imag. **2017**(1), 12–21 (2017)
114. Sudharsan, M., Brunton, S.L., Riley, J.J.: Lagrangian coherent structures and inertial particle dynamics. ArXiv e-prints (2015). Arxiv:1512.05733
115. Sujudi, D., Haimes, R.: Identification of swirling flow in 3D vector fields. Technical report, Department of Aeronautics and Astronautics, MIT (1995). AIAA Paper 95–1715
116. Sydney, A., Baharani, A., Leishman, J.G.: Understanding brownout using near-wall dual-phase flow measurements. In: Proceedings of the American Helicopter Society, 67th Annual Forum. Virginia Beach (2011)
117. Theisel, H.: Designing 2D vector fields of arbitrary topology. Comput. Graph. Forum (Proc. Eurographics) **21**(3), 595–604 (2002)
118. Theisel, H., Rössl, C., Seidel, H.P.: Compression of 2D vector fields under guaranteed topology preservation. Comput. Graph. Forum (Proc. Eurographics) **22**(3), 333–342 (2003)
119. Theisel, H., Seidel, H.P.: Feature flow fields. In: Proceedings of the Symposium on Data Visualisation, pp. 141–148 (2003)
120. Theisel, H., Weinkauf, T., Hege, H.C., Seidel, H.P.: Saddle connectors - an approach to visualizing the topological skeleton of complex 3D vector fields. In: Proceedings of the IEEE Visualization, pp. 225–232 (2003)
121. Theisel, H., Weinkauf, T., Hege, H.C., Seidel, H.P.: Grid-independent detection of closed stream lines in 2D vector fields. In: Vision, Modeling and Visualization, vol. 4, pp. 421–428 (2004)
122. Theisel, H., Weinkauf, T., Hege, H.C., Seidel, H.P.: Stream line and path line oriented topology for 2d time-dependent vector fields. In: Proceedings of the Conference on Visualization 2004, pp. 321–328. IEEE Computer Society (2004)
123. Tierny, J., Favelier, G., Levine, J.A., Gueunet, C., Michaux, M.: The topology toolkit. IEEE Trans. Visual Comput. Graphics **24**(1), 832–842 (2017)
124. Tricoche, X., Scheuermann, G., Hagen, H.: A topology simplification method for 2D vector fields. In: Proceedings of the Visualization, pp. 359–366 (2000). https://doi.org/10.1109/VISUAL.2000.885716
125. Tricoche, X., Scheuermann, G., Hagen, H.: Continuous topology simplification of planar vector fields. In: Proceedings of the Conference on Visualization 2001, pp. 159–166. IEEE Computer Society (2001)
126. Tricoche, X., Wischgoll, T., Scheuermann, G., Hagen, H.: Topology tracking for the visualization of time-dependent two-dimensional flows. Comput. Graph. **26**(2), 249–257 (2002)
127. Truesdell, C., Noll, W.: The nonlinear field theories of mechanics. In: Flugge, S., (ed.) Handbuch der Physik, Band III/3. Springer, Berlin (1965)
128. Truesdell, C., Rajagopal, K.R.: An Introduction to the Mechanics of Fluids. Springer, Boston (2010)
129. Üffinger, M., Sadlo, F., Ertl, T.: A time-dependent vector field topology based on streak surfaces. IEEE TVCG **19**(3), 379–392 (2013)
130. Wang, W., Wang, W., Li, S.: From numerics to combinatorics: a survey of topological methods for vector field visualization. J. Vis. **19**(4), 727–752 (2016)
131. Weinkauf, T.: Extraction of topological structures in 2D and 3D vector fields. Ph.D. thesis, University Magdeburg (2008)

132. Weinkauf, T., Hege, H.C., Theisel, H.: Advected tangent curves: a general scheme for characteristic curves of flow fields. Comput. Graph. Forum (Proc. Eurographics) **31**(2), 825–834 (2012)
133. Weinkauf, T., Sahner, J., Theisel, H., Hege, H.C.: Cores of swirling particle motion in unsteady flows. IEEE Trans. Vis. Comput. Graph. (Proc. Visualization) **13**(6), 1759–1766 (2007)
134. Weinkauf, T., Theisel, H.: Streak lines as tangent curves of a derived vector field. IEEE TVCG (Proc. Visualization) **16**(6), 1225–1234 (2010)
135. Weinkauf, T., Theisel, H., Hege, H.C., Seidel, H.P.: Boundary switch connectors for topological visualization of complex 3D vector fields. In: VisSym, pp. 183–192 (2004)
136. Weinkauf, T., Theisel, H., Hege, H.C., Seidel, H.P.: Topological construction and visualization of higher order 3D vector fields. Computer Graphics Forum (Proc. Eurographics) **23**(3), 469–478 (2004)
137. Westermann, R., Johnson, C., Ertl, T.: Topology-preserving smoothing of vector fields. IEEE Trans. Visual Comput. Graphics **7**(3), 222–229 (2001)
138. Wiebel, A., Chan, R., Wolf, C., Robitzki, A., Stevens, A., Scheuermann, G.: Topological flow structures in a mathematical model for rotation-mediated cell aggregation. In: Topological Data Analysis and Visualization: Theory, Algorithms and Applications, pp. 1–12 (2009)
139. Wiebel, A., Tricoche, X., Scheuermann, G.: Extraction of separation manifolds using topological structures in flow cross sections. In: Topology-Based Methods in Visualization II, pp. 31–43. Springer (2009)
140. Wiebel, A., Tricoche, X., Schneider, D., Janicke, H., Scheuermann, G.: Generalized streak lines: analysis and visualization of boundary induced vortices. IEEE Trans. Vis. Comput. Graph. **13**(6), 1735–1742 (2007). https://doi.org/10.1109/TVCG.2007.70557
141. Wilde, T., Rössi, C., Theisel, H.: Recirculation surfaces for flow visualization. IEEE Trans. Vis. Comput. Graph. (Proc. IEEE Scientific Visualization) **25**(1), 946–955 (2019). https://doi.org/10.1109/TVCG.2018.2864813
142. Wischgoll, T., Scheuermann, G.: Detection and visualization of closed streamlines in planar flows. IEEE Trans. Visual Comput. Graphics **7**(2), 165–172 (2001)
143. Wischgoll, T., Scheuermann, G.: Locating closed streamlines in 3D vector fields. Methods **16**, 19 (2002)
144. Zehner, B., Watanabe, N., Kolditz, O.: Visualization of gridded scalar data with uncertainty in geosciences. Comput. Geosci. **36**(10), 1268–1275 (2010)
145. Zhou, J., Adrian, R.J., Balachandar, S., Kendall, T.M.: Mechanisms for generating coherent packets of hairpin vortices in channel flow. J. Fluid Mech. **387**, 353–396 (1999)

An Overview of the Topology ToolKit

Talha Bin Masood, Joseph Budin, Martin Falk, Guillaume Favelier, Christoph Garth, Charles Gueunet, Pierre Guillou, Lutz Hofmann, Petar Hristov, Adhitya Kamakshidasan, Christopher Kappe, Pavol Klacansky, Patrick Laurin, Joshua A. Levine, Jonas Lukasczyk, Daisuke Sakurai, Maxime Soler, Peter Steneteg, Julien Tierny, Will Usher, Jules Vidal, and Michal Wozniak

Abstract This software paper gives an overview of the features supported by the Topology ToolKit (TTK), which is an open-source library for topological data analysis (TDA). TTK implements, in a generic and efficient way, a substantial collection of reference algorithms in TDA. Since its initial public release in 2017, both its user

T. B. Masood · M. Falk · P. Steneteg
Linköping University, Norrköping, Sweden
e-mail: talha.bin.masood@liu.se

M. Falk
e-mail: martin.falk@liu.se

P. Steneteg
e-mail: peter.steneteg@liu.se

J. Budin · J. Vidal
Sorbonne Université, Paris, France
e-mail: joseph.budin@sorbonne-universite.fr

J. Vidal
e-mail: jules.vidal@sorbonne-universite.fr

G. Favelier · A. Kamakshidasan
INRIA Saclay - Île-de-France, Palaiseau, France
e-mail: guillaume.favelier@inria.fr

A. Kamakshidasan
e-mail: adhitya.kamakshidasan@inria.fr

C. Garth · C. Kappe
TU Kaiserslautern, Kaiserslautern, Germany
e-mail: garth@cs.uni-kl.de

C. Kappe
e-mail: kappe@cs.uni-kl.de

C. Gueunet
Kitware, New York, USA
e-mail: charles.gueunet@kitware.com

P. Guillou
Sorbonne Université, Paris, France
e-mail: pierre.guillou@sorbonne-universite.fr

I. Hotz et al. (eds.), *Topological Methods in Data Analysis and Visualization VI*, Mathematics and Visualization, https://doi.org/10.1007/978-3-030-83500-2_16

and developer bases have grown, resulting in a significant increase in the number of supported features. In contrast to the original paper introducing TTK [40] (which detailed the core algorithms and data structures of TTK), the purpose of this software paper is to describe the list of features currently supported by TTK, ranging from image segmentation tools to advanced topological analysis of high-dimensional data, with concrete usage examples available on the TTK website [42].

1 Introduction

Topological data analysis (TDA) [10, 34, 38] is a vibrant field of study at the cross roads between mathematics and computer science, which considers the *structure* of complex data. In particular thanks to advanced concepts such as persistent homology [10], TDA provides theories and algorithms for the multi-scale representation and analysis of the structural features of interest present in the data. It has been shown to be useful in a variety of fields, ranging from machine learning [7] to geometry pro-

L. Hofmann
Heidelberg University, Heidelberg, Germany
e-mail: lutz.hofmann@iwr.uni-heidelberg.de

P. Hristov
University of Leeds, Leeds, UK
e-mail: p.hristov@leeds.ac.uk

P. Klacansky · W. Usher
SCI Institute, University of Utah, Salt Lake City, USA
e-mail: klacansky@sci.utah.edu

W. Usher
e-mail: will@sci.utah.edu

P. Laurin · M. Wozniak
ShapeShift3D, Montreal, Canada
e-mail: patrick.laurin@shapeshift3d.com

M. Wozniak
e-mail: michal.wozniak@shapeshift3d.com

J. A. Levine
University of Arizona, Tucson , USA
e-mail: josh@email.arizona.edu

J. Lukasczyk
Arizona State University, Phoenix, USA
e-mail: jl@jluk.de

D. Sakurai
Kyushu University, Fukuoka, Japan
e-mail: d.sakurai@computer.org

M. Soler
Total / Sorbonne Université, Paris, France
e-mail: soler.maxime@total.com

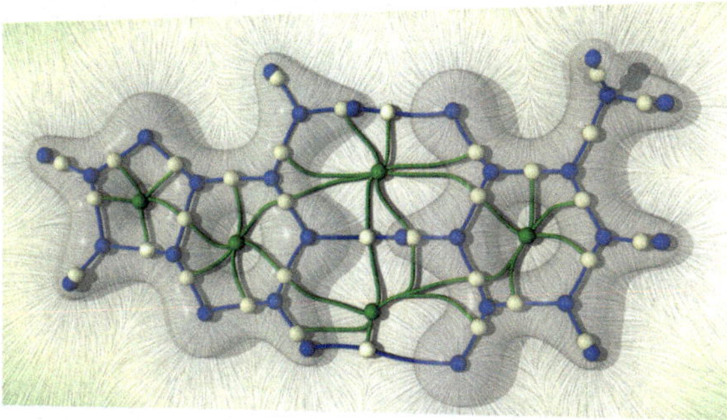

Fig. 1 Extraction of the covalent and non-covalent interactions in a molecular system with TTK. Covalent and hydrogen bonds are captured by the blue separatrices of the Morse-Smale complex, and steric effects (repulsive forces induced by the carbon cycles) are captured by saddle -saddle connectors (green)

cessing [49]. In scientific applications, TDA is particularly effective for the analysis of large-scale data sets [16]. The *Topology ToolKit* (TTK) [40] is an open-source library for TDA that has been released in 2017 under the permissive BSD license. It features a substantial collection of generic and efficient implementations of reference TDA algorithms. TTK is mostly written in C++ (~110k lines of code) to offer the best possible performance. To date, 15 institutions have contributed code to TTK, including 12 academic organizations (Arizona State University, CNRS, Heidelberg University, INRIA, Linkoping University, Los Alamos National Laboratory, Sorbonne Universite, TU Kaiserslautern, University of Arizona, University of Leeds, University of Utah, Zuse Institute Berlin) and 3 companies (Kitware, Total, ShapeShift3D). Since its initial release, TTK's website has collected more than 135k page-views, from more than 25k unique visitors, and its video tutorials have collected more than 16k Youtube views. TTK is accessible to developers through several APIs: C++, VTK/C++ or Python. For end users, TTK is directly accessible in the form of a plugin for ParaView [1] and an anaconda package [45]. Data can be provided to TTK in multiple forms: it can be sampled along 1D, 2D, or 3D regular grids (including periodic grids), or 1D, 2D, or 3D meshes (simplicial complexes). It can also be provided as point clouds of arbitrary dimension (Fig. 1).

The internal data structures and algorithms of TTK have already been presented in its companion paper [40], its end-user features have not been formally presented, other than in oral tutorials [12, 13, 15] or hackathons [28]. This software paper fills this gap by describing the high-level features of TTK through a list of concrete examples. Note that although the following examples will be discussed based on a usage of TTK with ParaView, the entire discussion holds for all TTK's APIs (C++, VTK/C++, Python) as each TTK *filter* in the presented ParaView pipelines (green box in the *pipeline browser*, top left of each screenshot) represents an individual TTK object. We also note that ParaView state files can be automatically exported to Python scripts. All the material necessary to reproduce the examples presented in

J. Tierny (✉)
CNRS / Sorbonne Université, Paris, France
e-mail: julien.tierny@sorbonne-universite.fr

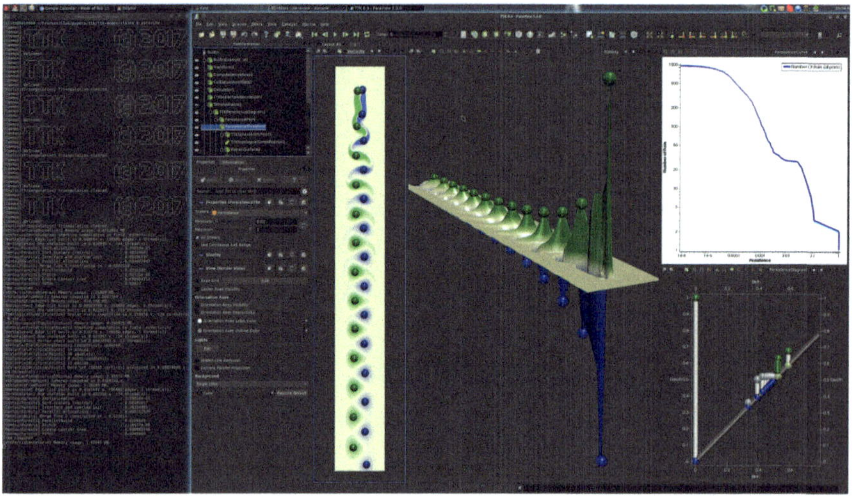

Fig. 2 Persistence-driven analysis pipeline applied to vortex tracking in computational fluid dynamics. Once TTK and its data package are installed, from the `ttk-data/` directory, run the following command to reproduce this example: "`paraview states/BuiltInExample1.pvsm`" (see [42] for further details)

this paper (data and ParaView state files) is available on the TTK website (section *Tutorials* [42]). Readers are invited to run these examples (see the caption of each figure) to further inspect interactively the content of each illustration.

2 Scalar Data

TTK supports the computation of a large number of topological abstractions for scalar data. In the applications, each topological abstraction supported by TTK serves a specific purpose. Critical points [5] extract points of interest. Merge/contour trees and Reeb graphs [18–21] estimate skeletons and meaningfully segment the input data along level sets. Persistence diagrams [10] visually represent the population of points of interest (critical points) as well as their salience (topological persistence). The Morse-Smale complex can be used to extract filament structures in data.

Typically, as illustrated in Fig. 2, to explore the data at multiple scales, the persistence diagram [10] is first computed to identify the main topological features present in the data and to discard the irrelevant features that correspond to noise. In particular, long vertical bars in the diagram (Fig. 2, bottom right) denote topologically salient features, while small bars near the diagonal correspond to spurious features. The persistence curve (Fig. 2, top right) provides a visual tool for inspecting relevant persistence thresholds, under which features should be interpreted as noise. In particular, these curves often exhibit in practice flat plateaus, which either separate features from noise (at persistence 0.07 in Fig. 2), or which separate group of features of different scales. Once a proper persistence threshold has been identified by the user, to reflect the corresponding noise removal in the original data, TTK's support for *topological* simplification [29, 41] can be used. Then, any topological object mentioned above (critical point, merge/contour tree, Reeb graph, Morse-Smale complex) computed after this data simplification step will subsequently be simplified, supporting multi-scale feature exploration. In Fig. 2, selecting the persistence threshold corresponding to the flat plateau located at the center of the persitence curve (persistence threshold of 0.07) enables the simplification of all spurious features, and robustly extracts the centers of the vortices of this fluid mechanic example (persistent extrema of the flow orthogonal curl component).

Figs. 3, 4, and 5 show further typical usage examples illustrating classical topological data analysis pipelines, where data is pre-simplified by preserving only the most persistent features (highlighted in the corresponding persistence diagrams).

In Fig. 3, the simplification is combined with the Morse complex (bottom) to extract cells in confocal microscopy (top left: input data). In this example, the segmentation is illustrated by representing the manifold of each local maximum with a distinct color. In particular, each maximum denotes the nucleus of a cell, its manifold the geometry of the cell, and the network of filament structures extracted from the separatrices of the Morse complex indicate the boundaries separating the cells. In this application, the removal of the small bars from the persistence diagram (top right) removes spurious maxima from the data and consequently resolves the possible over-segmentation provided by the Morse complex alone.

In Fig. 4, data pre-simplification is combined with the merge tree to extract bones in medical imaging. In particular, in this example, the user segmented the regions corresponding to each arc of the split tree containing a local maximum (which corresponds to locally dense regions). Here the level of persistence has been tuned to maintain only the five most persistent features (corresponding to the toes of the foot). Maintaining more features (in this example, for a persistence threshold of 150) would precisely segment the bones along each joint, which further illustrates the potential for multi-scale data exploration.

Note that TTK also offers functionality to design harmonic scalar fields by solving the Laplace equations subject to Dirichlet constraints [50] provided by the user at key locations (typically at extremities of prominent shape features). This is illustrated in Fig. 5 (left), which also illustrates skeleton extraction with the Reeb graph (right).

TTK also implements efficient algorithms [24, 33, 36] for the estimation of distances between persistence diagrams (such as the bottleneck and Wasserstein distances [10]). Recently, efficient and progressive algorithms [26, 48] have been integrated for the computation of barycenters of persistence diagrams [27, 46], which visually summarize the topological features of an ensemble data set, and which

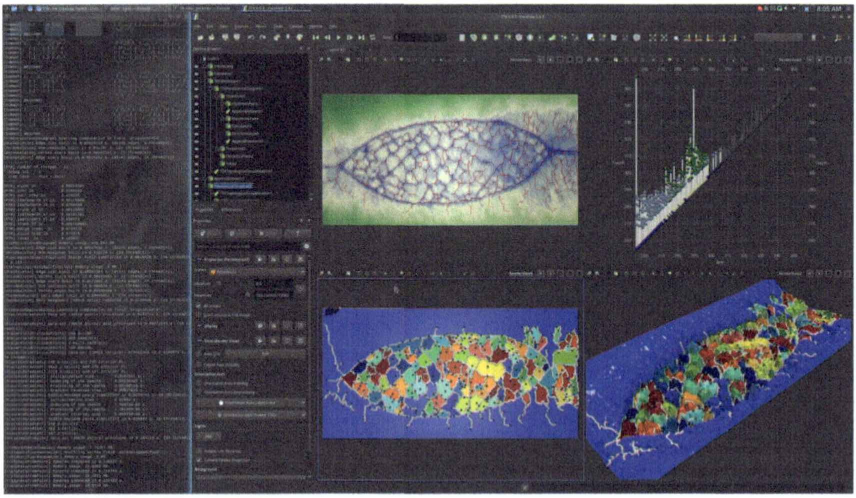

Fig. 3 Persistence-driven analysis pipeline, combined with Morse complex computation for cell enumeration in confocal microscopy (example reproduced from [10], page 217). Once TTK and its data package are installed, from the `ttk-data/` directory, run the following command to reproduce this example: "`paraview states/tribute.pvsm`" (see [42] for further details)

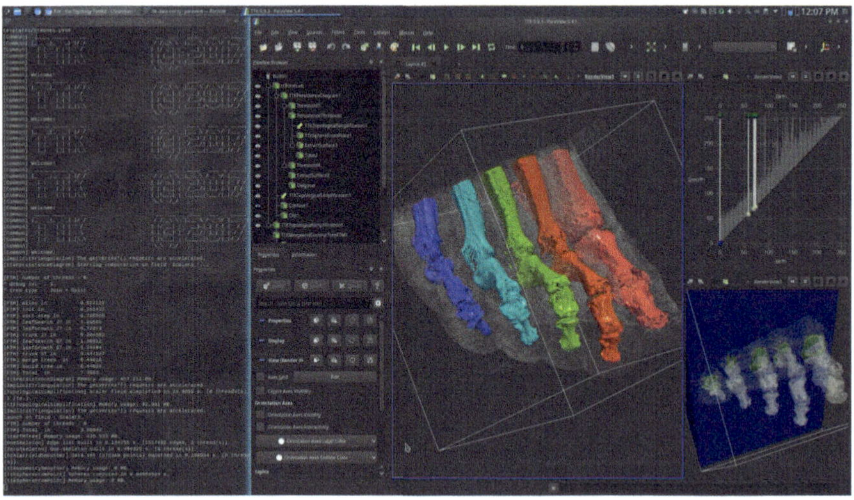

Fig. 4 Persistence-driven merge-tree based segmentation applied to bone extraction in medical imaging. Once TTK and its data package are installed, from the `ttk-data/` directory, run the following command to reproduce this example: "`paraview states/ctBones.pvsm`" (see [42] for further details)

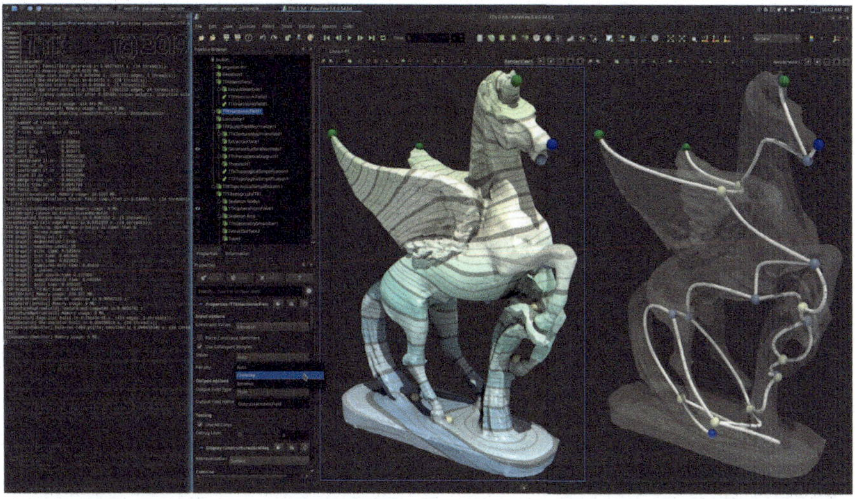

Fig. 5 Skeleton estimation from the Reeb graph [21] of a user designed harmonic field [50]. Once TTK and its data package are installed, from the `ttk-data/` directory, run the following command to reproduce this example: "`paraview states/harmonicSkeleton.pvsm`" (see [42] for further details)

enable an efficient clustering of the ensemble members based on their persistence diagram.

3 Bivariate Scalar Data

TTK supports the computation of several topological abstractions for bivariate data (where the data is characterized by two values defined at each vertex of the geometrical domain). TTK provides a fast implementation of continuous scatterplots [4], which can be interpreted as continuous histograms of bivariate data defined on volumes. They are particularly useful for understanding where and how volumetric data projects to the data range. Fiber surfaces [6, 25] extend the notion of isosurfaces to bivariate data and enable users to explore the regions in the volume corresponding to features of interest segmented manually in the continuous scatterplot. The Jacobi sets [9] are also implemented in TTK. They are the bivariate analog of critical points (points where both gradients are colinear), and they enable the extraction of filament structures in bivariate data. They correspond to *folds* of the volume when projecting it to the plane according to the bivariate data. TTK also supports the fast computation of Reeb spaces of bivariate data [39], which allows the *peeling* of the continuous scatterplot in regions that do not self-overlap during the projection of the volume induced by the bivariate data. These capabilities areillustrated in Fig. 6.

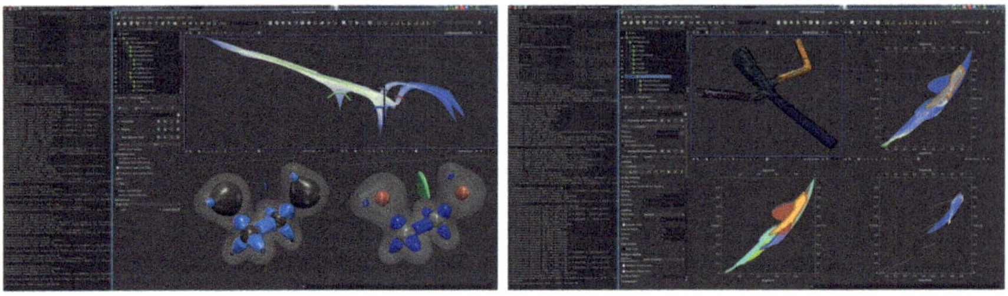

Fig. 6 Gallery of bivariate scalar data analysis. *Left:* Continuous scatterplot (top) of the electron density and reduced gradient of the ethane-diol molecule, some isosurfaces (bottom left) and fiber surfaces [6, 25] (bottom right) corresponding to the curves of matching color in the scatterplot. *Right:* Interactive continuous scatterplot peeling on fluid mechanics example (flow and curl magnitudes): a sheet of the simplified Reeb space [39] is selected by the user (orange), and its projection is independently isolated in the scatterplot for further individual inspection. Once TTK and its data package are installed (see [42] for further details), from the `ttk-data/` directory, run the following commands to reproduce these examples: "`paraview states/BuiltInExample2.pvsm`" (left) and "`paraview states/mechanical.pvsm`" (right)

In the left image, the user provides a few strokes on the main visual features of the continuous scatterplot (colored curves, top), and the corresponding structures in 3D are extracted as *fiber surfaces* (surfaces of matching colors, bottom). This feature definition captures subtle structures that are difficult to extract with the isosurfaces of either of the two fields of the bivariate data (bottom left). In the right image, the Reeb space segments the volume into regions that do not self-overlap when projected onto the plane given the bivariate data. Such regions can be isolated from the continuous scatterplot for further inspection. Furthermore, TTK also provides heuristics for persistence-like simplification mechanisms on bivariate Reeb spaces to enable multi-scale interactive exploration.

4 Uncertain Scalar Data

TTK supports the analysis of uncertain data, where the data is given as two scalar fields, representing the bounds of the interval of possible data values for each vertex of the domain. From this representation, mandatory critical points [17] can be extracted (Fig. 7). These objects correspond to regions where the appearance of at least one critical point is guaranteed for any realization of the uncertain data (i.e., for any scalar field randomly generated from the input intervals). This topological analysis enables the estimation of the structures that always occur despite the uncertainty as well as their geometrical variability. This construction can be used for instance to analyze ensemble data sets, in conjunction with clustering techniques, as illustrated by Favelier et al. [14].

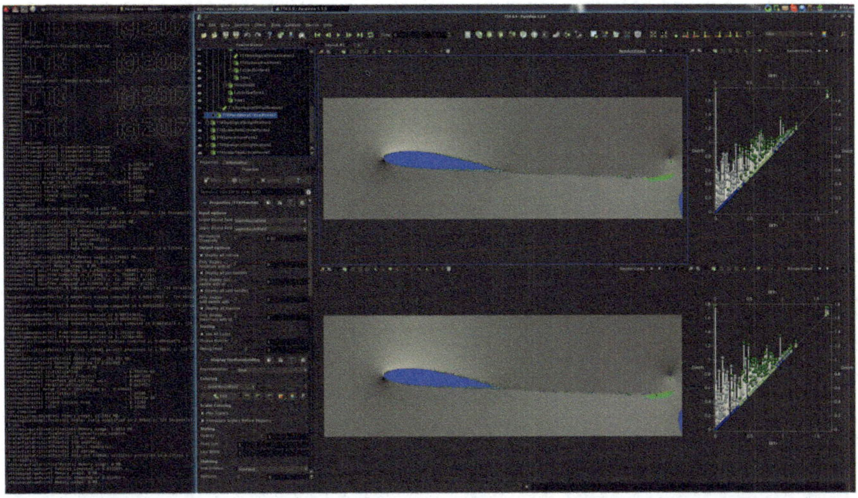

Fig. 7 Mandatory critical points [17] (colored regions) on the starting vortex example. Two members of the ensemble are shown, along with their persistence diagrams and their critical points in the domain. These critical points correspond to the vortices forming behind the wing. The most salient critical points land in the colored regions predicted by the algorithm. In this example, mandatory critical points (colored regions) help estimate visually the geometrical variability that can be expected in the locations of these vortices, given the uncertainty of the data. Once TTK and its data package are installed, from the `ttk-data/` directory, run the following command to reproduce this example: "`paraview states/uncertainStartingVortex.pvsm`" (see [42] for further details)

5 Time–Varying Scalar Data

TTK also provides several features for the analysis and visualization of time-varying data. The trajectory of critical points through time can be tracked with the Wasserstein matcher method introduced by Soler et al. [36]. This technique enables for instance to represent the path taken by vortices in computational fluid dynamics (Fig. 8, left). In addition, TTK supports the visualization and analysis of the topological evolution through time of features of interest, with the notion of nested tracking graphs [31] (Fig. 8, right). These graphs encode the temporal evolution of the connected components of sub-level sets, in the form of a nested hierarchy, where each hierarchy level (each shade of blue in Fig. 8, top right) correspond to a distinct isovalue (and hence a specific sub-level set).

6 High–Dimensional Point Cloud Data

TTK recently integrated the popular package *scikit-learn* [35], leveraging in particular its dimension reduction capabilities: Principal Component Analysis, Spectral Embedding, Locally Linear Embedding, Isomap, Multi-Dimensional Scaling,

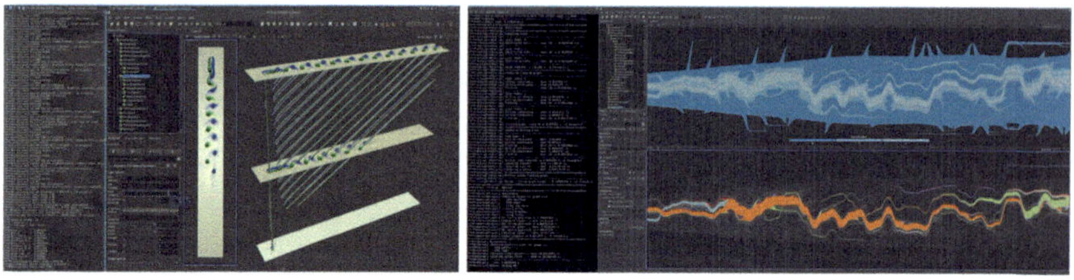

Fig. 8 Gallery of feature tracking in time-varying data. *Left:* critical point trajectory tracking with the Wasserstein matcher [36] (the height denote the temporal component). *Right:* Nested tracking graph [31] (viscous fingering data). Once TTK and its data package are installed (see [42] for further details), from the `ttk-data/` directory, run the following commands to reproduce these examples: "`paraview states/timeTracking.pvsm`" (left) and "`paraview states/nestedTrackingGraph.pvsm`"

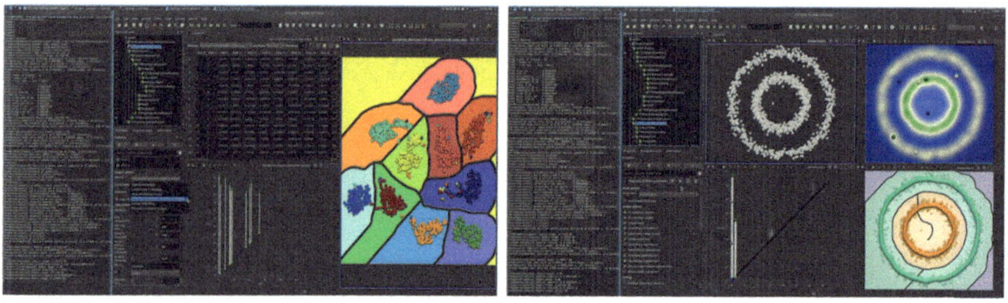

Fig. 9 Examples of topological analysis of high-dimensional point cloud data. *Left:* Persistence-driven clustering [7] of the *"mfeat"* data set (64 dimensions). The data is first projected to 2D with the t-SNE method (available from TTK's integration of scikit-learn [35]). Point colors indicate the ground-truth classification, whereas the clustering computed by TTK is reported by the background color (cells of the Morse complex). *Right:* Persistence-driven clustering [7] and beyond, on a toy point cloud example. In addition to the extraction of the correct clusters, TTK can also extract generators of the first homology group (1-dimensional cycles) with looping separatrices connecting saddles to maxima of density estimation (Gaussian kernel). Once TTK and its data package are installed (see [42] for further details), from the `ttk-data/` directory, run the following commands to reproduce these examples: "`paraview states/karhunenLoveDigits64Dimensions.pvsm`" (left) and "`paraview states/1manifoldLearningCircles.pvsm`" (right)

t-distributed Stochastic Neighbor Embedding. Then, high-dimensional point cloud data (typically in the form of a CSV file) can be processed by TTK. Typically, the data is first projected to 2D or 3D with one of the above dimension reduction methods (Fig. 9). Next, a density estimation (e.g., Gaussian kernel) is performed on a regular grid to describe the projection of the input point cloud (Fig. 9, top right). From this point, any tool of the TTK arsenal can be employed to further analyze, visualize, and explore the data. For instance, persistence-driven clustering [7] can easily be deployed with TTK [8]. The k most persistent features can be selected from the persistence diagram (Fig. 9) to drive a pre-simplification of the data, in order to control the number of clusters (where k is the number of desired clusters). Note that, in practice, a relevant value of k can often be visually inferred from the flat plateaus of the persistence curve (see Fig. 2, top right). Similarly to the notion of eigen gap

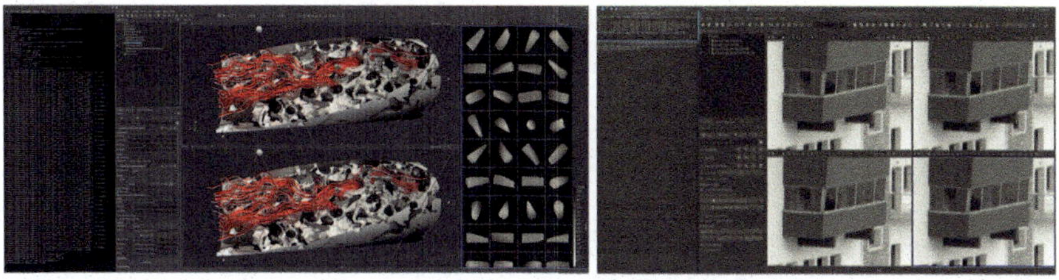

Fig. 10 Examples of in situ data reduction with TTK. *Left:* View-based surface approximation [30] (top: ground-truth, bottom: approximation). *Right:* Topology-controlled lossy compression [37]. Once TTK and its data package are installed (see [42] for further details), from the `ttk-data/` directory, run the following commands to reproduce these examples: "`paraview states/geometryApproximation.pvsm`" (left) and "`paraview states/persistenceDrivenCompression.pvsm`" (right)

[32] in spectral clustering, heuristics can be derived for an automatic selection [8]. Next, the Morse complex can be extracted to isolate each basin of attraction of each of the k remaining maxima (Fig. 9, bottom right, where two clusters are extracted, corresponding to the two *rings* present in the data). The final clustering can be projected from the cells of the Morse complex to the input point cloud with TTK's generic interpolator. Note that TTK enables topological explorations that go beyond simple clustering, such as the extraction of generators of homology groups, as illustrated in Fig. 9 (bottom, right), where looping separatrices linking saddles to maxima are used to extract such generators, hence visually conveying to the user additional information about the internal structure of each cluster. In particular, such generators, when mapped back onto the data, provide visual hints that enable users to identify to which cycle a given data point belongs to. Moreover, it also helps users appreciate the importance of a given group of cycles, given the size of its generator in the data. The left example of Fig. 9 further illustrates the clustering capabilities of TTK on the *mfeat* data set (64 dimensions, 2000 points). The ground-truth classification is given by the colors on the points, whereas the non-supervised classification obtained from the topological clustering is given by the background color (one color per cell of the Morse complex). This example nicely illustrates how TTK can effectively help visualize the intrinsic structure of high-dimensional data (Fig. 10).

7 In Situ Topological Analysis

TTK can be efficiently run in situ (i.e., directly from a simulation source code without storing data to disk) using the Catalyst API [3]. TTK's website reports a complete tutorial [44] with the open-source fluid mechanic simulation code *Code_Saturne* [11], where TDA capabilities are run on the file, without data storage, after each computation of a simulation time step.

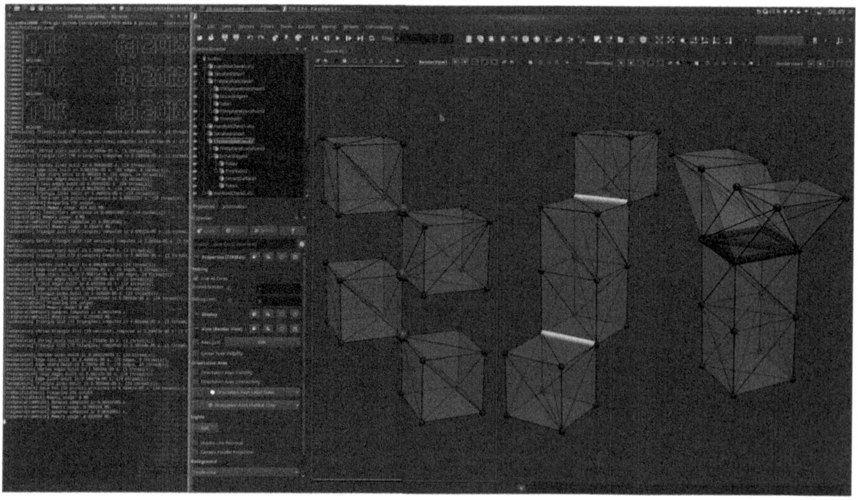

Fig. 11 Example of convenience TTK module: check for manifold-ness on several simplicial complexes. Non-manifold vertices, edges and triangles are reported in green, white, and blue respectively. Once TTK and its data package are installed, from the `ttk-data/` directory, run the following command to reproduce this example: "`paraview states/manifoldChecks.pvsm`" (see [42] for further details)

In addition, TTK offers lossy compression and data reduction tools, to allow the in situ storage of reduced information. In particular, regular grid data can be saved in the TTK file format (**.ttk*), which implements the topologically controlled compression framework by Soler et al. [37]. This framework enables to compress data in a lossy way while guaranteeing the exact preservation of the persistence diagrams of the most salient features. This methodology guarantees, in practice, that any topological analysis run on the compressed data is faithful to the original data. TTK also implements the award-winning image-based geometry approximation method by Lukasczyk et al. [30]. Additionally, TTK implements the latest specification of Cinema databases [2], which enables users to interactively explore large ensembles of data sets stored as Cinema databases and to apply specific analysis pipelines to selections of members, expressed with SQL queries on the meta-data of the members.

8 Convenience

Finally, TTK provides a number of features that make its deployment more convenient for users, including generic data interpolators (interpolating data from any type of object onto any type of object), convertors, mesh processing, and analysis (subdivision, point merging, manifold checks, Fig. 11, etc.).

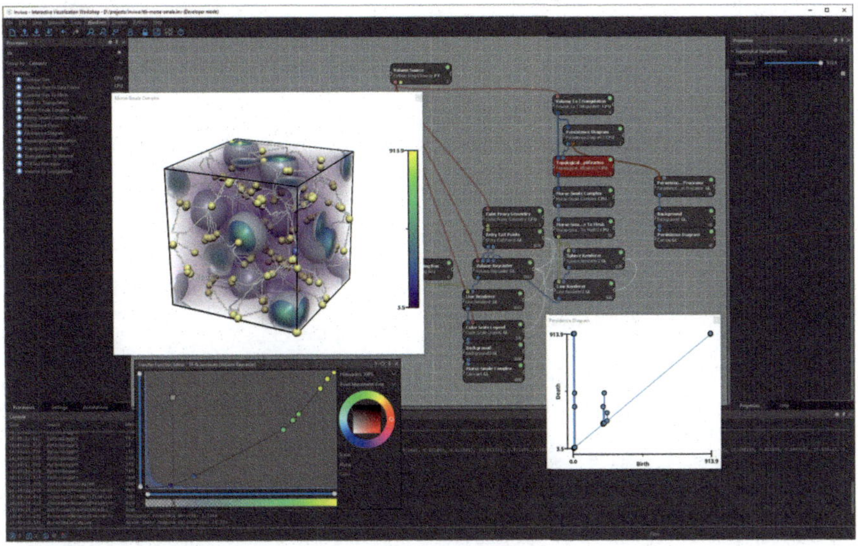

Fig. 12 Integration of TTK in a software ecosystem other than VTK/ParaView: Inviwo [23], a software framework for the rapid prototyping of visualizations, written in C++ and exploiting modern graphics hardware. This example shows the topological analysis with the Morse-Smale complex (with persistence-driven data pre-simplification) of charge densities in iron oxide [22]

9 Conclusion and Perspectives

This paper presented a brief overview of the main end-user features available in the Topology ToolKit (TTK) along with example application scenarios. The material that is necessary to reproduce these examples is available on the TTK website [42]. The data analysis pipelines presented in this paper can be easily reproduced with ParaView, with Python scripts (ParaView supports the automatic export of analysis pipelines to Python scripts), with VTK or direct C++ code. The examples illustrated in this paper ranged from basic image segmentation capabilities to the advanced topological analysis of high-dimensional point cloud data. We refer the reader to TTK's online user forum for further discussions and usage examples [43].

In the future, we are looking forward to further extending TTK's developer and user communities. We see TTK as an opportunity to grow as a community by federating our software engineering efforts, to make our research more accessible, reproducible and visible to others. In that regard, we warmly welcome any contributors (see TTK's contribution page: https://topology-tool-kit.github.io/contribute.html), especially with experience in vector and tensor data analysis. We will also work toward the improved integration of TTK in third-party data analysis and visualization tools, as done, for example, in collaboration with the Inviwo [23] development team (Fig. 12). Future directions of development of TTK include an improved support for statistical tasks based on topological data representations as well as an improved integration of TTK on supercomputers. Such improvements will be partially conducted in the context of the *VESTEC* project [47], which focuses on novel supercomputing methodologies for urgent decision making, and for which TTK is one of the core software technologies.

Acknowledgements We would like to thank the anonymous reviewers for their thoughtful remarks and suggestions. This work is partially supported by the European Commission grant H2020-FETHPC-2017 "VESTEC" (ref. 800904).

References

1. Ahrens, J., Geveci, B., Law, C.: Paraview: An End-User Tool for Large-Data Visualization. The Visualization Handbook. Elsevier, Amsterdam (2005)
2. Ahrens, J., Jourdain, S., O'Leary, P., Patchett, J., Rogers, D.H., Petersen, M.: An image-based approach to extreme scale in situ visualization and analysis. In: IEEE SuperComputing (2014)
3. Ayachit, U., et al.: Paraview catalyst: enabling in situ data analysis and visualization. In: In Situ Infrastructures for Enabling Extreme-Scale Analysis and Visualization (ISAV 2015) (2015)
4. Bachthaler, S., Weiskopf, D.: Continuous scatterplots. IEEE Transactions on Visualization and Computer Graphics (Proc. of IEEE VIS) (2008)
5. Banchoff, T.F.: Critical points and curvature for embedded polyhedral surfaces. Am. Math. Monthly **77**, 475–485 (1970)
6. Carr, H., Geng, Z., Tierny, J., Chattopadhyay, A., Knoll, A.: Fiber surfaces: generalizing isosurfaces to bivariate data. In: Computer Graphics Forum (Proc. of EuroVis) (2015)
7. Chazal, F., Guibas, L.J., Oudot, S.Y., Skraba, P.: Persistence-based clustering in Riemannian manifolds. J. ACM **60**, 1–38 (2013)
8. Cotsakis, R., Shaw, J., Tierny, J., Levine, J.A.: Implementing persistence-based clustering of point clouds in the topology ToolKit. In: TopoInVis Book (2020)
9. Edelsbrunner, H., Harer., J.: Jacobi Sets of Multiple Morse Functions. Cambridge Books Online, Cambridge (2004)
10. Edelsbrunner, H., Harer, J.: Computational Topology: An Introduction. AMS Press, Naga (2009)
11. EDF: Code_saturne. https://www.code-saturne.org/cms/
12. Falk, M., et al.: Topological data analysis made easy with the topology toolkit, What is new? In: Proceedings of IEEE VIS Tutorials (2020). https://topology-tool-kit.github.io/ieeeVis2020Tutorial.html
13. Falk, M., et al.: Topological data analysis made easy with the topology ToolKit, A sequel. In: Proceedings of IEEE VIS Tutorials (2019). https://topology-tool-kit.github.io/ieeeVis2019Tutorial.html
14. Favelier, G., Faraj, N., Summa, B., Tierny, J.: Persistence atlas for critical point variability in ensembles. In: IEEE Transactions on Visualization and Computer Graphics (Proc. of IEEE VIS) (2018)
15. Favelier, G., et al.: Topological data analysis made easy with the Topology ToolKit. In: Proceedings of IEEE VIS Tutorials (2018). https://topology-tool-kit.github.io/ieeeVis2018Tutorial.html
16. Favelier, G., Gueunet, C., Tierny, J.: Visualizing ensembles of viscous fingers. In: IEEE SciVis Contest (2016)
17. Guenther, D., Salmon, J., Tierny, J.: Mandatory critical points of 2D uncertain scalar fields. Computer Graphics Forum (Proc. of EuroVis) (2014)
18. Gueunet, C., Fortin, P., Jomier, J., Tierny, J.: Contour forests: fast multi-threaded augmented contour trees. In: Proceedings of IEEE Large Data Analysis and Visualization (2016)
19. Gueunet, C., Fortin, P., Jomier, J., Tierny, J.: Task-based augmented merge trees with fibonacci heaps. In: Proceeding of IEEE Large Data Analysis and Visualization (2017)

20. Gueunet, C., Fortin, P., Jomier, J., Tierny, J.: Task-based augmented contour trees with fibonacci heaps. IEEE Trans. Parall. Distrib. Syst. (2019)
21. Gueunet, C., Fortin, P., Jomier, J., Tierny, J.: Task-based augmented Reeb graphs with dynamic ST-trees. In: Eurographics Symposium on Parallel Graphics and Visualization (2019)
22. Jakobsson, E., Bin-Masood, T., Hotz, I., Abrikosov, I., Steneteg, P.: Topology-guided analysis and visualization of charge density fields : a case study (2019, Submitted manuscript)
23. Jönsson, D., et al.: Inviwo - a visualization system with usage abstraction levels. IEEE Trans. Visual. Comput. Graph. (2019)
24. Kerber, M., Morozov, D., Nigmetov, A.: Geometry helps to compare persistence diagrams. ACM J. Exp. Algorith. **22**, 1–20 (2016)
25. Klacansky, P., Tierny, J., Carr, H.A., Geng, Z.: Fast and exact fiber surfaces for tetrahedral meshes. IEEE Trans. Visual. Comput. Graph. (2017)
26. Kontak, M., Vidal, J., Tierny, J.: Statistical parameter selection for clustering persistence diagrams. In: Proceedings of SuperComputing Workshop on Urgent HPC (2019)
27. Lacombe, T., Cuturi, M., Oudot, S.: Large scale computation of means and clusters for persistence diagrams using optimal transport. In: NIPS (2018)
28. Lukasczyk, J., et al..: Report of the TopoInVis TTK Hackathon: experiences, lessons learned, and perspectives. In: TopoInVis (2019)
29. Lukasczyk, J., Garth, C., Maciejewski, R., Tierny, J.: Localized topological simplification of scalar data. In: IEEE Transactions on Visualization and Computer Graphics (Proc. of IEEE VIS) (2020)
30. Lukasczyk, J., Kinner, E., Ahrens, J., Leitte, H., Garth, C.: Voidga: A view-approximation oriented image database generation approach. In: Proceedings of IEEE Large Data Analysis and Visualization (2018)
31. Lukasczyk, J., Weber, G.H., Maciejewski, R., Garth, C., Leitte, H.: Nested tracking graphs. In: Computer Graphics Forum (Proc. of EuroVis) (2017)
32. von Luxburg, U.: A tutorial on spectral clustering. Stat. Comput. **17**, 395–416 (2007)
33. Morozov, D.: Dionysus (2010). http://www.mrzv.org/software/dionysus
34. Pascucci, V., Tricoche, X., Hagen, H., Tierny, J.: Topological data analysis and visualization: theory. In: Algorithms and Applications, Springer (2010)
35. Pedregosa, F., et al..: Scikit-learn: machine learning in python. J. Mach, Learn. Res. **12**, 2825–2830 (2011)
36. Soler, M., Plainchault, M., Conche, B., Tierny, J.: Lifted wasserstein matcher for fast and robust topology tracking. In: Proceedings of IEEE Large Data Analysis and Visualization (2018)
37. Soler, M., Plainchault, M., Conche, B., Tierny, J.: Topologically controlled lossy compression. In: Proceedings of PacificVis (2018)
38. Tierny, J.: Topological Data Analysis for Scientific Visualization. Springer, Cham (2018)
39. Tierny, J., Carr, H.: Jacobi fiber surfaces for bivariate reeb space computation. In: IEEE Transactions on Visualization and Computer Graphics (Proc. of IEEE VIS) (2016)
40. Tierny, J., Favelier, G., Levine, J.A., Gueunet, C., Michaux, M.: The topology ToolKit. In: IEEE Transactions on Visualization and Computer Graphics (2017). https://topology-tool-kit.github.io/
41. Tierny, J., Pascucci, V.: Generalized topological simplification of scalar fields on surfaces. In: IEEE Transactions on Visualization and Computer Graphics (Proc. of IEEE VIS) (2012)
42. TTK-Contributors: TTK Online Tutorials. https://topology-tool-kit.github.io/tutorials.html
43. TTK-Contributors: TTK User Forum. https://groups.google.com/forum/#!forum/ttk-users
44. TTK-Contributors: tutorial on in-situ topological data analysis with TTK and catalyst. https://topology-tool-kit.github.io/catalyst.html
45. TTK-Contributors: TTK Anaconda package (2019). https://anaconda.org/conda-forge/topologytoolkit
46. Turner, K., Mileyko, Y., Mukherjee, S., Harer, J.: Fréchet means for distributions of persistence diagrams. Disc. Compu, Geom. **52**, 44–70 (2014)
47. VECSTEC-Consortium: Visual exploration and sampling ToolKit for extreme computing. https://vestec-project.eu/

48. Vidal, J., Budin, J., Tierny, J.: Progressive Wasserstein Barycenters of persistence diagrams. In: IEEE Transactions on Visualization and Computer Graphics (Proc. of IEEE VIS) (2019)
49. Vintescu, A., Dupont, F., Lavoué, G., Memari, P., Tierny, J.: Conformal factor persistence for fast hierarchical cone extraction. In: Eurographics (short papers) (2017)
50. Xu, K., Zhang, H., Cohen-Or, D., Xiong, Y.: Dynamic harmonic fields for surface processing. Comput. Graph **33**, 391–398 (2009)

Implementing Persistence-Based Clustering of Point Clouds in the Topology ToolKit

Ryan Cotsakis, Jim Shaw, Julien Tierny, and Joshua A. Levine

Abstract We show how the scalar field topology features of the Topology ToolKit (TTK) can be leveraged in a pipeline for persistence-based clustering of point clouds. While TTK provides numerous features for computing topological structures of scalar fields on unstructured meshes, prior to this work, it allowed for only basic point cloud input. In this work, we implemented two new modules in TTK: one for sampling scalar fields based on either distance or density of the point cloud and a second for computing persistence-based clusters. Both modules provide heuristics for automatically specifying key thresholds so as to simplify user interaction. This document outlines the implementation details of the two modules and provides experimental results that demonstrate their modularity and utility.

1 Introduction

Clustering is an important tool used widely in data analysis. We consider the problem of clustering point clouds. The input dataset is an unstructured collection of points that are a discrete subset of $\mathbb{R}^d$. Clustering seeks to partition the points into a set of logical groups (clusters) such that points in the same group are more similar to each other than they are to points in the other groups. From a data analysis perspective, clustering is often an important module for other downstream tasks that either directly utilize the clusters, or study how many, how large, or how spread out clusters are.

R. Cotsakis and J. Shaw—Contributed equally to the work.

R. Cotsakis · J. Shaw
University of British Columbia, Vancouver, Canada

J. Tierny
CNRS/Sorbonne Université, Paris, France
e-mail: julien.tierny@sorbonne-universite.fr

J. A. Levine (✉)
University of Arizona, Tucson, USA
e-mail: josh@email.arizona.edu

I. Hotz et al. (eds.), *Topological Methods in Data Analysis and Visualization VI*,
Mathematics and Visualization, https://doi.org/10.1007/978-3-030-83500-2_17

When the data is unstructured, as is the case of point clouds, clustering typically requires building a model based on other prior knowledge or assumptions on the dataset. This knowledge can be used to infer distance and/or similarity measures or otherwise make decisions about how to define what properties delineate grouping. As a result there is no singular goal for clustering applicable in all settings.

On the other hand, segmenting a scalar field has similar goals to clustering and provides an interesting counterpoint. Compared to segmenting a point cloud, segmenting a scalar field has a variety of tools that naturally decompose the field based on features encoded in the scalar field. Common techniques for this include watersheds in image segmentation [5, 6] and topological segmentation via the Morse complex [12, 26]. A particular advantage of topological segmentation techniques is the ability to provide a hierarchy of segmentations that respect a notion of simplification. Specifically, topological persistence allows one to rank the importance of segments, jointly simplifying the scalar field while leading to a coarser segmentation [13].

Leveraging the strengths of using scalar field topology for segmentation, one approach for segmenting point clouds is to first compute a scalar field from a given input point cloud. This scalar field serves to provide the additional prior information and provide structure for the dataset. For example, one simple proposition is to use a scalar field that estimates the density of the points themselves, resulting in clusters that relate to those computed by density-based clustering [14]. Of course, the choice of scalar field is a free parameter to the clustering algorithm and other scalar fields could be use that encode distance or similarity. After segmenting the scalar field, one can then associate each point with the labels assigned to nearby regions of the domain of the scalar field, and then use this assignment to produce the final clustering.

1.1 Contributions

In this work we leverage the features of scalar field segmentation for point cloud segmentation. We primarily follow the methodology of Chazal et al.'s algorithm ToMATo [9] for designing the scalar field, except we replace the domain as represented by a neighborhood graph with a simple triangulated grid on which we sample the field. Our main contributions are the implementation details for adapting this approach into a large framework for topological analysis, the Topology ToolKit (TTK) [27]. Specifically, our contributions are:

- We describe the implementation for persistence-based clustering of point clouds in TTK. This involved implementing two new modules: one for computing a scalar field from a point and a second for clustering via persistence.
- As we aim for a non-parametric method, we design heuristics for automatically selecting parameters involved in both modules that work well in most settings.
- We report experimental details on the efficacy of our heuristics as well as discuss important design considerations.

2 Related Work

Clustering methods are well-studied over the past forty years, with common approaches that include k-means [19], density-based clustering like DBSCAN and its variants [14, 24], mode-seeking methods [18] such as mean shift [10], and spectral clustering [30]. These methods have been used in a wide variety of applications in data analysis, image processing, computer graphics, and statistics. Thus, a variety of practical tools implement clustering, thus making it desirable for the users of TTK as well.

Our work emphasizes using topological tools for clustering, via computing segmentations that can be simplified through topological persistence [13]. Topological analysis has led to a variety of tools for data analysis in general, and relies on the field of persistent homology to compute and rank features (see Edelsbrunner and Harer for a thorough introduction [11]). Chazal et al. were one of the pioneers of using topological persistence for guiding clustering of point clouds [9] and as mentioned previously our work is using a conceptually similar pipeline adapted to TTK [27]. This approach builds on key theoretical results where the same authors show one can recover structural information of scalar fields from samples [8].

Others have also pursued using topological analysis for clustering. Particularly, Beksi and Papnikolopoulos have shown the utility of clustering 3D point clouds [2, 4] and designing signatures for points [3]. Moon et al. design the persistence terrace to help provide a summary plot to guide the inference process [20]. These methods all focus on low-dimensional point sets (typically two- and three-dimensional data). Nevertheless, many applications of clustering focus on higher dimensional point clouds. Using topological analysis can also work in this setting, but there are certain challenges with computing and visualizing the topological structure of segments. Oesterling et al. show methods for capturing the topology of high-dimensional point cloud density fields [21] by constructing topological landscapes to provide a tangible metaphor [22].

3 Software Design Overview

We implemented persistence-based clustering for point clouds in TTK by creating two new modules. The first module, *ScalarFieldFromPointCloud*, computes a scalar field from a given input point cloud that will be used to analyze the structure of the point cloud. The second module, *PersistenceSimplification*, is used to simplify the scalar field in preparation for topological segmentation. Both modules enable a few user controls for segmentation, but can also be used in an automatic mode where all parameters are set by default heuristics.

To complete the clustering, we wrap these modules in a pipeline with five steps: (1) read the input point cloud, (2) compute a scalar field, (3) simplify this field with persistence, (4) perform topological segmentation with the Morse complex, and (5) map the segmentation labels from the scalar field domain to the input point clouds. Steps (1), (4), and (5) leverage existing modules in both TTK as well as standard methods that exist in the ParaView [1]. Steps (2) and (3) are developed as standard TTK modules that we also enable as ParaView plugins for experimentation.

4 Computing Scalar Fields from Point Clouds

As previously mentioned, ScalarFieldFromPointCloud takes a point cloud as an input and constructs a triangulated regular grid surrounding the input point cloud. The grid size as well as boundary padding may be specified by the user. The grid is the domain of the scalar field that ScalarFieldFromPointCloud outputs. As with Chazal et al. [9] we experimented with two difference options for calculating the scalar values on each grid point: a kernel density estimation (KDE) or a distance field.

For the distance field, the scalar values on the regular grid can simply be taken to be the distance to the closest point in the point cloud. In effect, this will yield very small values for points with nearby neighbors, whereas the KDE will yield the largest values in regions of high density. Subsequently, in this document we only report on experiments using density.

The scalar values for the KDE are calculated as follows: We construct identical Gaussian distributions centered at each point in the point cloud. A continuous scalar function $f : \mathbb{R}^d \rightarrow \mathbb{R}$ defined on the ambient space can be constructed by taking the sum of all of the Gaussian distributions. The standard deviation or width of the

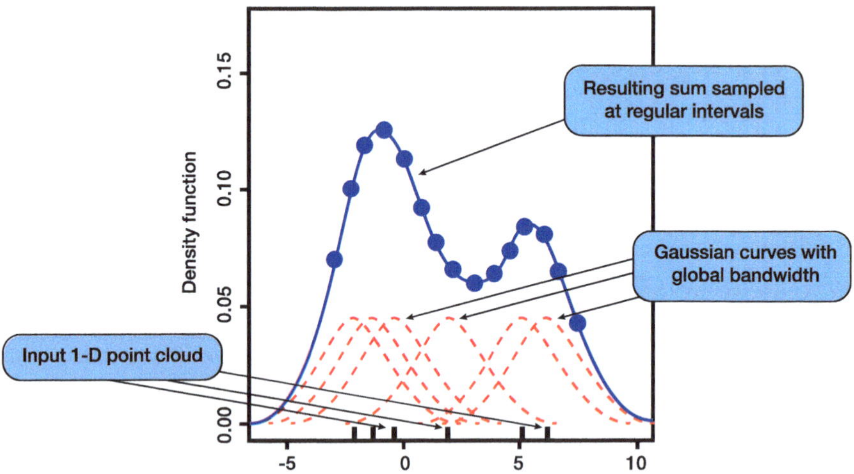

Fig. 1 Gaussian kernel density estimate. A Gaussian of a certain bandwidth is centered at each point. The Gaussians are added and the purple curve is the resulting probability distribution that is estimated from the samples. Image taken from Wikipedia

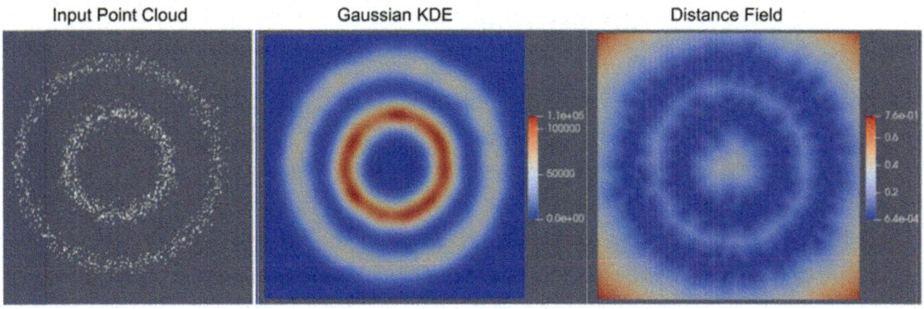

Fig. 2 The input to ScalarFieldFromPointCloud on the left, the output scalar field in the middle when using the Gaussian KDE option, and the output scalar field on the right when using the distance field option. For the KDE, the colour map is red at locations with a high density of points, and blue in sparse regions; in the distance field, the colour map is red far away from the points and blue near the points

Gaussians is denoted as the *bandwidth*, which is a parameter the user may choose. The value of the output scalar field for a particular vertex in the regular grid is determined to be the value of f at that point. See Fig. 1 for a graphic explanation (Fig. 2).

4.1 Parameter Setting

We also implemented a feature that automatically estimates what the bandwidth should be. For each point in the point cloud, we calculate the distance to the $k-$th closest neighbor. The parameter k is an integer, and we estimate it as follows

$$k = (0.587 \cdot n^{4/5})^{1/d}$$

and then rounding k up. n is the number of points in the point cloud, d is the dimension of the data (currently we allow $d = 2$ and $d = 3$). This method is modified from Equation 19 of [23] to include the $1/d$ scaling to the $n^{4/5}$ term. Finally, the mean distance to the $k-$th closest neighbor for every point in the point cloud is taken to obtain the final bandwidth which is used. This method of bandwidth selection can be used in ScalarFieldFromPointCloud module by selecting the "Automatic Bandwidth for KDE" option in ParaView.

We also considered a method that uses adaptively-sized bandwidths for the Gaussian kernels, proportional to the k-th nearest neighbor [7]. When exploring this variation, we found that the least persistent clusters would have an even smaller persistence with this algorithm, making it likely for PersistenceSimplification module to discard them as noise.

5 Persistence-Based Clustering

We briefly review the key concepts we used in persistent homology, for a full introduction we recommend Edelsbrunner and Harer [11]. Given a scalar field, $f : \mathbb{R}^d \to \mathbb{R}$, persistent homology can be used to study the evolution of the sublevel sets $f^{-1}(-\infty, a]$ for every real value a. Intuitively, as one sweeps through increasing values of a, the connectivity of the sublevel sets will change, and persistent homology helps to gauge both when and what types of changes occur. Specifically, one can observe when new components are created and destroyed, and how long (in terms of values for a) they exist. Such events will occur at values for a which correspond to critical points of f. These events can be grouped into what are commonly referred to as *persistence pairs* that correspond to a pair of critical points. For a persistence pair c_i and c_j such that $f(c_i) < f(c_j)$, we define the *persistence* as $f(c_j) - f(c_i)$, which corresponds to the span of function values for which the corresponding feature was present. The *persistence diagram* embeds these pairs into a useful tool for data analysis, as it describes all such pairs and conveniently arranges them so that pairs with higher persistence are made prominent.

Given the persistent diagram, a common task to filtering such features is to select a persistence threshold and retain all pairs whose persistence is above the threshold. The second module we implemented, PersistenceSimplification, performs this form of persistence simplification on an input scalar field. In our setting, simplification prepares the scalar field for topological segmentation and ultimately, clustering the point cloud. Tools for persistence simplification already exist in TTK, but our module encapsulates this process so as to make it easier for an end user. Specifically, our new module relies components from two existing modules: FTMTreePP [17] (which computes a merge tree that can be used to construct the persistence threshold for each persistence pair) and TopologicalSimplification [28] (which is used to compute a new, simplified scalar field that preserves the persistence pairs that are above threshold).

Previously, in TTK, a user would have to separately compute the persistence diagram, use thresholds to select a set of persistence pairs to preserve, and then run TopologicalSimplification to produce a simplified field. Our new module combines these features into a single module. First, the scalar field is given to the FTMTreePP object, and the persistence pairs are computed. The pairs are then sorted based on their persistence, and then a persistence threshold is decided based on user selected parameters (including automatic selection). Finally, TopologicalSimplification is called using the pairs with sufficiently high persistence as constraints on the input scalar field. Our module then computes this simplified field and returns it as output. Thus, unlike the separate components that exist in TTK, this module starts with a scalar field and produces a new, simplified one with no intermediate steps, greatly simplifying the process of simplification.

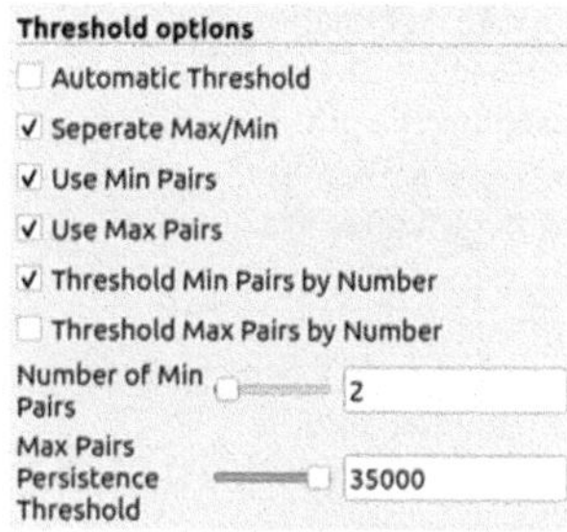

Fig. 3 GUI for PersistenceSimplification. By checking "Separate Max/Min", the user may select minimum-saddle pairs and maximum-saddle pairs in different ways. By ticking "Use Min/Max Pairs" the module will choose to retain minimum/maximum-saddle pairs. Since "Threshold Min Pairs by Number" is specified, we can choose the number of minimum-saddle pairs to retain. Alternatively, since "Threshold Max Pairs by Number" is not specified, we retain maximum-saddle pairs by thresholding persistence instead.

5.1 User Options

The PersistenceSimplification module makes this process easy by not exposing the persistence diagram to the user nor requiring manual pair selection. Instead, the user is given the option to have PersistenceSimplification choose which points are sufficiently persistent automatically as described below. Nevertheless, we consider a number of use cases for the PersistenceSimplification module. Although we are focused on clustering applications, we have made the module as versatile as possible. The user can override the automatic threshold by interacting with the GUI in ParaView to choose various threshold options:

- The user has the option to manually threshold which persistence pairs (min-saddle, max-saddle, or both) are used to simplify the scalar field based on a known persistence threshold.
- Alternatively, if a threshold is not known, the user can instead threshold based on the number of pairs they would like to simplify with. The most persistent points will be chosen. This is conceptually similar to setting the number of clusters, k, in k-means.

In addition the user can choose to simplify the scalar field using the maximum-saddle pairs and the minimum-saddle pairs separately, or both together. For density-based clustering, we expect maxima to delineate centroids for clustering, but this module provides more flexibility capabilities to be used in other settings too (Fig. 3).

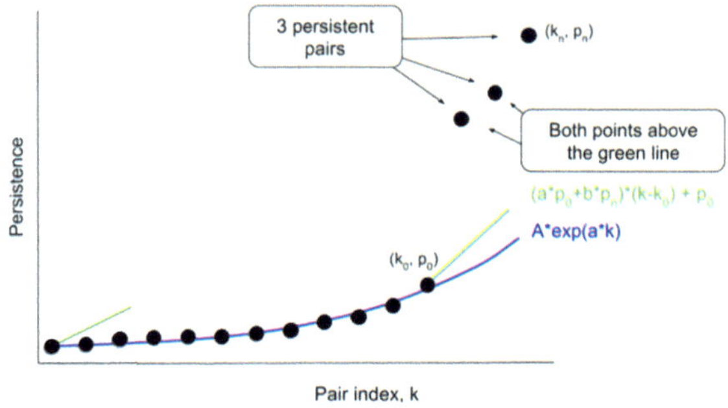

Fig. 4 Automatic persistence thresholding. Each black point is a critical point pair. We sort the pairs by persistence. We assume the noise assumes an exponential shape. $(a \cdot p_0 + b \cdot p_n)(k - k_0) + p_0$ is the tangent line to the exponential plus some slope $b \cdot p_n$, where p_n is the persistence of the most persistent critical pair. The algorithm essentially looks point by point to see if the next point is above this green line. If the next two points lie above the line, then all further points are persistent enough

5.2 *Automatic Parameter Setting*

We developed our automatic threshold mechanism based on the concept of finding large jumps in persistence that separate topological noise (with relatively low persistence) from topological signal (with relatively high persistence). We define such jumps by first sorting all pairs by their index and looking for locations where the tangent of the curve does a poor job of predicting the persistence of the next highest pair, relative to how much this curve increases overall. For an illustration, see Fig. 4.

Specifically, let p_n be the persistence of the most persistent critical pair in our dataset. To identify gaps automatically, we examine the persistence of each pair in increasing order. Let $a = 0.2$ and $b = 0.025$ be two fitting parameters (a describes the exponential growth in persistence pairs and b weights relative to the maximum). For a critical pair with persistence p_0, we ask if (1) the next most persistent pair has persistence greater than $(1 + a)p_0 + bp_n$ and if (2) the pair after that has persistence greater than $(1 + 2a)p_0 + 2bp_n$. If the answer to both questions is yes, then the persistence threshold is determined to be $(1 + a)p_0 + bp_n$. If the answer to either question is no, then we remove the initial pair and begin the analysis again on the next pair, until a persistence threshold is decided.

This progressive approach is based on the observation that noisy persistent pairs fit an exponential function quite well. Thus, the two parameters we select model this process. Specifically, a is chosen to be the parameter that describes the exponential growth of persistence in Fig. 4. The parameter b is chosen as a value that captures discontinuities in these curves. Ideally, one would directly estimate a and b using data fitting, or let the user vary them. As heuristics, we found they performed quite well for our data even with fixed parameters.

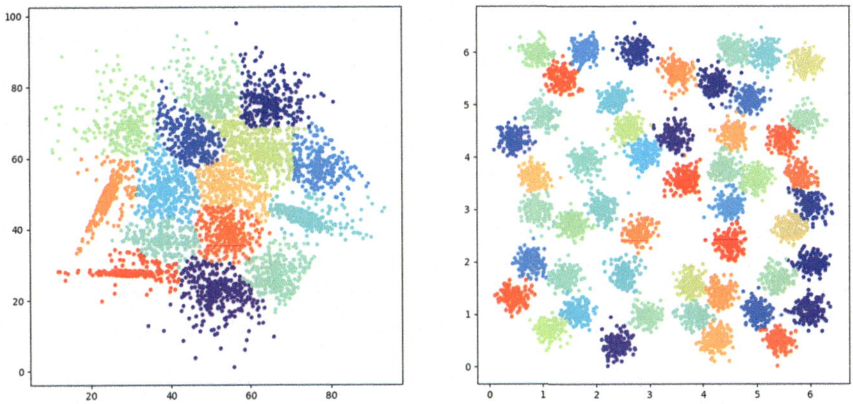

Fig. 5 Our automatic persistence clustering, applied to two datasets from the k-means clustering suite. Left: dataset S4, Right: dataset A3. Each colour indicates a different cluster, as labeled by our algorithm

6 Experimental Results

We test the effectiveness of our automatic parameter selection technique and analyze where it fails. For the following experiments, we only cluster with parameters selected through our automatic selection technique. We used a collection of different datasets and benchmarks. For evaluating automatic feature detection efficacy, we used the Ultsch's Fundamental Clustering Problems Suite (FCPS) [29][1] and Fränti and Sieranoja's k-means clustering testing suite [16][2]. For comparisons against other clustering methods, we evaluated on the scikitlearn clustering datasets [25][3].

6.1 Automatic Feature Detection

Results for our clustering method with automatic parameter selection are shown in Figs. 5 and 6. On the two k-means datasets, our method was quite effective at separating individual clusters of both different sizes and shapes.

For the FCPS datasets we chose, our technique encountered additional challenges that stem from the automatic selection of persistence thresholds. While generally reasonable, we found our thresholds to produce a few additional clusters than necessary, particularly for the TwoSuns (we end up with 4 instead of 2 clusters) and Lsun (we end up with 5 instead of 3 clusters). Note that if we manually set the number

[1] Downloaded from https://www.uni-marburg.de/fb12/arbeitsgruppen/datenbionik/data.

[2] Downloaded from http://cs.joensuu.fi/sipu/datasets/.

[3] Extracted directly from `sklearn.datasets`.

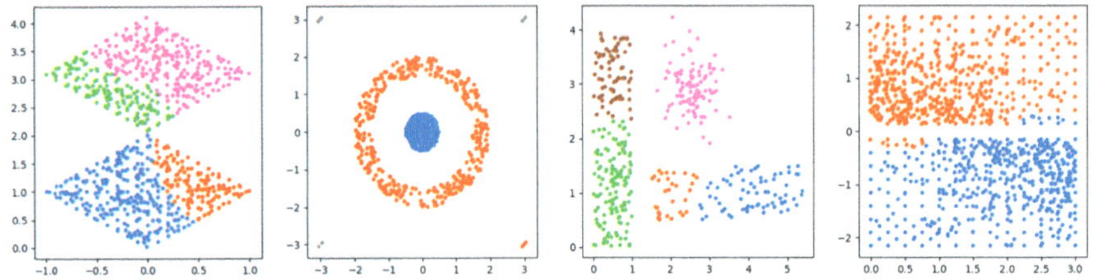

Fig. 6 Our automatic persistence clustering, applied to four datasets from FCPS. From left-to-right, we evaluated the TwoSuns, Target, Lsun, and Wingnut datasets. Each colour indicates a different cluster, as labeled by our algorithm

of clusters to the appropriate number, or method produced correct clusterings for both. The Wingnut dataset proved a bit challenging when the data sparsity made it challenging to separate the top from the bottom, as shown in the few points that are not separated well. We discuss this issue further in Sect. 6.2.

6.2 Comparison Against Other Clustering Methods

We also evaluated our clustering method against a variety of other techniques[4], as shown in Fig. 7. Note that the data generation process in scikitlearn involves randomness, which explains the minor discrepancies between Fig. 7 and figures on their website.

In the datasets, we are given the ground truth for what the clusters should be. This means we can apply the Fowlkes-Mallows score for clustering [15]. The Fowlkes-Mallow score is a goodness of clustering score from 0 to 1, with 1 being a perfect clustering and smaller values being bad clusterings. We have labeled each method and each dataset with a Fowlkes-Mallows score and summed up the scores for all datasets and each algorithm in Table 1.

All algorithms in Fig. 7 except ours which is outlined in red and denoted “Persistence” needs to have parameters specified. For these other algorithms, we used the same parameters as in https://scikit-learn.org/stable/modules/clustering.html, which were picked to perform well on these datasets. The best algorithms according to the sum of scores is Persistence and DBSCAN. DBSCAN, however, does not classify all points as seen by the black outliers in the figure.

In Fig. 7 we see that the only data set for which our automatic clustering method fails drastically is the last data set on the bottom row, which has only 1 cluster. The data itself is not ideal for our method when compared to the other clusters in Fig. 7. Particularly, it suffers from the fact that it is sampled from a uniform distribution on a square and thus there are no major topological variations in the resulting density

---

[4] See scikit-learn.org/stable/modules/clustering.html for information on the clustering methods we compared to.

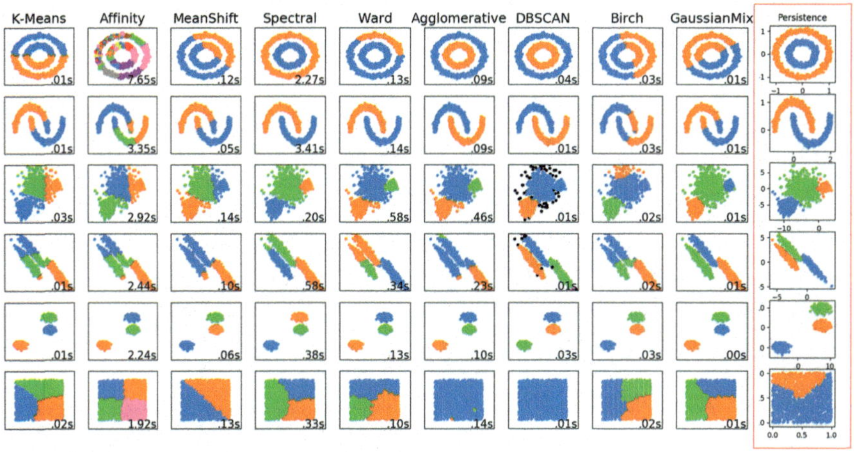

Fig. 7 A comparison of clustering methods. Each colour indicates a different labeling according to the algorithm. Our method, "Persistence", is outlined in red.

Table 1 The Fowlkes-Mallows score, a metric for goodness of clustering, for the corresponding algorithm (column) and dataset (row). The metric ranges from 0 to 1, with 0 being a bad clustering and 1 being a perfect clustering. The datasets correspond in order to the datasets in Fig. 7

	K-Means	Affinity	Mean shift	Spectral	Ward	Agglom.	DBSCAN	Birch	Gauss. mix	Persistence
Circles	0.50	0.36	0.42	1.0	0.66	1.0	1.0	0.51	0.50	**1.0**
Moons	0.74	0.67	0.77	1.0	1.0	1.0	1.0	0.57	0.75	**1.0**
Varied density	0.87	0.88	0.90	0.96	0.95	0.95	0.75	0.74	**0.98**	0.96
Anisotropic	0.73	0.74	0.75	0.97	0.79	0.69	0.98	0.72	1.0	**1.0**
Blobs	1.0	1.0	1.0	1.0	1.0	1.0	1.0	1.0	1.0	**1.0**
No structure	0.58	0.50	0.71	0.58	0.59	0.99	**1.0**	0.59	0.58	0.77
Sum of scores	*4.42/6.0*	*4.15/6.0*	*4.55/6.0*	*5.51/6.0*	*4.99/6.0*	*5.63/6.0*	*5.73/6.0*	*4.13/6.0*	*4.81/6.0*	***5.73/6.0***

field. This behavior is also byproduct of estimating density by summed Gaussians, which would naturally weight dense areas higher than sparse areas. As a result, the one singular cluster has a large area and is ultimately undersampled.

These two conditions (uniform sampling and summing Gaussians) result in "patchy-ness" in the density estimation. Figure 8 illustrates the clustering procedure for the final data set. Despite the relatively low quality density estimation, our automatic threshold for persistence almost worked. In this test, a in the automatic persistence algorithm was set to 0.2. In reality, by fitting an exponential curve to the data, we get that a should be 0.33 instead. This leads to a more appropriate clustering.

It turns out that the TwoSuns datasets in Fig. 6 fails to automatically cluster for the same reasons as above. The same patchiness in the scalar field results leading to an incorrect thresholding. Interestingly, fitting an exponential curve to the persistence curve yields $a = 0.53$, which we found would indeed lead to the correct two clusters.

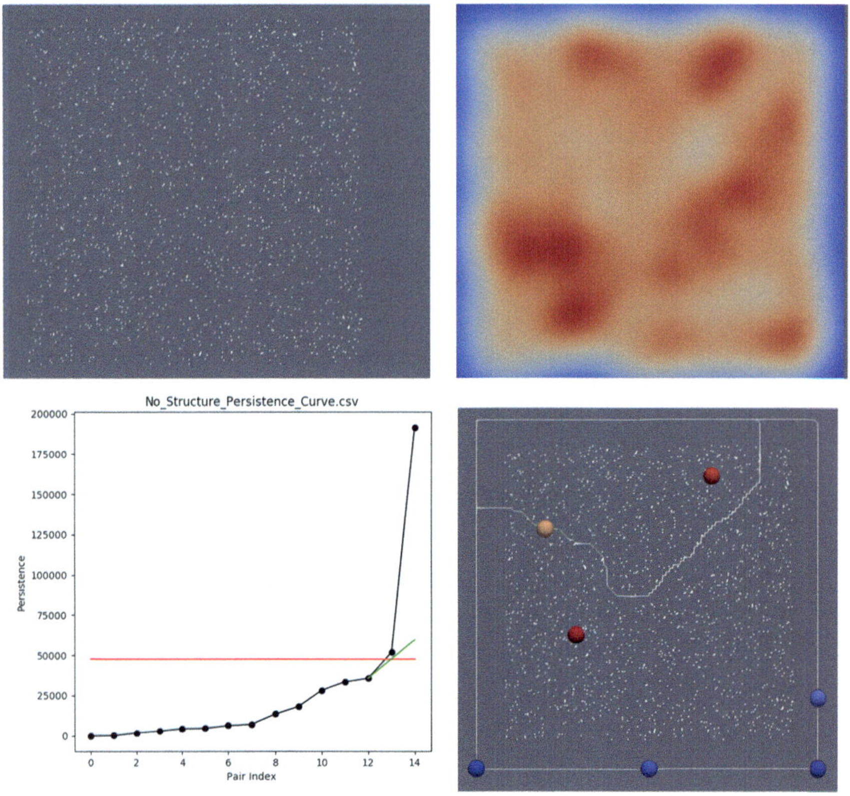

Fig. 8 Clustering of a data set which is sampled from a uniform distribution on a square. Top Left—Original data set. Top Right—Kernel density estimation; red is high, blue is low. Bottom Left–Automatic persistence thresholding, with the red line representing the persistence at which we threshold. Each black dot represents a maximum-saddle pair. We order the pairs by persistence. The point just above the red threshold lies above the green line, hence the algorithm terminates. Bottom Right—Segmented domain, with each segment representing a cluster. The red spheres are maxima, blue minima, and yellow saddle

It seems for both data sets that the bandwidth is too small for the density estimation. However, we noticed that the automatic bandwidth selection on the data sets in Fig. 5 produced bandwidths that could not be increased without yielding incorrect clusterings. This observation points out some of the challenges with setting parameters, but in general our automatic method worked well as a heuristic for the data sets we experimented with.

7 Discussion

Our work demonstrates the effectiveness of using scalar field topology for clustering point clouds. Moreover, we also discussed implementing this technique in TTK, creating new modules to both ease the process for a user as well as designing user heuristics for automatically specifying parameters. In general, clustering is often an extremely useful first step in the data analysis pipeline, but as the problem is ill-constrained there is no single solution that works in all settings.

Our method allows for both distance fields and KDE to building the scalar field from point clouds. While KDE is a natural choice, its use does restrict the clustering approach to be sensitive to the density of the input point cloud, similar to DBSCAN. Of course, there is further control in terms of the persistence and other user-defined parameters. In some contexts, this control is desirable, but we leave it to future work to explore what other scalar fields might be better suited in different contexts.

On the computational side, we made an up front choice to limit our approach to working with point clouds in low-dimensions. This report demonstrates the efficacy on mainly two-dimensional datasets, but we have also experimented with both three-dimensional point clouds and point clouds sampled from surfaces embedded in three dimensions. Our results do extend, although our method for automatically specifying parameters in both modules does have a dependence on the underlying dimension of the data. We leave it to future work to specify these parameters (k, a, and b) in terms of the dimension of the data. For higher dimensional data, utilizing a similar pipeline is also work in progress, but it has significant challenges with visualizing the resulting clustering. One method that shows promise is to first project the input data to a lower dimensional manifold, as is frequently done in other settings.

Since we sample our scalar field on a triangulated grid, the computational workload will ultimately be dependent on the resolution of this grid. TTK is optimized for fast access on triangulations, which is advantageous compared to other alternative simplicial complexes we could use such as Rips complexes. While we require a sampling density sufficient for capturing the separation between clusters, it would be interesting to consider specifying a non-uniform triangulation that adapts to the data so as to reduce the computational burden. It is future work as how best to balance clustering accuracy with computational cost.

Finally, our implemented modules are available for download, as well as the ParaView state files used to produce the results presented in this paper[5].

Acknowledgements This work is partially supported by the European Commission grant ERC-2019-COG *"TORI"* (ref. 863464). This material is based upon work supported by the U.S. Department of Energy, Office of Science, Office of Advanced Scientific Computing Research, under Award Number(s) DE-SC-0019039.

[5] https://github.com/bluenote-1577/ttk.

References

1. Ahrens, J., Geveci, B., Law, C.: Paraview: an end-user tool for large-data visualization. In: The Visualization Handbook, pp. 717–731 (2005)
2. Beksi, W.J., Papanikolopoulos, N.: 3D point cloud segmentation using topological persistence. In: 2016 IEEE International Conference on Robotics and Automation (ICRA), pp. 5046–5051. IEEE (2016)
3. Beksi, W.J., Papanikolopoulos, N.: Signature of topologically persistent points for 3D point cloud description. In: 2018 IEEE International Conference on Robotics and Automation (ICRA), pp. 1–6. IEEE (2018)
4. Beksi, W.J., Papanikolopoulos, N.: A topology-based descriptor for 3D point cloud modeling: theory and experiments. Image Vis. Comput. **88**, 84–95 (2019)
5. Bertrand, G.: On topological watersheds. J. Math. Imag. Vis. **22**(2-3), 217–230 (2005)
6. Beucher, S.: Watersheds of functions and picture segmentation. In: ICASSP 1982. IEEE International Conference on Acoustics, Speech, and Signal Processing, vol. 7, pp. 1928–1931. IEEE (1982)
7. Breiman, L., Meisel, W., Purcell, E.: Variable kernel estimates of multivariate densities. Technometrics **19**(2), 135–144 (1977)
8. Chazal, F., Guibas, L.J., Oudot, S.Y., Skraba, P.: Scalar field analysis over point cloud data. Discrete Comput. Geometry **46**(4), 743 (2011)
9. Chazal, F., Guibas, L.J., Oudot, S.Y., Skraba, P.: Persistence-based clustering in Riemannian manifolds. J. ACM (JACM) **60**(6), 41 (2013)
10. Comaniciu, D., Meer, P.: Mean shift: a robust approach toward feature space analysis. IEEE Trans. Pattern Anal. Mach. Intell. **24**(5), 603–619 (2002)
11. Edelsbrunner, H., Harer, J.: Computational Topology: An Introduction. American Mathematical Society (2009)
12. Edelsbrunner, H., Harer, J., Natarajan, V., Pascucci, V.: Morse-smale complexes for piecewise linear 3-manifolds. In: Proceedings of the Nineteenth Annual Symposium on Computational Geometry, pp. 361–370. ACM (2003)
13. Edelsbrunner, H., Letscher, D., Zomorodian, A.: Topological persistence and simplification. In: Proceedings 41st Annual Symposium on Foundations of Computer Science, pp. 454–463. IEEE (2000)
14. Ester, M., Kriegel, H., Sander, J., Xu, X.: A density-based algorithm for discovering clusters in large spatial databases with noise. In: Proceedings of Knowledge Discovery and Data Mining, pp. 226–231 (1996)
15. Fowlkes, E.B., Mallows, C.L.: A method for comparing two hierarchical clusterings. J. Am. Stat. Assoc. **78**(383), 553–569 (1983)
16. Fränti, P., Sieranoja, S.: K-means properties on six clustering benchmark datasets. Appl. Intell. **48**(12), 4743–4759 (2018)
17. Gueunet, C., Fortin, P., Jomier, J., Tierny, J.: Task-based augmented merge trees with Fibonacci heaps. In: 2017 IEEE 7th Symposium on Large Data Analysis and Visualization (LDAV), pp. 6–15. IEEE (2017)
18. Koontz, W.L.G., Narendra, P.M., Fukunaga, K.: A graph-theoretic approach to nonparametric cluster analysis. IEEE Trans. Comput. **25**(9), 936–944 (1976)
19. Lloyd, S.: Least squares quantization in PCM. IEEE Trans. Inf. Theory **28**(2), 129–137 (1982)
20. Moon, C., Giansiracusa, N., Lazar, N.A.: Persistence terrace for topological inference of point cloud data. J. Comput. Graph. Stat. **27**(3), 576–586 (2018)
21. Oesterling, P., Heine, C., Janicke, H., Scheuermann, G., Heyer, G.: Visualization of high-dimensional point clouds using their density distribution's topology. IEEE Trans. Vis. Comput. Graph. **17**(11), 1547–1559 (2011)
22. Oesterling, P., Heine, C., Weber, G.H., Scheuermann, G.: Visualizing nd point clouds as topological landscape profiles to guide local data analysis. IEEE Trans. Vis. Comput. Graph. **19**(3), 514–526 (2012)

23. Orava, J.: K-nearest neighbour kernel density estimation, the choice of optimal k. Tatra Mountains Math. Publ. **50**(1), 39–50 (2011)
24. Sander, J., Ester, M., Kriegel, H.P., Xu, X.: Density-based clustering in spatial databases: the algorithm GDBSCAN and its applications. Data Min. Knowl. Disc. **2**(2), 169–194 (1998)
25. Scikitlearn: Clustering (2019). https://scikit-learn.org/stable/modules/clustering.html. Accessed 02 Jan 2019
26. Thom, R.: Sur une partition en cellules associée à une fonction sur une variété. Comptes Rendus Hebdomadaires des Seances de l Academie des Sciences **228**(12), 973–975 (1949)
27. Tierny, J., Favelier, G., Levine, J.A., Gueunet, C., Michaux, M.: The topology toolkit. IEEE Trans. on Visualization and Computer Graphics (Special Issue IEEE VIS 2017: SciVis) **24**(1), 832–842 (2018)
28. Tierny, J., Pascucci, V.: Generalized topological simplification of scalar fields on surfaces. IEEE Trans. Vis. Comput. Graph. **18**(12), 2005–2013 (2012)
29. Ultsch, A.: Clustering with SOM: U * C. In: Proceedings Workshop on Self-Organizing Maps, pp. 75–82 (2005)
30. Von Luxburg, U.: A tutorial on spectral clustering. Stat. Comput. **17**(4), 395–416 (2007)

Report of the TopoInVis TTK Hackathon: Experiences, Lessons Learned, and Perspectives

Jonas Lukasczyk, Jakob Beran, Wito Engelke, Martin Falk, Anke Friederici, Christoph Garth, Lutz Hofmann, Ingrid Hotz, Petar Hristov, Wiebke Köpp, Talha Bin Masood, Małgorzata Olejniczak, Paul Rosen, Jan-Tobias Sohns, Tino Weinkauf, Kilian Werner, and Julien Tierny

Abstract This paper documents the organization, the execution, and the results of the Topology ToolKit (TTK) hackathon that took place at the TopoInVis 2019 conference. The primary goal of the hackathon was to promote TTK in our research community as a unified software development platform for topology-based data analysis algorithms. To this end, participants were first introduced to the structure and capabilities of TTK, and then worked on their own TTK-related projects while being mentored by senior TTK developers. Notable outcomes of the hackathon were first steps towards Python and Docker packages, further integration of TTK in Inviwo, the extension of TTK with new algorithms, and the discovery of current limitations of TTK as well as future development directions.

J. Lukasczyk (✉) · C. Garth · J.-T. Sohns · K. Werner
Technische Universität Kaiserslautern, Kaiserslautern, Germany
e-mail: jl@jluk.de

C. Garth
e-mail: garth@cs.uni-kl.de

J.-T. Sohns
e-mail: j_sohns12@cs.uni-kl.de

K. Werner
e-mail: kwerner@cs.uni-kl.de

J. Beran
Stockholm University, Stockholm, Sweden
e-mail: jakob.beran@misu.su.se

W. Engelke · M. Falk · I. Hotz · T. B. Masood
Linköping University, Norrköping, Sweden
e-mail: wito.engelke@liu.se

M. Falk
e-mail: martin.falk@liu.se

I. Hotz
e-mail: ingrid.hotz@liu.se

T. B. Masood
e-mail: talha.bin.masood@liu.se

I. Hotz et al. (eds.), *Topological Methods in Data Analysis and Visualization VI*, Mathematics and Visualization, https://doi.org/10.1007/978-3-030-83500-2_18

1 Introduction

The Topology ToolKit (TTK) [17] is an open-source software library for topological data analysis (TDA) and scientific visualization. At the time of writing, TTK consists of more than 60 modules—contributed from various researchers from several institutions—that provide efficient algorithms to compute contour trees [7], Reeb graphs [8], persistence diagrams [3, 4], topological simplifications [18], Morse-Smale complexes [15], nested tracking graphs [11, 13], fiber surfaces [10], image-based geometry reconstructions [12], and many more TDA products. The core feature of TTK is its efficient and unified approach to topological data representation that makes it possible to coherently chain these different algorithms. Therefore, developers can contribute new algorithms to TTK in a modular fashion, and end-users can utilize TTK as a production tool for interactive TDA (Fig. 1). Furthermore, this unified approach makes it possible for researchers to reproduce results, benchmark algorithms, and develop new algorithms in an existing interrelated software environment.

However, TTK has two major limitations: a) TTK is difficult to use and extend due to its extensive capabilities and complex software architecture; and b) TTK is not easily accessible since it requires manual compilation and requires several dependencies to utilize all features (which makes it especially difficult to install TTK on non UNIX systems). These limitations often discourage new developers and end-users to utilize TTK in their projects.

A. Friederici · W. Köpp · T. Weinkauf
KTH Royal Institute of Technology, Stockholm, Sweden
e-mail: ankef@kth.se

W. Köpp
e-mail: wiebkek@kth.se

T. Weinkauf
e-mail: weinkauf@kth.se

L. Hofmann
Heidelberg University, Heidelberg, Germany
e-mail: lutz.hofmann@iwr.uni-heidelberg.de

P. Hristov
University of Leeds, Leeds, UK
e-mail: mm16pgh@leeds.ac.uk

M. Olejniczak
University of Warsaw, Warsaw, Poland
e-mail: malgorzata.olejniczak@cent.uw.edu.pl

P. Rosen
University of South Florida, Tampa, USA
e-mail: prosen@usf.edu

J. Tierny
CNRS, Sorbonne Universite, Paris, France
e-mail: julien.tierny@sorbonne-universite.fr

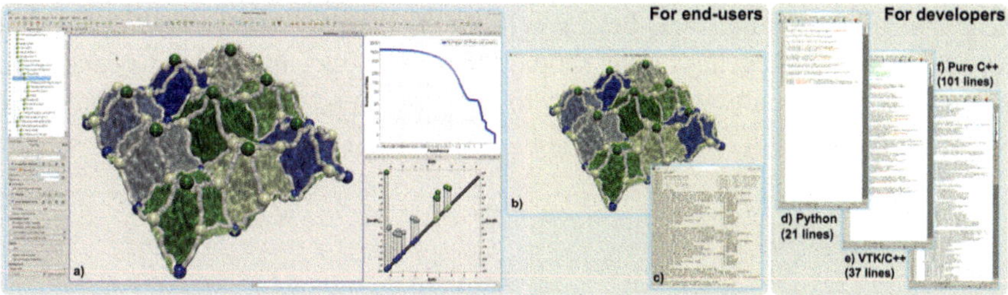

Fig. 1 TTK is a software platform for topological data analysis and scientific visualization. It is both accessible to end-users—via ParaView plugins **a**, VTK-based generic GUIs **b**, and command-line programs **c**—and to developers—via Python **d**, VTK/C++ **e** or dependence-free C++ **f** bindings. TTK provides an efficient and unified approach to topological data representation and simplification, which enables in this example a discrete Morse-Smale complex **a** to comply to the level of simplification dictated by a piecewise linear persistence diagram (bottom-right linked view, a). Code snippets are provided (**d**–**f**) to reproduce this pipeline

In an effort to overcome these limitations and simultaneously grow TTK's user and developer base, two senior TTK developers (Julien Tierny and Jonas Lukasczyk) organized a hackathon as an event in which they would be able to directly interact with people from the topology-based visualization research community, understand their practical needs, and address the concrete problems they face during integration of TTK in their own projects. This event would therefore be fundamentally different to previous TTK related workshops, such as the TTK tutorials at IEEE VIS in 2018 [6] and 2019 [5] that followed a teacher-centered teaching approach. At the same time, the TopoInVis conference series was experimenting with novel initiatives to increase interest and collaboration within the same community. Thus, the hackathon became part of said initiative, as it seemed—and proved to be—beneficial for both the conference organizers and conference participants to organize the hackathon as a co-located conference event. However, the hackathon organizers did not have first-hand experience about organizing and conducting such an event. Therefore, this work documents

- the organizational aspects of the hackathon that can be used to successfully conduct similar events in the future (Sect. 2); and
- the practical results of the hackathon that actually advanced TTK (Sect. 3).

2 Organization

The hackathon organizers had no prior experience about hosting hackathons; including getting people interested, coping with different backgrounds of the participants, determining a schedule, and setting reasonable goals for such an event. It was only clear that the hackathon will be a single day event that will be co-located with TopoInVis, and that the hackathon should focus on the specific problems developers and end-users encounter when they try to integrate TTK within their own projects.

2.1 Preparation

To get an initial sense on the number, the different experience levels, and the interests of potential participants, the organizers provided a five-minute online survey that consisted of ten multiple choice questions. Thirteen people completed the survey and they all registered for the hackathon. Participants were first asked about their familiarity with TDA and TTK, as well as their programming skills (Fig. 2), which indicated that the hackathon program needs to reflect the diverse backgrounds of the participants. Next, each participant was asked to select two of the following suggested hackathon topics she or he is interested in (total number of votes are shown in parenthesis):

- (6) TTK integration into an existing system
- (5) Actual Data Analysis with TTK
- (5) TTK support for vector fields
- (4) TTK support for periodic grids
- (4) TTK packaging
- (2) TTK support for tensor fields

Two participants also used the option to suggest a new topic: the portation of an existing algorithm to TTK. This poll also indicated that participants are interested in largely different topics, so the hackathon program should not focus on a single subject. The remaining questions covered the types of datasets participants commonly

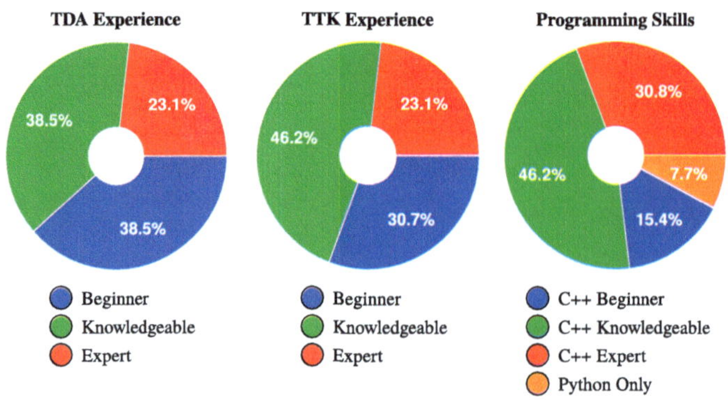

Fig. 2 Experience levels of the 13 participants in regard to topological data analysis **a**, TTK **b**, and programming languages **c**

08:00 - 10:00	Introduction
10:30 - 12:30	Workgroups
14:00 - 16:00	Workgroups
16:30 - 18:00	Conclusions

Fig. 3 Schedule of the TTK hackathon: After providing an overview of TTK's structure and usage, the participants split into groups that worked on different tasks, and subsequently presented their results in a concluding session

have to process, if they plan to bring these datasets to the hackathon, and minor organisational issues.

2.2 *Program*

The hackathon was designed as a single-day event that consisted of four sessions (Fig. 3) led by both organizers. To address the diverse backgrounds of the participants (Sect. 2.1), the organizers presented in the first session the internal structure of TTK and demonstrated its capabilities on several examples, where it was possible for participants to follow along if they already had TTK installed.

The introduction provided a common basis for the rest of the hackathon where participants split into small groups consisting of 4–5 people to work on a specific, self-determined coding project based on the topics indicated in the survey. The overall goal of the coding projects was to help participants with initial steps towards first results, and to teach them how to continue these projects themselves later on, as it was already anticipated by the organizers beforehand that most projects can not be completed in one day. Initially, the organizers intended to provide a more extensive introduction that would cover the first two sessions, but due to the experience level and interest of the participants the introduction was cut short to have more time for the actual coding projects.

In the last session, a representative of each group summarized their corresponding achievements of the day, the encountered problems, and the next steps they have to take in order to complete their projects. To this end, every representative had to give a ten minute presentation, followed by a discussion.

3 Results

This section presents the most prominent results of the workgroups including improved accessibility, ported algorithms, the integration of TTK in the production tool Inviwo [9], and the support of periodic grids.

3.1 Packaging

An often problematic aspect of scientific data analysis and visualization is the maintenance of software environments, i.e. creating a working installation of an analysis package and all its dependencies. Since TTK is shipped with a rich API (C++, VTK, Python) and a plugin for the production visualization platform ParaView, its full-option compilation can be challenging for novice users. Moreover, this process differs substantially between platforms. The resulting complexity of getting dependencies right for a TTK installation is overwhelming to users, and is thus a significant obstacle towards making TTK's methods available for a large user base.

3.1.1 Docker

For many scenarios (including scientific workloads), container technologies such as Docker [14] have proven to be a viable solution. In brief, a container captures not only an application, but also all of its dependencies in an otherwise self-contained, minimal install of a base operating system. Execution of containers is achieved via virtualization technologies built into all major operating systems. One goal of the packaging project was therefore to create a build process for Docker containers, which would expose TTK to a large user base.

The actual Docker packaging concept of TTK that was brought forth by the hackathon is based on the client-server model already provided by ParaView. Specifically, the docker container only contains a ParaView server build, and a full-option build of TTK, which is integrated into the ParaView server via its common plugin mechanism. Once this container is running, a vanilla ParaView client downloaded from Kitware's website can then be used to connect to the server running inside the container. Hence, all TTK algorithms are executed on the server within the container, and the results are send to the client.

During the hackathon, the workgroup was able to produce a minimum viable prototype. The discussions between hackathon participants were essential in deriving this server-client based container solution. It could be confirmed that the resulting containers are surprisingly easy to use and incur few limitations. Among the latter is a limitation to server-side software rendering, since hardware accelerated graphics within containers require complex vendor-specific setups that reduce the portability of the containers. Since the hackathon, the corresponding implementation underwent several improvements towards robustness and usability, and is now included in TTK. Currently, publicly available Docker containers for different ParaView versions can be downloaded from the DockerHub container registry [2].

3.1.2 Anaconda

The packaging workgroup also explored another approach to make TTK more accessible by providing TTK as an Anaconda [1] package. Anaconda is a cross-platform open-source distribution platform for Python-based data science software that includes its own package manager. Similar to a docker image, each Anaconda package specifies a build procedure and a list of dependencies; in this case to other Anaconda packages. VTK is already available as an Anaconda package, so the workgroup investigated if the same build structure can be adapted to the VTK layer of TTK. To this end, it was necessary to make use of custom CMake functions that are already included in VTK 9, but were undocumented at the time of the hackathon. The workgroup spend most of its time on including the various dependencies of TTK such as Sqlite, GraphViz, and Eigen. Based on the initial steps taken during the hackathon, the VTK-layer of TTK is now available as the Anaconda package TOPOLOGYTOOLKIT from the CONDA- FORGE channel [1].

3.2 *Vector Field Robustness Module*

Due to the limited availability of vector field topology algorithms in TTK, this workgroup set out to implement 2D vector field robustness calculation [19] and simplification [16] techniques.

3.2.1 Vector Field Robustness

In brief, robustness is a metric for pairing and canceling critical points using minimum perturbation of the L^∞ norm of vector magnitudes. The algorithm itself consist of three phases:

1. computation and classification of critical points, i.e., sources, sinks, and saddles;
2. construction of a specialized merge tree that considers the vector magnitude field and the previously calculated critical point locations and types; and
3. if the user intends to apply topological simplification, regions of the vector field—specified as sublevel set regions of the vector magnitude field—are numerically perturbed.

3.2.2 Implementation

The process of implementing this module can essentially be split into two parts: 1) building the module infrastructure, and 2) implementing the algorithm. The workgroup faced different struggles at each of these steps and required a lot of assistance from the organizers.

At the infrastructure level, there are multiple layers of code that need to be written, including the ParaView level, the VTK level, and finally the TTK level. Configuring these is non-trivial without extensive experience in at least VTK, but to some extent TTK as well. First, the selection and usage of data types at each level is non-trivial, especially if the module is supposed to support vector fields of different data types. Second, across all layers many definitions have to be repeated and consistent, which can easily result in errors, i.e., at the Paraview layer the user interface controls class members of the VTK layer, which are then passed to the base layer as parameters.

Implementing the algorithm portion of the module has its own mix of challenges. First is the used programming language. Fortunately, the original code was already written in c++, but if it would not have been, then the code would need to be translated first. Next, the code required translating existing mesh definitions into those used by TTK/VTK. Again, the workgroup members pointed out that it was fortunate that TTK's mesh specification is fairly easy to use, which made this code transition trivial. The final challenge was minimizing the use of external libraries, since ideally plugins should be self-contained. This was especially problematic for this particular module since the first phase of the algorithm—the computation and classification of critical points—was handled by an external library. Moreover, at the time of writing, TTK was only able to compute and classify critical points of scalar fields, so removing this dependency would require a lot of additional effort.

3.2.3 Results

The workgroup spend most of its time on the module infrastructure, leaving the actual implementation of the different phases of the algorithm to be completed after the hackathon. Yet, the participants felt confident to finish the module by themselves.

3.3 Extending the Integration of TTK in Inviwo

Inviwo [9] is a rapid prototyping framework for visualizing spatial and abstract data. The network editor of Inviwo provides the functionality for building a visualization pipeline out of individual blocks, or processors. A basic subset of TTK's functionality has been part of Inviwo for some time in form of various processors for creating TTK triangulations, topological simplification, and persistence diagrams. During the course of the hackathon, this subset was to be extended to also expose the Morse-Smale complex computation.

3.3.1 Implementation

There were some initial difficulties in understanding the extensive TTK infrastructure and the plethora of function arguments and results, which might partially have been caused by API requirements of VTK as TTK uses VTK to wrap algorithms. This is essentially the same limitation also reported by the vector field topology workgroup (Subsect. 3.2). These difficulties were, however, quickly resolved through interactions and discussions with the hackathon organizers. One design decision was to decouple the computation from the output required for the subsequent visualization, i.e. the geometric information of critical points and saddle point connections. Thus, two processors were created. The first accepts a TTK triangulation, performs the Morse-Smale calculations, and outputs the results. The second processor then generates geometric primitives from the results.

3.3.2 Results

At the end of the hackathon it was possible to compute the Morse-Smale Complex in Inviwo using the underlying TTK functionality and to visualize the results. Figure 4

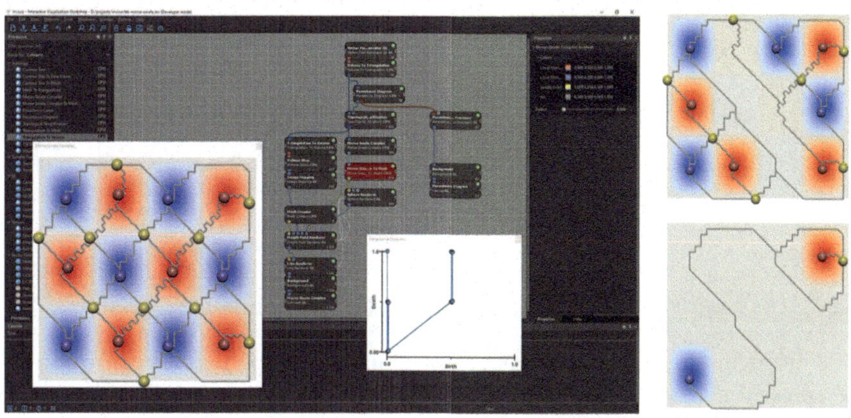

Fig. 4 Results of the TTK-based Morse-Smale complex computation in Inviwo. The figure shows the underlying visualization pipeline (background), and Morse-Smale complexes for different simplification thresholds (foreground/right)

depicts the Inviwo network computing the Morse-Smale complexes for a synthetic dataset. In the future, Inviwo will be extended to support an even larger part of the TTK functionality as well as incorporate periodic boundaries; a topic covered by another workgroup (see Sect. 3.4).

3.4 *Periodic Grids*

Periodic boundary conditions are often used to model a very large (or infinite) system using a small representative part called a *unit cell*. Many physical, chemical, and biological systems exhibit repetitive symmetric patterns. These systems are ideal for study and analysis based on simulations on small unit cells with periodic boundary conditions. Other examples where periodic boundaries are used include the study of metals, crystal lattices, and bulk solvents in molecular dynamics simulations.

Topological algorithms that are already implemented in TTK—such as the Morse-Smale complex or merge tree computation—are able to process any domain that can be represented via a simplicial complex. However, TTK's original domain data structures—i.e., implicit and explicit triangulations—did not support periodicity. Therefore, this workgroup set out to extend the underlying triangulation data structures to support periodicity boundary conditions.

3.4.1 Implementation

At the base layer, TTK provides a unified triangulation interface that is capable of representing any domain, and this interface is used by all topological algorithms available in TTK.The TTK triangulation class structure is shown in Fig. 5.

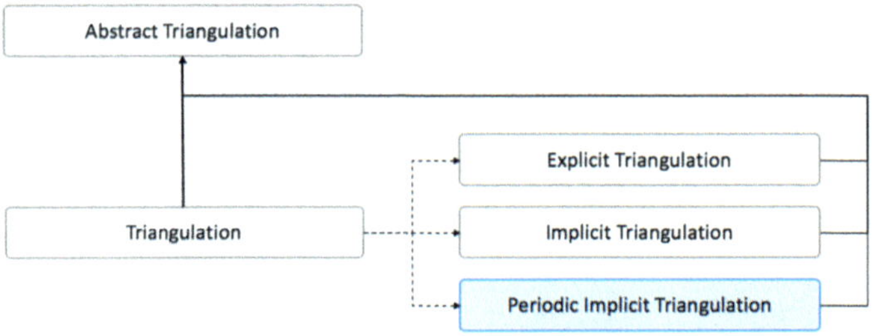

Fig. 5 The inheritance structure of the different triangulation classes in TTK

TTK distinguishes between two types of triangulations: `ExplicitTriangulation` to represent a triangulation where the cell connectivity is provided explicitly, and `ImplicitTriangulation` to provide a highly memory efficient triangulation for structured grids. `ImplicitTriangulation`, however, assumes that the boundary is not periodic.

The workgroup decided to implement a new triangulation class for structured grids with periodic boundaries called `PeriodicImplicitTriangulation`. Figure 6 illustrates the concept behind this implementation for 2D grids, where additional edges and triangles are created between boundary vertices to establish periodicity in all coordinate directions. This concept directly translates to 3D grids. Based on the design decisions discussed during the hackathon, the workgroup completed the implementation of the new class after the hackathon and integrated it into the TTK master branch.

Fig. 6 The periodic triangulation in 2D is accomplished by adding the dashed edges and triangles to the triangulation shown with solid edges

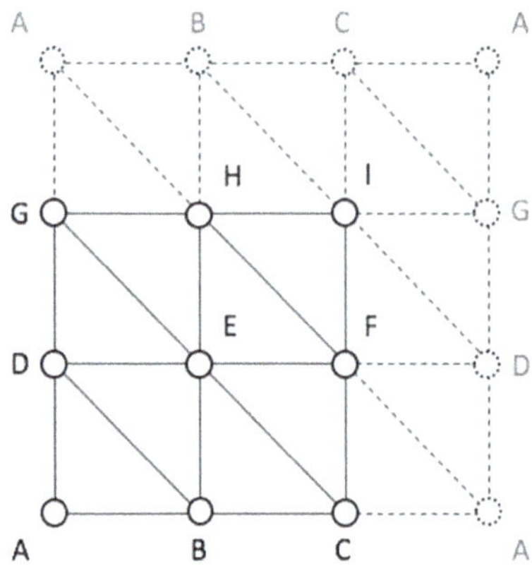

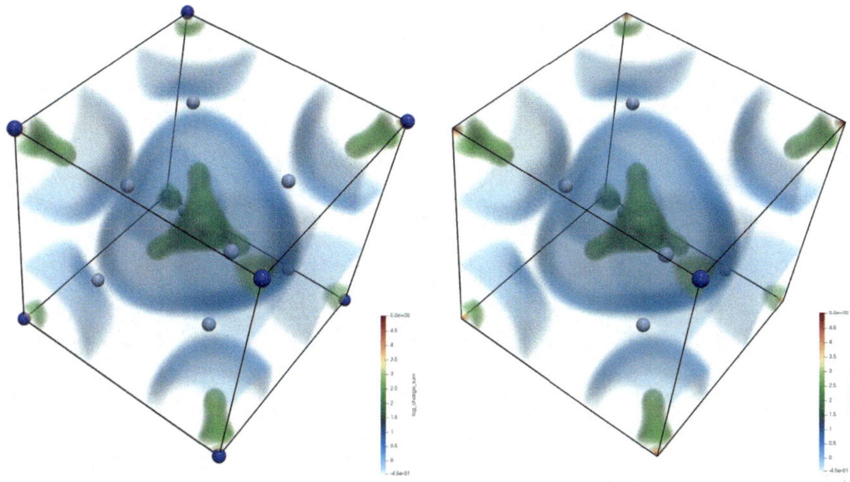

Fig. 7 Results of the Morse-Smale complex computation for a methane crystal modeled via a fixed unit cell that contains two methane molecules with non-periodic (left) and periodic (right) boundary conditions. The images show a volume rendering of the underlying scalar field and the location of the identified critical points, where maxima and saddles are represented as dark and light blue spheres, respectively. Note that without periodic boundaries eight maxima are obtained at the corners of the grid, however with periodic boundaries a single maximum is obtained

3.4.2 Results

A working example of the Morse-Smale complex computation on a periodic domain is shown in Figs. 7 and 8. Currently, `PeriodicImplicitTriangulation` establishes periodic boundaries in all coordinate directions, but in the future this class can be extended to toggle periodicity for individual directions.

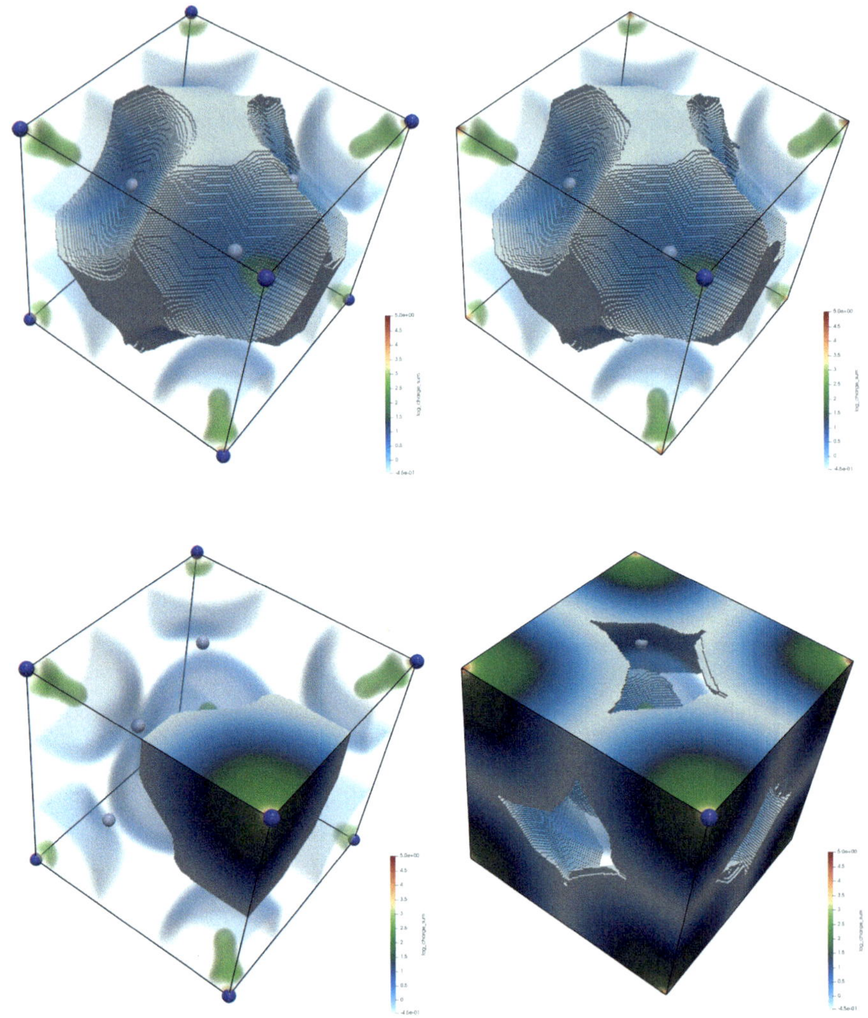

Fig. 8 Morse-Smale complex-based segmentation of a methane crystal unit cell containing two methane molecules with a non-periodic (left) and a periodic (right) boundary. (First row) One methane molecule is located in the middle of the domain and hence it is correctly identified as an ascending manifold of a maximum in the Morse-Smale complex with or without periodic boundaries. (Second row) However, for the molecule located at the corner of the cell, the ascending manifold with periodic boundary correctly identifies the region corresponding to the second molecule. The ascending manifold without periodic boundary fails to correctly segment the volume into two methane molecules

4 Conclusion

Overall, the participants and organizers consider the hackathon a success, and the majority of attendees stated that they would participate again in a future hackathon. Besides the practical outcomes of the hackathon (Sect. 4.1), the senior developers also learned about organizing such an event (Sect. 4.2), and derived future development directions based on feedback of the participants (Sect. 4.3).

4.1 Workgroup Results

The hackathon initiated several significant coding projects that were later completed and added to TTK's source code. Specifically, due to the hackathon, TTK is now also shipped via a Docker image and an Anaconda package (Sect. 3.1), is further integrated into Inviwo (Sect. 3.3), and supports periodic grids (Sect. 3.4).

The coding project that worked on a new vector field robustness module (Sect. 3.2) is not yet completed, but yielded valuable insight into the practical challenges new developers face while adding new algorithms to TTK. Based on that insight, TTK's internal API is now getting revised (Sect. 4.3).

4.2 Organizational Aspects

A major factor for the success of the TTK hackathon was to organize it as a co-located event at the TopoInVis conference. This was advantageous to both the hackathon organizers as well as the participants, as conference participants correspond to TTK's target user base, and participants were able to simply extend their stay by one day to attend the hackathon.

Although the organizers originally intended to spend the same amount of time on introducing TTK and working on the actual coding projects, it turned out to be better to spend more time on the actual coding sessions. In future hackathons, the organizers recommend to use the revised schedule (Fig. 3), and to adjust the content of the introduction session based on the attendees. To this end, the initial questionnaire prior to the hackathon was essential in assessing the different experience levels and interests of the participants. A shortcoming of the organization was the missed opportunity to collect a formal feedback of the participants via another questionnaire. The organizers strongly recommend to do so in future events.

4.3 TTK Development Directions

Already at the beginning of the hackathon it became apparent that many participants struggled with installing TTK on their system. In fact, many potential users are currently discouraged to even try out TTK because it is very difficult to get it up and running. The hackathon sprout two valuable approaches to this problem by making TTK accessible via a Docker container and an Anaconda package. Yet, the Docker container requires superuser privileges and only supports software rendering, and the Anaconda package does not feature ParaView as a front end. Besides advancing these approaches, it also seems valuable to improve and simplify the general build process of TTK in the future.

Moreover, it appeared that TTK's internal API could be greatly simplified in the interest of readability and accessibility to new developers, without changes in performance or in its core design principles. Based on feedback from the participants, the senior developers are currently revising TTK's internal API to make the code base more transparent, simpler, and easier to modify, in order to drastically reduce the amount of effort users and developers have to spend on learning and extending TTK.

In conclusion, it is the belief of the organizers that the hackathon was essential for growing TTK's user and developer base, for addressing current limitations, and deriving future development directions.

Acknowledgements We thank all participants for their significant contributions and valuable feedback. This work was partially supported by the European Commission grant ERC-2019-COG "TORI" (ref. 863464), and the German research foundation (DFG) within the IRTG 2057 "Physical Modeling for Virtual Manufacturing Systems and Processes".

References

1. Anaconda Software Distribution (2016). https://anaconda.com
2. TTK Docker Container Repository (2019). https://hub.docker.com/u/topologytoolkit
3. Edelsbrunner, H., Harer, J.: Computational Topology: an Introduction. American Mathematical Society, Providence (2010)
4. Edelsbrunner, H., Letscher, D., Zomorodian, A.: Topological persistence and simplification. Discrete Comput. Geom. **28**(4), 511–533 (2002)
5. Falk, M.,et al..: Topological data analysis made easy with the topology toolkit, a sequel (2019). https://topology-tool-kit.github.io/ieeeVisTutorial.html
6. Favelier, G., et al.: Topological data analysis made easy with the topology toolkit (2018). https://topology-tool-kit.github.io/ieeeVisTutorial.html
7. Gueunet, C., Fortin, P., Jomier, J., Tierny, J.: Task-based augmented merge trees with fibonacci heaps. In: IEEE Symposium on Large Data Analysis and Visualization (2017)
8. Gueunet, C., Fortin, P., Jomier, J., Tierny, J.: Task-based augmented reeb graphs with dynamic ST-Trees (2019)
9. Jönsson, D., et al.: Inviwo-a visualization system with usage abstraction levels. IEEE Trans. Visual. Comput. Graph. **26**(11), 3241–3254 (2019)
10. Klacansky, P., Tierny, J., Carr, H., Geng, Z.: Fast and exact fiber surfaces for tetrahedral meshes. IEEE Trans. Visual Comput. Graph. **23**(7), 1782–1795 (2016)

11. Lukasczyk, J., Garth, C., WeberWeber, G., Biedert, T., Maciejewski, R., Leitte, H.: Dynamic nested tracking graphs. IEEE Trans. Visual. Comput. Graph. **26**, 249–258 (2019)
12. Lukasczyk, J., Kinner, E., Ahrens, J., Leitte, H., Garth, C.: VOIDGA: a view-approximation oriented image database generation approach. In: IEEE 8th Symposium on Large Data Analysis and Visualization (LDAV) (2018)
13. Lukasczyk, J., Weber, G., Maciejewski, R., Garth, C., Leitte, H.: Nested tracking graphs. Comput. Graph. Forum **36**, 12–22 (2017)
14. Saha, P., Beltre, A., Uminski, P., Govindaraju, M.: Evaluation of docker containers for scientific workloads in the cloud. CoRR **abs/1905.08415** (2019). http://arxiv.org/abs/1905.08415
15. Shivashankar, N., Natarajan, V.: Parallel computation of 3D Morse-Smale complexes. Comput. Graph. Forum **31**, 965–974 (2012)
16. Skraba, P., Wang, B., Chen, G., Rosen, P.: 2D vector field simplification based on robustness. In: IEEE Pacific Visualization Symposium (PacificVis), pp. 49–56. IEEE (2014)
17. Tierny, J., Favelier, G., Levine, J.A., Gueunet, C., Michaux, M.: The topology ToolKit. IEEE Trans. Visual. Comput. Graph. (2017). https://topology-tool-kit.github.io/
18. Tierny, J., Pascucci, V.: Generalized topological simplification of scalar fields on surfaces. IEEE Trans. Visual. Comput. Graph. **18**(12), 2005–2013 (2012)
19. Wang, B., Rosen, P., Skraba, P., Bhatia, H., Pascucci, V.: Visualizing robustness of critical points for 2D time-varying vector fields. Comput. Graph. Forum, **32**, 221–230 (2013)

Printed by Printforce, the Netherlands